# Methods in Neurosciences

Volume 27

# Measurement and Manipulation of Intracellular Ions

Methods in Neurosciences

Editor-in-Chief

P. Michael Conn

Methods in Neurosciences

Volume 27

# Measurement and Manipulation of Intracellular Ions

Edited by

Jacob Kraicer

*Department of Physiology*
*Health Sciences Centre*
*The University of Western Ontario*
*London, Ontario, Canada*

*and*

*Human Frontier Science Program*
*Strasbourg, France*

S. Jeffrey Dixon

*Division of Oral Biology and Department of Physiology*
*Health Sciences Centre*
*The University of Western Ontario*
*London, Ontario, Canada*

ACADEMIC PRESS

San Diego New York Boston London Sydney Tokyo Toronto

*Front cover photograph*: Fura-2 stained neuron. Neuron from the nucleus accumbens in cell culture stained with fura-2. The image was obtained using Axon Imaging Workbench by digitizing an integrated frame (500 msec) from a CCD camera (Cohu). The cell was illuminated at 380 nm. Courtesy of J. Plant and B. A. MacVicar, University of Calgary.

This book is printed on acid-free paper. ♾



Academic Press, Inc.
A Division of Harcourt Brace & Company
525 B Street, Suite 1900, San Diego, California 92101-4495

*United Kingdom Edition published by*
Academic Press Limited
24-28 Oval Road, London NW1 7DX

International Standard Serial Number: 1043-9471

International Standard Book Number: 0-12-185297-0

Printed and bound in the United Kingdom
Transferred to Digital Printing, 2011

*To my mentors, John Logothetopoulos, Claude Fortier, and Don Hatcher*

JACOB KRAICER

*To my parents, Michael and Helen*

S. JEFFREY DIXON

# Table of Contents

# Contributors to Volume 27

*Article numbers are in parentheses following the names of contributors. Affiliations listed are current.*

JULIO ALTAMIRANO (19), Departamento de Neurobiología, Instituto Mexicano de Psiquiatría, México 14370 D. F., México, and Department of Physiology and Biophysics, University of Texas Medical Branch, Galveston, Texas 77555

FRANCISCO J. ALVAREZ-LEEFMANS (19), Departamento de Neurobiología, Instituto Mexicano de Psiquiatría, México 14370 D. F., México, and Departamento de Farmacología y Toxicología, Centro de Investigation y de Estudios Avanzados del IPN, 07000 D.F., México

HERMAN S. BACHELARD (8), Magnetic Resonance Center, Department of Physics, University of Nottingham, Nottingham NG7 2RD, United Kingdom

SUSAN E. BATES (7), Department of Pharmacology, United Medical and Dental Schools, St. Thomas' Hospital, London SE1 7EH, United Kingdom

MARK O. BEVENSEE (12), Department of Cellular and Molecular Physiology, Yale University School of Medicine, New Haven, Connecticut 06510

JOACHIM BIWERSI (17), Cardiovascular Research Institute, University of California, San Francisco, San Francisco, California 94143

WALTER F. BORON (12), Department of Cellular and Molecular Physiology, Yale University School of Medicine, New Haven, Connecticut 06510

JONATHAN A. COLES (18), Neurobiologie Intégrative, INSERM U.394, F-33077 Bordeaux, Cedex, France

WILLIAM E. CROWE (19), Department of Physiology, Medical College of Pennsylvania and Hahnemann University, Philadelphia, Pennsylvania 19129

CHARLES M. DEBER (4), Research Institute and Department of Biochemistry, The Hospital for Sick Children, University of Toronto, Toronto, Ontario, Canada M5G 1X8

JOACHIM W. DEITMER (14), Abteilung für Allgemeine Zoologie, FB Biologie, Universität Kaiserslautern, D-67653 Kaiserslautern, Germany

J. DEMPSTER (6), Department of Physiology and Pharmacology, University of Strathclyde, Glasgow, Scotland, United Kingdom

S. JEFFREY DIXON (9), Division of Oral Biology and Department of Physiology, Faculty of Dentistry, The University of Western Ontario, London, Ontario, Canada N6A 5C1

J. KEVIN FOSKETT (13), Division of Cell Biology, The Hospital for Sick Children, Toronto, Ontario, Canada M5G 1X8

KENNETH A. GIULIANO (1), Department of Biological Sciences, Center for Light Microscope Imaging and Biotechnology, Carnegie Mellon University, and Department of Neurological Surgery, University of Pittsburgh School of Medicine, Presbyterian University Hospital, Pittsburgh, Pennsylvania 15213

ALISON M. GURNEY (7), Department of Pharmacology, United Medical and Dental Schools, St. Thomas' Hospital, London SE1 7EH, United Kingdom

HERBERT R. HALVORSON (16), Department of Pathology, Division of Biochemical Research, Henry Ford Health Sciences Center, Detroit, Michigan 48202

J. A. HELPERN (16), Division of Medical Physics, Nathan S. Kline Institute for Psychiatric Research, Orangeburg, New York 10962

JOHN HOYLAND (6), Laboratory of Neural and Secretory Signalling, Department of Neurobiology, BBSRC Babraham Institute, Babraham, Cambridge CB2 4AT, United Kingdom

STEPHEN J. KARLIK (3), Department of Radiology, University Hospital, London, Ontario, Canada N6A 5A5

RICHARD P. KRAIG (10), Departments of Neurology, and Pharmacological and Physiological Sciences, The University of Chicago, Chicago, Illinois 60637

CHRISTOPHER D. LASCOLA (10), Department of Neurology, The University of Chicago, Chicago, Illinois 60637

P. M. LLEDO (6), Institut Alfred Fessard, CNRS, F-91198 Gif-sur-Yvette, France

ROBERT E. LONDON (15), Laboratory of Molecular Biophysics, National Institute of Environmental Health Sciences, National Institutes of Health, Research Triangle Park, North Carolina 27709

W. T. MASON (6), Laboratory of Neural and Secretory Signalling, Department of Neurobiology, BBSRC Babraham Institute, Babraham, Cambridge CB2 4AT, United Kingdom

MARY E. MORRIS (2), Departments of Anesthesia, Medicine and Pharmacology, University of Ottawa, Ottawa, Ontario, Canada K1H 8M5

THOMAS MUNSCH (14), Institut für Physiologie, Medizinische Fakultät, Universität Magdeburg, D-39120 Magdeburg, Germany

ELIZABETH MURPHY (15), Laboratory of Molecular Biophysics, National Institute of Environmental Health Sciences, National Institutes of Health, Research Triangle Park, North Carolina 27709

OGNEN A. C. PETROFF (11), Department of Neurology, Yale University School of Medicine, New Haven, Connecticut 06510

DOUGLAS R. PFEIFFER (4), Department of Medical Biochemistry, Ohio State University, Columbus, Ohio 43210

JAMES W. PRICHARD (11), Department of Neurology, Yale University School of Medicine, New Haven, Connecticut 06510

MARLI A. ROBERTSON (13), Division of Gastroenterology and Nutrition, The Hospital for Sick Children, Toronto, Ontario, Canada M5G 1X8

GERRY A. SMITH (8), Department of Biochemistry, University of Cambridge, Cambridge CB2 1QW, United Kingdom

D. LANSING TAYLOR (1), Department of Biological Sciences, Center for Light Microscope Imaging and Biotechnology, Carnegie Mellon University, Pittsburgh, Pennsylvania 15213

A. S. VERKMAN (17), Departments of Medicine and Physiology, University of California, San Francisco, San Francisco, California 94143

DAVID A. WILLIAMS (5), Muscle and Cell Physiology Laboratory, Department of Physiology, The University of Melbourne, Parkville, Victoria 3052, Australia

JOHN X. WILSON (9), Department of Physiology, The University of Western Ontario, London, Ontario, Canada N6A 5C1

G. ANDREW WOOLLEY (4), Department of Chemistry, University of Toronto, Toronto, Ontario, Canada M58 1A1

R. ZOREC (6), Institute of Pathophysiology, School of Medicine, University of Ljubljana, Ljubljana, Slovenia

# Preface

Neuroscientists, as do other biologists, recognize the crucial importance of intracellular ions in cellular physiology and pathology. Under resting conditions, cytosolic ion concentrations are closely regulated by intracellular buffers and membrane transport systems. Changes in ion concentrations contribute to the regulation of numerous cell functions, including metabolism, secretion, motility, membrane permeability, cell proliferation, and apoptosis. Phenomena such as long-term potentiation and depression and excitotoxic cell death are thought to be mediated through changes in intracellular ion concentrations. To understand these processes, it is important that experimental biologists monitor changes in intracellular ion levels in response to various physiological and pathological stimuli. Similarly, it is crucial to experimentally manipulate intracellular ion concentrations and observe the resulting effects on cell function.

The purpose of this volume is to provide the reader with expert guidance on practical methods for measuring and manipulating cytosolic ionic activities. The first four chapters provide an overview of the use of optical probes, ion-selective microelectrodes and magnetic resonance spectroscopy for measuring ion levels, and the use of ionophores for manipulating intracellular ion concentrations. These are followed by chapters presenting relevant techniques, grouped according to the ion of interest ($Ca^{2+}$, $H^+$, $Na^+$, $Mg^{2+}$, and $Cl^-$). We have not attempted to include chapters on all available techniques or even all biologically relevant ions. Emphasis has been placed on widely studied ions, such as $Ca^{2+}$, and techniques that can be easily modified to measure related ionic species. The concluding chapter presents methods for the measurement of cell volume, the control of which is intimately linked to regulation of intracellular ion levels. In many cases, the techniques presented can be applied only to *in vitro* preparations; however, where possible, techniques for *in vivo* measurement of ion levels have also been included. Although many of the biological preparations described in this volume are of neural origin, most techniques are readily applicable to other tissues. Therefore, we hope that this volume will be of use to all cell biologists with interests in the role of cytosolic ions in cellular function, regulation, and pathology.

We thank the authors for their excellent contributions. Their enthusiasm and cooperation have made our role as editors a pleasant task. We especially thank Drs. Mary E. Morris and Stephen J. Karlik for their invaluable assistance in editing the chapters on ion-selective microelectrodes and NMR spectroscopy, respectively. Last, we thank Dr. P. Michael Conn, Editor-in-Chief of *Methods in Neurosciences*, and Shirley Light of Academic Press for their assistance and encouragement.

JACOB KRAICER
S. JEFFREY DIXON

# Methods in Neurosciences

Volume 1 Gene Probes
*Edited by P. Michael Conn*

Volume 2 Cell Culture
*Edited by P. Michael Conn*

Volume 3 Quantitative and Qualitative Microscopy
*Edited by P. Michael Conn*

Volume 4 Electrophysiology and Microinjection
*Edited by P. Michael Conn*

Volume 5 Neuropeptide Technology: Gene Expression and Neuropeptide Receptors
*Edited by P. Michael Conn*

Volume 6 Neuropeptide Technology: Synthesis, Assay, Purification, and Processing
*Edited by P. Michael Conn*

Volume 7 Lesions and Transplantation
*Edited by P. Michael Conn*

Volume 8 Neurotoxins
*Edited by P. Michael Conn*

Volume 9 Gene Expression in Neural Tissues
*Edited by P. Michael Conn*

Volume 10 Computers and Computations in the Neurosciences
*Edited by P. Michael Conn*

Volume 11 Receptors: Model Systems and Specific Receptors
*Edited by P. Michael Conn*

Volume 12 Receptors: Molecular Biology, Receptor Subclasses, Localization, and Ligand Design
*Edited by P. Michael Conn*

Volume 13 Neuropeptide Analogs, Conjugates, and Fragments
*Edited by P. Michael Conn*

Volume 14 Paradigms for the Study of Behavior
*Edited by P. Michael Conn*

Volume 15 Photoreceptor Cells
*Edited by Paul A. Hargrave*

Volume 16 Neurobiology of Cytokines (Part A)
*Edited by Errol B. De Souza*

Volume 17 Neurobiology of Cytokines (Part B)
*Edited by Errol B. De Souza*

Volume 18 Lipid Metabolism in Signaling Systems
*Edited by John N. Fain*

# [1] Light-Optical-Based Reagents for the Measurement and Manipulation of Ions, Metabolites, and Macromolecules in Living Cells

Kenneth A. Giuliano and D. Lansing Taylor

The chemical processes that comprise living organisms are the result of the precise orchestration of ions, metabolites, macromolecules, macromolecular assemblies, and organelles in time and space. Regulation of free ion concentration is crucial to processes that include maintenance of membrane potential, control of biochemical energetics, and mediation of signal transduction. Hence, we begin this survey by describing the progress made in the design and use of fluorescent indicators of the major intracellular free ion concentrations. Next, the development of fluorescent analogs of macromolecules, fluorogenic substrates, and protein-based optical biosensors is discussed. Then, we describe how free ion indicators are coupled with other fluorescence-based reagents in the same living cell to provide an understanding of the complex temporal and spatial relationships between molecular processes. Finally, we discuss the progress and the prospects of what we believe to be an exciting application of optical-based reagents—the manipulation of specific molecular events using photomodulated ions, metabolites, and macromolecules. The use of photomodulated reagents to change the chemical and molecular properties of cells coupled with the use of reagents that detect specific chemical and molecular events in living cells will propel us toward the dissection of the myriad of chemical reactions that define cellular physiology. Figure 1 portrays in graphical form how cells can be used as living cuvettes for the measurement and manipulation of ions, metabolites, macromolecules, macromolecular assemblies, and organelles.

## Fluorescent Indicators of Intracellular Ion Concentration: Powerful Tools for Describing Cellular Physiology

### *Rationale for Using Fluorescent Indicators in Living Cells*

Cells contain endogenous fluorescent compounds such as the nicotinamide adenine dinucleotides, the flavins, and the fluorophoric amino acids trypto-

*Methods in Neurosciences, Volume 27*

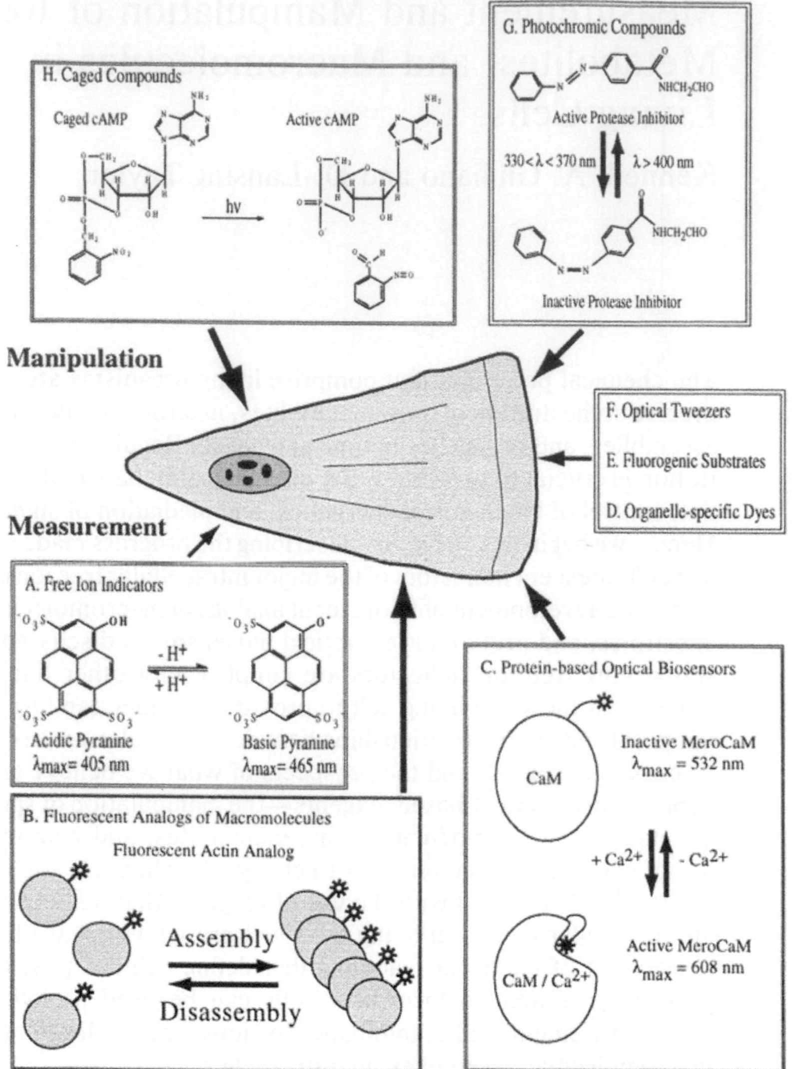


FIG. 1 Cells as living microcuvettes: Measurement and manipulation of chemical and molecular activities in single living cells. The potential to measure and manipulate several specific molecular processes in the same cell is being realized. Optical-based methods for the measurement (A–E) and manipulation (F–H) of cellular components are depicted surrounding a living cell. Examples of several classes of optical reagents are shown. These include (A) pyranine as a fluorescent indicator of $pH_i$ [K. A. Giuliano and R. J. Gillies, *Anal. Biochem.* **167**, 362 (1987)]; (B) fluorescent analogs of actin to measure the dynamics of the actin cytoskeleton [Y. L. Wang and D. L.

phan, tyrosine, and phenylalanine (1, 2). Because endogenous fluorophores are integral components of a vast number of metabolites and macromolecules within the cell, it is difficult to isolate the fluorescence associated with a single molecular process when exciting fluorescence in the ultraviolet (UV) and blue wavelengths. Nevertheless, maps of endogenous NADH fluorescence lifetimes have been described (3). Significant advances have been made by developing probes that fluoresce at the longer wavelengths of visible and near-infrared light (4–6). Once introduced into the appropriate cellular compartment, these reagents report the dynamics of ions, metabolites, and macromolecules in living cells.

## *Advantages of Fluorescence-Based Detection of Ions in Living Cells*

Fluorescence-based detection of ions in living cells is a relatively simple, flexible, and sensitive technique. Living specimens range from a single cell that yields a temporally and spatially resolved map of intracellular ion concentration to a large number of cells whose total fluorescence signal reports the physiological state of the population. The surface layer of cells in a perfused organ system can also be used to monitor the intracellular chemistry of intact tissue (7, 8). Intracellular ion concentrations can be measured using flow cytometry (9–12), spectrofluorometry (13–15), and fluorescence microscopy (16).

Fluorescence-based intracellular ion detection is now characterized by the increasing number of commercially available reagents, the ease with which the reagents can be loaded into and detected within living cells, and the

Taylor, *J. Histochem. Cytochem.* **28,** 1198 (1980); R. L. DeBiasio, L. L. Wang, G. W. Fisher and D. L. Taylor, *J. Cell Biol.* **107,** 2631 (1988)]; (C) a protein-based optical biosensor that detects the activation of calmodulin (CaM) [K. Hahn, R. DeBiasio, and D. L. Taylor, *Nature* (*London*) **359,** 736 (1992); K. M. Hahn, A. S. Waggoner, and D. L. Taylor, *J. Biol. Chem.* **265,** 20335 (1990)]; (D) organelle-specific dyes [R. E. Pagano, *Methods Cell Biol.* **29,** 75 (1989)]; (E) fluorogenic enzyme substrates [B. Rotman, J. A. Zderic, and M. Edelstein, *Proc. Natl. Acad. Sci. U.S.A.* **50,** 1 (1963); Z. Huang, *Biochemistry* **30,** 8535 (1991)]; (F) optical tweezers for the micromanipulation of macromolecules and macromolecular assemblies [S. C. Kuo and M. P. Sheetz, *Science* **260,** 232 (1993); S. M. Block, *Nature* (*London*) **360,** 493 (1992)]; and (H) a photoactivatable caged cAMP molecule for the manipulation of intracellular enzyme cascades [S. R. Adams and R. Y. Tsien, *Annu. Rev. Physiol.* **55,** 755 (1993)]. Details of these methods and reagents are found within the text.

important improvements made in the affinity and selectivity toward specific ionic species. Reagents for measuring intracellular pH ($pH_i$) and intracellular free calcium ion concentration ($pCa_i$) have evolved the most. Early generations of $pH_i$ (15, 17) and $pCa_i$ (18) detection reagents had suboptimal ion dissociation constants for many cytoplasmic measurements or suboptimal fluorescence characteristics that hindered their detection in living cells, or both. Reagents for the measurement of $pH_i$ and $pCa_i$ available today have ion dissociation constants that match closely the concentration of these ions within the cytoplasm of many cells. In addition, many ion reagents are covalently linked to inert carrier macromolecules to aid in their intracellular incorporation, to minimize interaction with intracellular components, and to maximize their retention in the cell (16, 19).

Recent developments include improvements in reagents for measuring $pH_i$ and $pCa_i$, and the introduction of new fluorescent reagents for the measurement of other intracellular ions (20). Reagents with selectivity for the metal ions $Mg^{2+}$ (21), $Na^+$ (22–24), and $K^+$ (23) have been introduced. A fluorescent reagent for the measurement of chloride, an important intracellular metal counterion, is also available (25, 26). The fluorescence measurement of cytosolic $Ca^{2+}$, pH, $Na^+$, $Mg^{2+}$, and $Cl^-$ is discussed in detail in subsequent chapters.

## Methodology for Introducing Fluorescent Indicators into Living Cells, Data Acquisition, and Image Processing

### *Cell Loading*

Fluorescent reagents range in size from low molecular mass fluorescent indicators of intracellular ion concentration to substantially larger, fluorescently labeled macromolecules. Methods for the introduction of these reagents into living cells have evolved during the past 5 years and are well established. Most of the fluorescent ion indicators are available in versions that contain organic acid moieties that reversibly bind either the metal ion of interest or intracellular protons. It is this interaction with the ion of interest that results in a measurable fluorescence change in the reagent. A population of adherent or suspended eukaryotic cells is usually incubated with an esterified form of these reagents that diffuses easily across the plasma membrane into the cytoplasm, where endogenous esterases hydrolyze the reagents back into their active free acid forms, which are much less likely to leak out of (9, 27) or be extruded from (28) the cell. This technique avoids mechanical perturbation of the cells and requires no specialized equipment. Disadvantages include improper compartmentalization of the probes within the cell (19, 29, 30), incomplete ester hydrolysis that can lead to erroneous fluorescence signals (2, 31), changes in $pH_i$ resulting from ester hydrolysis (32), and a measurable leakage of the indicator out of the cell. Therefore, the use of ion indicators coupled to neutral carriers has been recommended whenever possible (19, 29).

Fluorescent reagents that do not easily diffuse across the plasma membrane into the cytoplasm because of size or charge characteristics must be introduced using other methods. Microinjection uses only small amounts of fluorescent reagent and good quantitation of loading is possible (33, 34). However, microinjection also requires specialized equipment and careful manipulation of the cells, and yields relatively small numbers of loaded cells. Methods for loading populations of cells include scrape loading (35), glass-bead loading (36), syringe loading (37), and osmotic lysis (38, 39). These methods result in large numbers of labeled cells, but one needs to use relatively large volumes of concentrated fluorescent reagents, and selection of suitably labeled cells requires additional methods such as flow cytometry (35). Mechanical methods of cell loading have been reviewed in detail (35, 40). Two other membrane-perturbing cell-loading techniques are electroporation (41–44) and cell fusion (45); both are used to load macromolecules into large numbers of cells. Electroporation uses specialized equipment and more of the material to be loaded than microinjection uses. For cell-fusion-based loading, membrane ghosts of one cell type, usually red blood cells, are used to entrap the reagent to be loaded, and then are fused with the target cells (45). After fusion of the loaded carrier particles with living cells, the contents of the carriers are released into the cytoplasm. This leaves the living cell with patches of foreign membrane, with unknown consequences. The cell type to be loaded will usually dictate the loading method to be used. For example, relatively harsh methods (scrape loading or syringe loading) would be unsuitable for neuronal cells containing fragile processes; microinjection into the cell body would be more appropriate.

## Data Acquisition

Fluorescent reagents report the dynamics of ions, metabolites, and macromolecules in living cells through measurable changes in their fluorescence distribution or intensity or shifts in their fluorescence spectra or other spectroscopic parameters, such as fluorescence lifetime, fluorescence anisotropy, or resonance energy transfer. For example, indicators of intracellular ion concentration that change fluorescence intensity at a single wavelength on ion binding report changes in ion concentration instantaneously, but calibration of the signals from these indicators is problematic (46). Fluorescence intensity changes are also dependent on cellular pathlength, leakage of indicator from the cell, and photobleaching of the indicator (2, 19). On the other hand, reagents that exhibit shifts in either their excitation or emission spectra can be used in a fluorescence ratio mode. In this mode, leakage of fluorescent reagents out of cells, changes in cellular pathlength, and photobleaching are corrected for by measuring the ratio of fluorescence intensity at two wavelengths (9, 19, 47). It is also possible to cointroduce, for example, ion-sensitive and -insensitive fluorescence reagents into the same cell and ratio the fluorescence of the two reagents (48). If the cellular distribution, rates of photobleaching, and leakage are similar for each indicator, the ratio of their intensities describes the intracellular ion concentration independent of pathlength, photobleaching, and intracellular indicator concentration. The price one must pay for using any type of fluorescence ratio technique is the overhead of added instrumentation and data processing. A pair of data points must be collected as close together in time as required by the rate of physiological changes. The kinetics of the process to be measured will dictate the speed and precision of the instrumentation. This often means that fluorescence filter sets or monochrometers must be changed rapidly or that two separate detectors, one for each fluorescence wavelength, have to be available.

## *Fluorescence Imaging Instrumentation*

The instrumentation used to measure the dynamics of intracellular fluorescent indicators is based on either the flow cytometer, spectrofluorometer, or fluorescence microscope. The flow cytometer (10–12, 49) and the spectrofluorometer (50, 51) have the potential to measure fluorescence changes in a population of living cells with high time resolution, but give virtually no information about spatially resolved molecular events. High temporal and spatial resolution are obtainable using the light microscope as a microspectrofluorometer and cells as living cuvettes. Fluorescent indicators can be studied in single cells for long periods of time when the cellular environment is properly controlled (19, 52). Furthermore, the automated selection of several modes of light microscopy and computerized acquisition and processing of data have been integrated into a single instrument (53). The multimode microscope has extended the versatility of microspectrofluorometry by allowing one to measure the interplay of several biological processes in a single living cell at one time (54, 55).

## Use of Fluorescent Reagents to Measure Specific Molecular Processes in Living Cells: Entering a New Era

### *Development of New Fluorescence-Based Reagents and Optical Biosensors: Augmenting the Present Generation of Intracellular Free-Ion-Concentration Indicators*

Changes in intracellular free ion concentration usually control a cascade of other molecular events that are likely to include macromolecules and macromolecular assemblies. By measuring the temporal and spatial dynamic distribution of specific macromolecules, cellular events can be dissected into their molecular components (55, 56). Ingenious design has already given us a wide array of fluorescent ion indicators and many fluorescence-based indicators of metabolite and macromolecular dynamics. The introduction of functional fluorescent analogs of macromolecules into living cells, or fluorescent analog cytochemistry, is used to measure the dynamic distribution of specific macromolecules in the presence of thousands of other molecular species (56–59). For example, fluorescent analogs of actin and myosin II (60), calmodulin (48, 61, 62), gelsolin (63), and even the glycolytic enzyme aldolase (64) have been used to describe the role that the dynamic actin/myosin II cytoskeleton plays in the physiology of nonmuscle cells. Furthermore, fluorescent analogs of other proteins (65), lipids (66, 67), and nucleic acids (68) are now available.

The localization of enzymes and their activities within living cells is essential in describing the dynamics of specific molecular processes. To this end, the mobilities of fluorescent analogs of two glycolytic enzymes have been measured in living cells using fluorescence photobleaching and fluorescence ratio imaging (64, 69, 70). The results suggest that there is a dynamic partitioning of some of the enzymes between the solid and fluid phases in the cytoplasm and that this partitioning regulates, in part, glycolysis and most likely other metabolic pathways (71). Other enzyme activities, namely, aminopeptidase (72) and protein kinase C (73), have been detected using new fluorogenic substrates. The development of new fluorogenic substrates and analogs of enzymes will accelerate our knowledge of enzyme dynamics *in vivo*.

Protein-based optical biosensors are designed to sense changes in the environment, conformation, posttranslational modification, or activity of specific proteins in living cells (74). The expanding number of modes of fluorescence microscopy available on modern instruments enhances their use (53). In one case, two fluorescent analogs of calmodulin are used to measure calmodulin activation and binding to target molecules using fluorescence ratio (75) and fluorescence anisotropy (62) techniques. In another case, fluorescence energy transfer is used to measure the cytoplasmic cAMP-dependent dissociation of the catalytic and regulatory subunits of cAMP-dependent protein kinase (76, 77). The reagent therefore senses the presence of intracellular cAMP. Finally, a fluorescent analog of a genetically engineered smooth muscle myosin II is being developed to measure the phosphorylation level of the regulatory light chain (74). This reagent will be used to map the assembly and motor activity of myosin II in living cells whose actin cytoskeleton is known to play an important role in cell motility and physiology (77a).

To begin illustrating other applications of fluorescence-based reagents, an electroluminescent lamp-based phase fluorometer has been designed to detect oxygen concentration (78). Fluorescent indicators of other biologically relevant oxygen-containing species—hydrogen peroxide (79) and superoxide and hydroxyl radicals (80)—have been reported. Using fluorescence lifetime imaging, Lakowicz *et al.* (3) describe the free and bound states of the endogenous fluorophore and mediator of many intracellular oxidation–reduction reactions, NADH. Final in this short survey of metabolic reagents is a fluorescent lipid analog whose fluorescence emission is used to photoactivate a radioactive azide compound that, in turn, reacts with and labels nearby proteins involved in the intracellular transport and metabolism of lipids (81).

Nearly all protein-based fluorescent reagents are prepared by covalently labeling a purified protein with a reactive fluorescent dye and chemically characterizing the activity of the modified protein *in vitro* before it is introduced into a living cell (56). An elegant way to introduce an optically based reagent into a large population would be to take advantage of the cellular genetic machinery and have the cells of interest express the reagent and distribute it within the cytoplasm. Button and Brownstein (82) describe the expression of apoaequorin, a jellyfish protein whose luminescence is triggered by $Ca^{2+}$, in cultured mammalian cells. Transfected cells treated with the coelenterate luciferin, coelenterazine, report $pCa_i$ as a result of the $Ca^{2+}$-induced luminescence of aequorin (82). Another genetically based strategy is to use biosynthetic methods to introduce unnatural fluorescent amino acids site specifically into protein-based reagents (83). In this method, a suppressor tRNA, chemically aminoacylated with a fluorescent amino acid analog, is constructed to recognize an *amber* nonsense codon within the protein of interest during *in vitro* transcription and translation of the protein-based reagent (83). The potential of this technique lies in the ability to label a protein at a specific site with a specific fluorescent dye.

The refinement of fluorescent indicators of intracellular free ion concentration over the last few years coupled with new technology for their detection in living cells played an important role in forming a new vision of light microscopy (55). From these small and relatively simple ion indicators, a new generation of more complex macromolecule-based reagents has arisen. The potential for future generations of these reagents appears to be limited only by our ability to ask the appropriate biochemical and molecular questions.

## *Determination of Several Biochemically Active Species in Single Living Cells Is Now Possible*

Fluorescent indicators of intracellular ions, metabolites, and macromolecules are beginning to fill the spectrum of visible light, so it would be a natural extension to measure the dynamics of two or more species in the same living cell using the appropriately designed indicators. For example, commercially available indicators of $pCa_i$ and $pH_i$ now span a wide range of visible wavelengths, so a logical choice would be to measure $pCa_i$ and $pH_i$ in the same cell with two nonoverlapping fluorescent ion indicators. Indeed, Davies *et al.* (84) and Martinez-Zaguilan *et al.* (51) and have simultaneously measured physiologically induced changes in $pH_i$ and $pCa_i$ in living cells and have found temporal relationships between the two. This suggests that the signaling pathways mediated by two distinct second messengers are coupled at some point along their paths.

Fluorescence-based multicolor measurements in living cells have also been described using labeled macromolecular components of the cytoplasm (85, 86). The potential of this methodology for dissecting intracellular signaling pathways is already being realized. Hahn *et al.* (48), using a fluorescent indicator of $pCa_i$ and a calmodulin-based optical biosensor in the same living cell, showed that calmodulin activation correlates both temporally and spatially with decreased in $pCa_i$, a finding that until now has only been suggested to occur *in vivo* using evidence obtained *in vitro*. Gough and Taylor (62), using fluorescence anisotropy imaging microscopy of another calmodulin-based optical biosensor, went on to show that intracellularly activated calmodulin binds to target proteins, some of which act to reorganize that actin cytoskeleton in the same cell. A myosin II fluorescent analog was incorporated into cells together with the calmodulin analog in this case. These examples point out that the careful design and detection of complementary fluorescence-based reagents will help us define complex interrelationships in living cells.

## Reagents Possessing Photomodulated Activity Complement Existing Fluorescence-Based Reagents

### *Demonstration of Photomodulation of Ions, Metabolites, and Macromolecules in Several Systems*

The history and prospects for biologically relevant photomodulation reagents parallel the development of fluorescence indicators of chemical and molecular activities. We can begin to test mechanisms of live cell function at a new level of sophistication by being able to manipulate specific chemical and molecular activities in time and space.

Just as the early fluorescence-based intracellular indicators were small molecules for the detection of free ion concentrations, a generation of small photomodulated calcium ion and nucleotide reagents appeared first (87). Some of these reagents belong to a class of "caged" calcium ion reagents. These reagents are calcium ion chelators whose ability to bind metal ions is determined by the presence of a blocking molecule bound to the chelator through a covalent yet photolabile bond. Photolysis of the covalent bridge releases the blocking molecule, which can also include part of the chelator, and causes a conformational change in the chelator that can result in a several order increase in the calcium ion dissociation constant (87). Hence, uncaging releases a bolus of free calcium ions at the site of photolysis. The blocking molecules, usually *o*-nitrobenzylic compounds, have been adapted from the wide array of photocleavable functional group protecting reagents designed for organic syntheses (88). Many classes of molecules have been caged and include neurotransmitters, nucleotides, amino acids free in solution and at the active sites of enzymes, and nitric oxide (87, 89). Macromolecules are also beginning to be caged using photocleavable reagents. A critical sulfhydryl group at the site of protein–protein interaction of the actin molecule has been caged, thus permitting the photomodulation of actin monomer assembly into filamentous polymers *in vitro* (90). Caging macromolecules is a challenging yet exciting new area. The potential to turn on enzymes, induce assembly of subunits, and initiate protein synthesis or gene expression will stimulate new chemical approaches to caging, and thus new experimental opportunities.

Another macromolecular photomodulation strategy, called photoactivation of fluorescence, is to cage the fluorescence of a labeled macromolecule, rather than its activity (91). Photoactivation of fluorescence has been used to measure the dynamics of labeled actin and tubulin in living cells (92). Irreversible photocleavage of covalent bonds has therefore become the hallmark of nearly all the biological photomodulation reagents available today.

Photochromic dyes belong to a class of reversibly photomodulated molecules that have recently found much attention as molecular electronic devices for use in optical recording media (93). A number of photochromic dyes have been studied (93) and a few have been applied to the photomodulation of biological macromolecules (94). Photochromic substrates for proteins (95, 96) and enzymes (94, 97, 98) have recently appeared, and these can be reversibly modulated with different wavelengths of light. In one example, a transition-state analog inhibitor of cysteine and serine proteases was rendered photochromic by conjugating it with an azobenzene (98). Irradiation of the substrate with $330 < \lambda < 370$-nm light resulted in a trans → cis isomerization of the azobenzene and a 500% increase in the activity of papain. Irradiation with $\lambda > 400$-nm light converted the inhibitor back into its trans configuration, thus restoring its ability to inhibit the protease. Other potentially useful photomodulation reagents for biological systems are photochromic synthetic polymers (99, 100) and crown ethers (101). Because of their reversible photomodulation and flexible spectral properties, photochromic reagents show great promise for the modulation of intracellular ions, metabolites, and macromolecules.

## *Light-Energy-Based Manipulation of Specific Cellular Processes Coupled with Sensitive Physiological Detection Reagents: Opportunities in Biological Research*

Light energy can be used both to induce a specific event and to measure the effect of the event on the physiology of a single living cell. Using a set of caged calcium ion and inositol triphosphate reagents and a fluorescent indicator of $pCa_i$, Kao *et al.* (102) induced and measured cytosolic calcium and inositol trisphosphate pulses in cultured fibroblasts. This work was possible because the absorption spectra of the reagents were designed so that excitation of the fluorescent calcium ion indicator did not cause photolysis of the calcium ion caging reagent and thus release an unwanted pulse of calcium ions. These experiments demonstrate the potential of an approach that couples photochemistry and photophysics to describe cellular physiology. The realm of optical-based discovery in cell biology will be realized by designing fluorescent detection and photomodulation reagents that span the UV, visible, and near-IR regions and combining them with the recent advances in machine-vision light microscopy. These advances include new types of illuminating light sources (such as laser diodes) that can be used to excite a new generation of fluorescent dyes with one or more photons (87, 103, 104), automated multimode microscopes (53, 54), and high-resolution charge-coupled detection devices (105). The convergence of mechanical, electronic, and computing technologies with biotechnology has sparked the rebirth of light microscopy as a tool for studying intracellular dynamics at the molecular level (55).

## Acknowledgments

This research was supported in part by National Science Foundation Science and Technology Center Grant MCB-8920118, Award 2044BR2 from the Council for Tobacco Research—United States, and IRG-58-34 from the American Cancer Society.

## References

1. D. L. Taylor and E. D. Salmon, *Methods Cell Biol.* **29,** 207 (1989).
2. E. D. W. Moore, P. L. Becker, K. E. Fogarty, D. A. Williams, and F. S. Fay, *Cell Calcium* **11,** 157 (1990).
3. J. R. Lakowicz, H. Szmacinski, K. Nowaczyk, and M. L. Johnson, *Proc. Natl. Acad. Sci. U.S.A.* **89,** 1271 (1992).
4. A. S. Waggoner, *in* "Applications of Fluorescence in the Biomedical Sciences" (D. L. Taylor, A. S. Waggoner, R. F. Murphy, F. Lanni, and R. R. Birge, eds.), p. 3. Alan R. Liss, New York, 1986.
5. R. B. Mujumdar, L. A. Ernst, S. R. Mujumdar, C. J. Lewis, and A. S. Waggoner, *Bioconjugate Chem.* **4,** 105 (1993).
6. P. L. Southwick, L. A. Ernst, E. W. Tauriello, S. R. Parker, R. B. Mujumdar, S. R. Mujumdar, H. A. Clever, and A. S. Waggoner, *Cytometry* **11,** 418 (1990).
7. R. Brandes, V. M. Figueredo, S. A. Camacho, B. M. Massie, and M. W. Weiner, *Am. J. Physiol.* **263,** H972 (1992).
8. S. A. Camacho, V. M. Figueredo, R. Brandes, and M. W. Weiner, *Am. J. Physiol.* **265,** H114 (1993).
9. G. Grynkiewicz, M. Poenie, and R. Y. Tsien, *J. Biol. Chem.* **260,** 3440 (1985).
10. I. Kurtz and R. S. Balaban, *Biophys. J.* **48,** 499 (1985).
11. R. J. Gillies, J. Cook, M. H. Fox, and K. A. Giuliano, *Am. J. Physiol.* **253,** C121 (1987).
12. R. F. Murphy and M. Roederer, *in* "Applications of Fluorescence in the Biomedical Sciences" (D. L. Taylor, A. S. Waggoner, R. F. Murphy, F. Lanni, and R. R. Birge, eds.), p. 545. Alan R. Liss, New York, 1986.
13. K. A. Giuliano and R. J. Gillies, *Anal. Biochem.* **167,** 362 (1987).
14. R. Y. Tsien, *Methods Cell Biol.* **30,** 127 (1989).
15. J. A. Thomas, P. C. Kolbeck, and T. A. Langworthy, *in* "Intracellular pH: Its Measurement, Regulation, and Utilization in Cellular Functions" (R. Nuccitelli and D. W. Deamer, eds.), p. 105. Alan R. Liss, New York, 1982.
16. G. R. Bright, G. W. Fisher, J. Rogowska, and D. L. Taylor, *Methods Cell Biol.* **30,** 157 (1989).

17. J. M. Heiple and D. L. Taylor, *in* "Intracellular pH: Its Measurement, Regulation, and Utilization in Cellular Functions" (R. Nuccitelli and D. W. Deamer, eds.), p. 21. Alan R. Liss, New York, 1982.
18. P. H. Cobbold and T. J. Rink, *Biochem. J.* **248,** 313 (1987).
19. G. R. Bright, G. W. Fisher, J. Rogowska, and D. L. Taylor, *J. Cell Biol.* **104,** 1019 (1987).
20. R. Haugland, *in* "Fluorescent and Luminescent Probes for Biological Activity" (W. T. Mason, ed.), p. 34. Academic Press, San Diego, 1993.
21. B. Raju, E. Murphy, L. A. Levy, R. D. Hall, and R. E. London, *Am. J. Physiol.* **256,** C540 (1989).
22. S. Kongasamut and D. A. Nachshen, *Biochim. Biophys. Acta* **940,** 241 (1988).
23. A. Minta and R. Y. Tsien, *J. Biol. Chem.* **264,** 19449 (1989).
24. A. T. Harootunian, J. P. Y. Kao, B. K. Eckert, and R. Y. Tsien, *J. Biol. Chem.* **264,** 19458 (1989).
25. N. P. Illsley and A. S. Verkman, *Biochemistry* **26,** 1215 (1987).
26. R. Krapf, C. A. Berry, and A. S. Verkman, *Biophys. J.* **53,** 955 (1988).
27. J. A. Thomas, R. N. Buchsbaum, A. Zimniak, and E. Racker, *Biochemistry* **18,** 2210 (1979).
28. L. Homolya, Z. Hollo, U. A. Germann, I. Pastan, M. M. Bottesman, and B. Sarkadi, *J. Biol. Chem.* **268,** 21493 (1993).
29. G. R. Bright, J. E. Whitaker, R. P. Haugland, and D. L. Taylor, *J. Cell. Physiol.* **141,** 410 (1989).
30. J. Hoyland, *in* "Fluorescent and Luminescent Probes for Biological Activity" (W. T. Mason, ed.), p. 223. Academic Press, San Diego, 1993.
31. M. W. Roe, J. J. Lemasters, and B. Herman, *Cell Calcium* **11,** 63 (1990).
32. D. C. Spray, J. Nerbonne, A. Campos De Carvalho, A. L. Harris, and M. V. L. Bennett, *J. Cell Biol.* **99,** 174 (1984).
33. P. A. Amato, E. R. Unanue, and D. L. Taylor, *J. Cell Biol.* **96,** 750 (1983).
34. P. McNeil, R. F. Murphy, F. Lanni, and D. L. Taylor, *J. Cell Biol.* **98,** 1556 (1984).
35. P. L. McNeil, *Methods Cell Biol.* **29,** 153 (1989).
36. P. L. McNeil and E. Warder, *J. Cell Sci.* **88,** 669 (1987).
37. M. S. F. Clarke and P. L. McNeil, *J. Cell Sci.* **102,** 533 (1992).
38. C. Y. Okada and M. Rechsteiner, *Cell (Cambridge, Mass.)* **29,** 33 (1982).
39. P. Rothenberg, L. Glaser, P. Schlesinger, and D. Cassel, *J. Biol. Chem.* **258,** 12644 (1983).
40. W. D. Richardson, *J. Cell Sci.* **91,** 319 (1988).
41. L. M. Mir, H. Banoun, and C. Paoletti, *Exp. Cell Res.* **175,** 15 (1988).
42. H. Lambert, R. Pankov, J. Gauthier, and R. Hancock, *Biochem. Cell Biol.* **68,** 729 (1990).
43. D. C. Chang and T. S. Reese, *Biophys. J.* **58,** 1 (1990).
44. J. C. Weaver, *J. Cell. Biochem.* **51,** 426 (1993).
45. T. Uchida, *Exp. Cell Res.* **178,** 1 (1988).
46. D. A. Williams and F. S. Fay, *Cell Calcium* **11,** 75 (1990).
47. L. Tanasugarn, P. McNeil, G. T. Reynolds, and D. L. Taylor, *J. Cell Biol.* **98,** 717 (1984).
48. K. Hahn, R. DeBiasio, and D. L. Taylor, *Nature (London)* **359,** 736 (1992).

49. T. M. Chused, H. A. Wilson, B. E. Seligmann, and R. Tsien, *in* "Applications of Fluorescence in the Biomedical Sciences" (D. L. Taylor, A. S. Waggoner, R. F. Murphy, F. Lanni, and R. R. Birge, eds.), p. 531. Alan R. Liss, New York, 1986.
50. S. Ohkuma and B. Poole, *Proc. Natl. Acad. Sci. U.S.A.* **75,** 3327 (1978).
51. R. Martinez-Zaguilan, G. M. Martinez, F. Lattanzio, and R. J. Gillies, *Am. J. Physiol.* **260,** C297 (1991).
52. N. M. McKenna and Y.-L. Wang, *Methods Cell Biol.* **29,** 195 (1989).
53. D. L. Farkas, G. Baxter, R. L. DeBiasio, A. Gough, M. A. Nederlof, D. Pane, J. Pane, D. R. Patek, K. W. Ryan, and D. L. Taylor, *Annu. Rev. Physiol.* **55,** 785 (1993).
54. K. A. Giuliano, M. A. Nederlof, R. DeBiasio, F. Lanni, A. S. Waggoner, and D. L. Taylor, *in* "Optical Microscopy for Biology" (B. Herman and K. Jacobson, eds.), p. 543. Wiley-Liss, New York, 1990.
55. D. L. Taylor, M. A. Nederlof, F. Lanni, and A. S. Waggoner, *Am. Sci.* **80,** 322 (1992).
56. Y. L. Wang and D. L. Taylor, *J. Histochem. Cytochem.* **28,** 1198 (1980).
57. Y.-L. Wang, *Methods Cell Biol.* **29,** 1 (1989).
58. D. L. Taylor, P. A. Amato, K. Luby-Phelps, and P. McNeil, *Trends Biochem. Sci.* **9,** 88 (1984).
59. D. L. Taylor and Y. L. Wang, *Proc. Natl. Acad. Sci. U.S.A.* **75,** 857 (1978).
60. R. L. DeBiasio, L. L. Wang, G. W. Fisher, and D. L. Taylor, *J. Cell Biol.* **107,** 2631 (1988).
61. K. Luby-Phelps, F. Lanni, and D. L. Taylor, *J. Cell Biol.* **101,** 1245 (1985).
62. A. H. Gough and D. L. Taylor, *J. Cell Biol.* **121,** 1095 (1993).
63. J. A. Cooper, D. J. Loftus, C. Frieden, J. Bryan, and E. L. Elson, *J. Cell Biol.* **106,** 1229 (1988).
64. L. Pagliaro and D. L. Taylor, *J. Cell Biol.* **107,** 981 (1988).
65. E. D. Salmon and P. Wadsworth, *in* "Applications of Fluorescence in the Biomedical Sciences" (D. L. Taylor, A. S. Waggoner, R. F. Murphy, F. Lanni, and R. R. Birge, eds.), p. 377. Alan R. Liss, New York, 1986.
66. R. E. Pagano and O. C. Martin, *Biochemistry* **27,** 4439 (1988).
67. R. E. Pagano, *Methods Cell Biol.* **29,** 75 (1989).
68. K. Ainger, D. Avossa, F. Morgan, S. J. Hill, C. Barry, E. Barbaese, and J. H. Carson, *J. Cell Biol.* **123,** 431 (1993).
69. L. Pagliaro, K. Kerr, and D. L. Taylor, *J. Cell Biol.* **94,** 333 (1989).
70. L. Pagliaro and D. L. Taylor, *J. Cell Biol.* **118,** 859 (1992).
71. L. Pagliaro, *News Physiol. Sci.* **8,** 219 (1993).
72. S. F. Bronk, S. P. Powers, and G. J. Gores, *Anal. Biochem.* **210,** 219 (1993).
73. C. S. Chen and M. Poenie, *J. Biol. Chem.* **268,** 15812 (1993).
74. K. Hahn, J. Kolega, J. Montibeller, R. DeBiasio, P. Post, J. Myers, and D. L. Taylor, *in* "Fluorescent and Luminescent Probes for Biological Activity" (W. T. Mason, ed.), p. 349. Academic Press, San Diego, 1993.
75. K. M. Hahn, A. S. Waggoner, and D. L. Taylor, *J. Biol. Chem.* **265,** 20335 (1990).
76. S. R. Adams, A. T. Harootunian, Y. J. Buechler, S. S. Taylor, and R. Y. Tsien, *Nature (London)* **349,** 694 (1991).

77. P. J. Sammak, S. R. Adams, A. T. Harootunian, M. Schliwa, and R. Y. Tsien, *J. Cell Biol.* **117,** 57 (1992).
77a. P. L. Post, K. M. Trybus, and D. L. Taylor, *J. Biol. Chem.* **269,** 12880 (1994).
78. K. W. Berndt and J. R. Lakowicz, *Anal. Biochem.* **201,** 319 (1992).
79. J. A. Royall and H. Ischiropoulos, *Arch. Biochem. Biophys.* **302,** 348 (1993).
80. S. Pou, Y. I. Huang, A. Bhan, V. S. Bhadti, R. S. Hosmane, S. Y. Wu, G. L. Cao, and G. M. Rosen, *Anal. Biochem.* **212,** 85 (1993).
81. A. G. Rosenwald, R. E. Pagano, and Y. Raviv, *J. Biol. Chem.* **266,** 9814 (1991).
82. D. Button and M. Brownstein, *Cell Calcium* **14,** 663 (1993).
83. J. Ellman, D. Mendel, S. Anthony-Cahill, C. J. Noren, and P. G. Schultz, *in* "Methods in Enzymology" (J. J. Langone, ed.), Vol. 202, p. 301. Academic Press, San Diego, 1991.
84. T. A. Davies, G. J. Weil, and E. R. Simons, *J. Biol. Chem.* **265,** 11522 (1990).
85. A. Waggoner, R. DeBiasio, R. Conrad, G. R. Bright, L. Ernst, K. Ryan, M. Nederlof, and D. Taylor, *Methods Cell Biol.* **30,** 449 (1989).
86. R. DeBiasio, G. R. Bright, L. A. Ernst, A. S. Waggoner, and D. L. Taylor, *J. Cell Biol.* **105,** 1613 (1987).
87. S. R. Adams and R. Y. Tsien, *Annu. Rev. Physiol.* **55,** 755 (1993).
88. V. N. R. Pillai, *in* "Organic Photochemistry" (A. Padwa, ed.), Vol. 9, p. 225. Dekker, New York, 1987.
89. W. Chamulitrat, S. J. Jordan, R. P. Mason, K. Saito, and R. G. Cutler, *J. Biol. Chem.* **268,** 11520 (1993).
90. G. Marriott, H. Miyata, and K. Kinosita, Jr., *Biochem. Int.* **26,** 943 (1992).
91. B. R. Ware, L. J. Brvenik, R. T. Cummings, R. H. Furukawa, and G. A. Krafft, *in* "Applications of Fluorescence in the Biomedical Sciences" (D. L. Taylor, A. S. Waggoner, R. F. Murphy, F. Lanni, and R. R. Birge, eds.), p. 141. Alan R. Liss, New York, 1986.
92. K. E. Sawin, J. A. Theriot, and T. J. Mitchison, *in* "Fluorescent and Luminescent Probes for Biological Activity" (W. T. Mason, ed.), p. 405. Academic Press, San Diego, 1993.
93. J. Seto, *in* "Infrared Absorbing Dyes," (M. Matsuoka, ed.), p. 71. Plenum, New York, 1990.
94. N. H. Wasserman and B. F. Erlanger, *in* "Molecular Models of Photoresponsiveness" (G. Montagnoli and B. F. Erlanger, eds.), Vol. 68, p. 269. Plenum, New York, 1982.
95. I. Willner, S. Rubin, and Y. Cohen, *J. Am. Chem. Soc.* **115,** 4927 (1993).
96. I. Willner, S. Rubin, J. Wonner, F. Effenberger, and P. Bauerle, *J. Am. Chem. Soc.* **114,** 3150 (1992).
97. I. Willner, S. Rubin, and T. Xor, *J. Am. Chem. Soc.* **113,** 4013 (1991).
98. P. R. Westmark, J. P. Kelly, and B. D. Smith, *J. Am. Chem. Soc.* **115,** 3416 (1993).
99. M. Irie, *in* "Molecular Models of Photoresponsiveness" (G. Montagnoli and B. F. Erlanger, eds.), Vol. 68, p. 291. Plenum, New York, 1982.
100. A. Fissi, O. Pieroni, F. Ciardelli, D. Fabbri, G. Ruggeri, and K. Umezawa, *Biopolymers* **33,** 1505 (1993).

101. S. Shinkai, *in* "Molecular Models of Photoresponsiveness" (G. Montagnoli and B. F. Erlanger, eds.), Vol. 68, p. 325. Plenum, New York, 1982.
102. J. P. Y. Kao, A. T. Harootunian, and R. Y. Tsien, *J. Biol. Chem.* **264,** 8179 (1989).
103. G. Patonay and M. D. Antoine, *Anal. Chem.* **63,** 321A (1991).
104. T. Imasaka, A. Tsukamoto, and N. Ishibashi, *Anal. Chem.* **61,** 2285 (1989).
105. R. Aikens, *in* "Fluorescent and Luminescent Probes for Biological Activity" (W. T. Mason, ed.), p. 277. Academic Press, San Diego, 1993.

# [2] Use of Ion-Selective Microelectrodes for Recording Intracellular Ion Levels

Mary E. Morris

## Introduction

The earliest measurements of intracellular ion levels used solid glass ion-sensitive electrodes (tip size, >100 $\mu$m) to record pH in crab muscle fibers and squid giant fibers (1). Subsequently smaller electrodes (tip size, ≥20 $\mu$m) were designed and used by Hinke in the first measurements of intracellular $Na^+$ and $K^+$, which were made in crab and lobster muscle (2). A few years later much smaller electrodes (≤1-$\mu$m tips) were developed by Lev and Bushinsky (3) and were used to record pH. $Na^+$, and $K^+$ in frog sartorius muscle and neurons of giant molluscs (4, 5). Although these electrodes were highly accurate, their fabrication required considerable technical expertise and they could be used only for measurements inside relatively large cells. When the technique was later modified by Thomas (6, 7) so that the ion-sensitive glass, instead of being exposed at the tip, was protected by being inserted into an insulating insensitive glass capillary (Fig. 1), measurements could then be made in much smaller cells.

However, when glass capillary microelectrodes were designed, with tips (≤1 m) containing a liquid ion exchanger membrane (8, 9), the fabrication and use of microelectrodes to measure both extracellular and intracellular ion levels became much more possible and available. Since that time techniques of construction have been improved and standardized in a variety of ways, multiple electrode assemblies have been combined and greatly miniaturized (10), and a large number of highly selective ion membrane phases that allow the measurement of $Na^+$, $K^+$, $H^+$, $Li^+$, $NH_4{}^+$, $Ca^{2+}$, $Mg^{2+}$, $Cl^-$, and other ions have been synthesized (11). The first intracellular ion recordings with liquid ion-selective microelectrodes were made in *Aplysia*—measurements of intracellular $K^+$ ($K_i^+$) and $Cl^-$ ($Cl_i^-$), which elegantly demonstrated how different levels of $Cl^-$ activity and the $Cl^-$ gradient determined whether cells responded to $CO_2$ by depolarization or hyperpolarization (12). However, later recordings were successfully made in *in vitro* preparations of amphibian dorsal root ganglion cells (13), spinal motoneurons (14, 15), sympathetic ganglion neurons (16), as well as in other tissue—skeletal, cardiac, and epithelial. The earliest successes with intracellular $Ca^{2+}$ ($Ca_i^{2+}$) measurements were with electrodes having fairly large tips in neurons

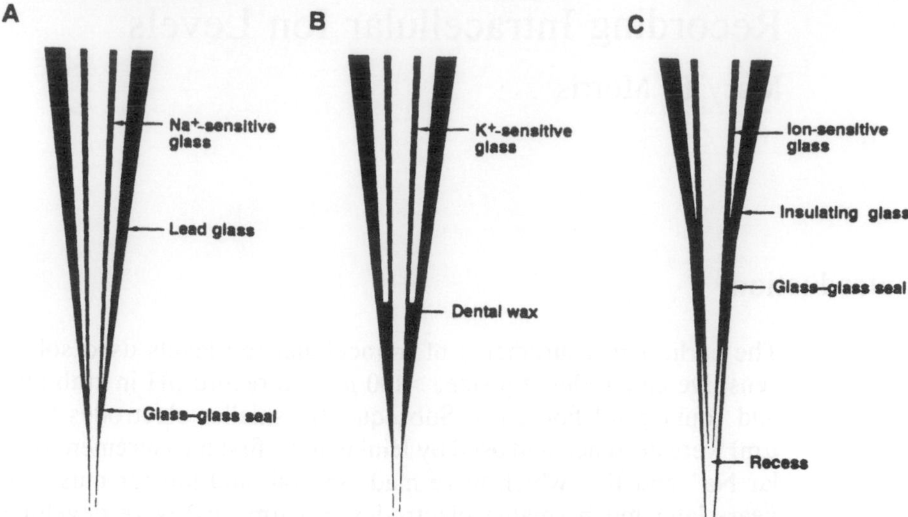


FIG. 1 Terminal portions of ion-sensitive glass membrane electrodes. (A and B) Protruding pointed tips (≈20 μm), with glass insulation of $Na^+$ and $K^+$ measuring electrodes designed by Hinke (2). (C) Recessed-tip design of Thomas (6). [Redrawn from Thomas (7) with permission.]

of *Aplysia* (17, 18) and *Helix pomatia* (19). Although few *in vivo* measurements of intracellular levels in mammalian neurons were attempted, in spite of the difficulties of constructing small enough dual-channel $Ca^{2+}$ electrodes with sufficient sensitivity in the low measuring range, stimulus-evoked increases in $Ca^{2+}$ were successfully measured in spinal motoneurons (20) (Fig. 2A) and with more limited success in the hippocampus (21) (Fig. 2B). More recently, exceptionally successful measurements of $Ca_i^{2+}$ and other ions in mammalian cortical, hippocampal, and thalamic neurons have been made by Silver and Erecińska (22, 23).

## Membrane Phase Sensors

The ion-selective sensors currently used in microelectrodes are either classical ion exchangers (e.g., Corning 477317 $K^+$ and 477315 $Cl^-$ compounds) or the more recently developed and commonly used neutral carriers [e.g., valinomycin $K^+$ sensor (Fluka 60398), ETH 1001 (Fluka 21048), and ETH 129 (Fluka 21196) for $Ca^{2+}$], which have been designed in the Laboratory of Organic Chemistry, Eidgenossiche Technische Höchschüle in Zürich (11, 24) and are available from Fluka Chemie AG (CH-9470 Buchs, Switzerland;

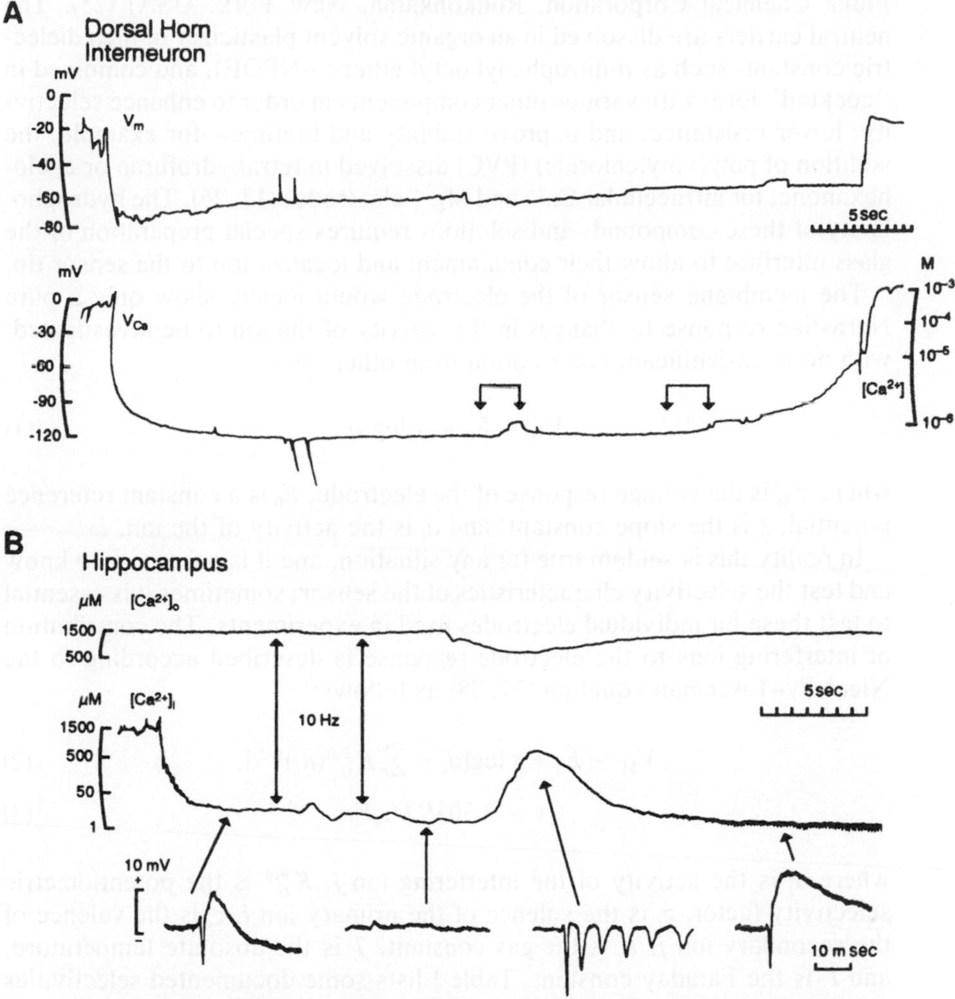


FIG. 2 Intracellular recordings of $Ca^{2+}$ *in vivo* in mammalian neurons. (A) Interneuron in dorsal horn of cat spinal cord: membrane potential ($V_m$) and calcium signal ($V_{Ca}$); arrows mark period of 20-Hz stimulation of dorsal root. [Redrawn and reprinted from *Neuroscience* **14,** Morris *et al.* (20), with kind permission from Elsevier Science Ltd., The Boulevard, Langford Lane, Kidlington OX5 1GB, UK]. (B) Pyramidal neuron in rat hippocampus: traces of extracellular ($[Ca^{2+}]_o$) and intracellular ($[Ca^{2+}]_i$) levels of $[Ca^{2+}]$. A large delayed increase in $[Ca^{2+}]_i$ and a decrease in $[Ca^{2+}]_o$ following a 10-Hz fimbrial stimulation were associated with repetitive bursts of population spikes (illustrated in oscilloscope traces below). [Redrawn from Krnjević *et al.* (21), with permission.]

Fluka Chemical Corporation, Ronkonkama, New York, USA) (25). The neutral carriers are dissolved in an organic solvent plasticizer of high dielectric constant, such as *o*-nitrophenyl octyl ether (*o*-NPOE), and combined in "cocktail" form with various other components in order to enhance selectivity, lower resistance, and improve stability and lifetime—for example, the addition of poly(vinylchloride) (PVC) dissolved in tetrahydrofuran or cyclohexanone, for intracellular $Ca^{2+}$ and $Mg^{2+}$ electrodes (17, 26). The hydrophobicity of these compounds and solutions requires special preparation of the glass interface to allow their containment and localization to the sensor tip.

The membrane sensor of the electrode would ideally show only a pure Nernstian response to changes in the activity of the ion to be investigated, with no or insignificant contribution from other ions:

$$V_M = E_0 + s \log a_i, \tag{1}$$

where $V_M$ is the voltage response of the electrode, $E_0$ is a constant reference potential, $s$ is the slope constant, and $a_i$ is the activity of the ion, $i$.

In reality this is seldom true for any situation, and it is necessary to know and test the selectivity characteristics of the sensor; sometimes it is essential to test these for individual electrodes used in experiments. The contribution of interfering ions to the electrode response is described according to the Nicolsky–Eisenman equation (27, 28) as follows:

$$V_M = E_0 + s \log[a_i + \sum K_{ij}^{\mathrm{Pot}}(a_j)^{z_i/z_j}], \tag{2}$$

$$s = 2.303RT/z_iF, \tag{3}$$

where $a_j$ is the activity of the interfering ion $j$, $K_{ij}^{\mathrm{Pot}}$ is the potentiometric selectivity factor, $z_i$ is the valence of the primary ion $i$, $z_j$ is the valence of the secondary ion $j$, $R$ is the gas constant, $T$ is the absolute temperature, and $F$ is the Faraday constant. Table I lists some documented selectivities for different interfering ions for a number of different available ionophores (11, 26, 29–37).

## Selectivity Measurement

Two methods are commonly used to measure electrode sensitivity—either the separate solution or the fixed interference technique (11). In the separate solution method (SSM), electrode responses to single-electrolyte solutions of the primary ion *i*—usually 0.1 *M* of a salt (e.g., chloride for cations)—and the interfering ion *j* are separately measured. This is the simplest method

TABLE I Selectivity Factors ($\log K_{ij}^{\mathrm{Pot}}$) of Microelectrodes

| Ion ($i^{z_i}$) | Membrane[a] (wt%) | | $\log K_{ij}^{\mathrm{Pot}}$ for $j^{z_j}$ | | | | | | Ref. |
|---|---|---|---|---|---|---|---|---|---|
| | | | $Na^+$ | $K^+$ | $H^+$ | $Li^+$ | $Mg^{2+}$ | $Ca^{2+}$ | |
| $Na^+$ | 10.0% | ETH 227 | 0 | −1.7 | −0.9 | 0.4 | −2.7 | −0.7 | Steiner *et al.* (29) |
| | 0.5% | NaTPB | | | | | | | |
| | 89.5% | *o*-NPOE | | | | | | | |
| $K^+$ | 5.0% | Valinomycin | −3.9 | 0 | — | −4.2 | −4.6 | −4.9 | Ammann *et al.* (30) |
| | 93.0% | 1,2-DM3NB | | | | | | | |
| | 2.0% | K-tet-4CPB | | | | | | | |
| $H^+$ | 6.0% | ETH 1907 | −9.6[b] | −8.8[b] | 0 | — | — | — | Chao *et al.* (31) |
| | 93.0% | *o*-NPOE | | | | | | | |
| | 1.0% | K-tet-4CPB | | | | | | | |
| $Li^+$ | 0.5% | Li ionophore V (Fluka) | −0.9[b] | −0.2[b] | 0.3[b] | 0 | −3.8[b] | −3.5[b] | Wilson *et al.* (32) |
| | 69.0% | TIN-TM | | | | | | | |
| | 30.5% | PVC | | | | | | | |
| $Li^+$ | 9.7% | ETH 149 | −1.3 | — | −0.5 | 0 | −2.7 | −0.6 | Ammann (11); Thomas *et al.* (33); Oehme (34) |
| | 85.5% | TEHP | | | | | | | |
| | 4.8% | NaTPB | | | | | | | |
| $Mg^{2+}$ | 10.0% | ETH 7025 | −3.2[c] | −2.7[c] | −2.3[c] | −3.4[c] | 0 | 0.8[c] | Schaller *et al.* (26) |
| | 80.4% | *o*-NPOE | | | | | | | |
| | 8.6% | K-tet-4CPB | | | | | | | |
| | 1.0% | ETH 500 | | | | | | | |
| $Ca^{2+}$ | 10.0% | ETH 1001 | −5.5[b,d] | −5.4[b,d] | −0.1[b,d] | — | −4.9[b,d] | 0 | Oehme *et al.* (35); Lanter *et al.* (36) |
| | 89.0% | *o*-NPOE | | | | | | | |
| | 1.0% | NaTPB | | | | | | | |
| $Ca^{2+}$ | 1.0% | ETH 129 | −3.7 | −4.0 | −1.6 | −3.3 | −4.9 | 0 | Schefer *et al.* (37) |
| | 65.6% | *o*-NPOE | | | | | | | |
| | 0.6% | K-tet-4CPB | | | | | | | |
| | 32.8% | PVC | | | | | | | |

[a] NaTPB, Sodium tetraphenyl borate; *o*-NPOE, 2-nitrophenyl octyl ether; 1-2-DM3NB, 1,2-dimethyl-3-nitrobenzene; K-tet-4CPB, potassium tetrakis (4-chlorophenyl) borate; TIN-TM, triisononyl trimellitate; PVC, poly(vinylchloride), high molecular weight; TEHP, tris(2-ethylhexyl) phosphate.

[b] Fixed interference method (FIM); all other data in this table were obtained using the separate solution method.

[c] Data from tests using microelectrodes containing 88% $Mg^{2+}$ ionophore cocktail/12% PVC.

[d] Data from tests using microelectrodes containing 86% $Ca^{2+}$ ionophore cocktail/14% PVC.

and is especially useful when a high degree of selectivity is present. With this method, the selectivity coefficient is calculated from Eq. (4).

$$K_{ij}^{\mathrm{Pot}} = 10^{(V_j - V_i)/s}[a_i/(a_j)^{z_i/z_j}], \tag{4}$$

where $V_j$ and $V_i$ are the measured voltage responses for single-electrolyte solutions, $a_j$ and $a_i$ are the activities and $z_j$ and $z_i$ are the valences for secondary (interfering) and primary ions, $j$ and $i$, respectively.

In the fixed interference method (FIM), measurements are made of electrode responses to varying activity and concentration of the primary ion $i$ in solutions containing a fixed background concentration of the interfering ion $j$. This method usually yields higher values of interference than the SSM and is the most highly recommended method (38). Figure 3A illustrates the graphical application of the FIM method, which allows description of the detection limit and its utilization in the calculation of the selectivity coefficient (11, 39). The observed voltage responses are plotted for different activities of the primary ion $i$. The point of intersection of the extrapolated linear parts of the curve, which correspond to $V_i$ and $V_j$, gives the value of $a_i$, which is in fact the practical limit of detection, and because here $V_i = V_j$, can be used to calculate the selectivity factor as follows:

$$K_{ij}^{\mathrm{Pot}} = a_i/(a_j)z_i/z_j, \tag{5}$$

or alternatively written,

$$K_{ij}^{\mathrm{Pot}} = a_i^{\text{detection limit}}/(a_j^{\text{background}})^{z_i/z_j}. \tag{6}$$

A third technique—the fixed primary method (FPM), which is essentially the reverse of the FIM and measures responses to a fixed activity and concentration of the primary ion in the presence of varying levels of the interfering ion—is used less often (11). This method is often employed to determine stability under varying pH conditions, or to test for the effects of anticipated changes in other interfering ions that may occur under experimental conditions (as in the example of Fig. 3B, which shows the voltage change of a $Ca^{2+}$ electrode in the absence of $Ca^{2+}$ in response to an increase in $Na^+$ concentration comparable to that predicted to occur during tetanic activation of a neuron). Whatever technique is used, it is important to state the exact methods used.

The level of acceptable interference from a secondary ion is usually arbitrarily defined as 1%. The following equation describes the limits of the selectivity factors (11), with the assumption that typical activity or concentration is being measured, and can be used to calculate the highest tolerable selectivity coefficient $K_{ij,\max}^{\mathrm{Pot}}$:

$$K_{ij,\max}^{\mathrm{Pot}} = [a_{i,\min}/(a_{j,\max})^{z_i/z_j}](p_{ij}/100), \tag{7}$$

where $a_{i,\min}$ is the lowest expected activity for the ion $i$ to be measured, $a_{j,\max}$

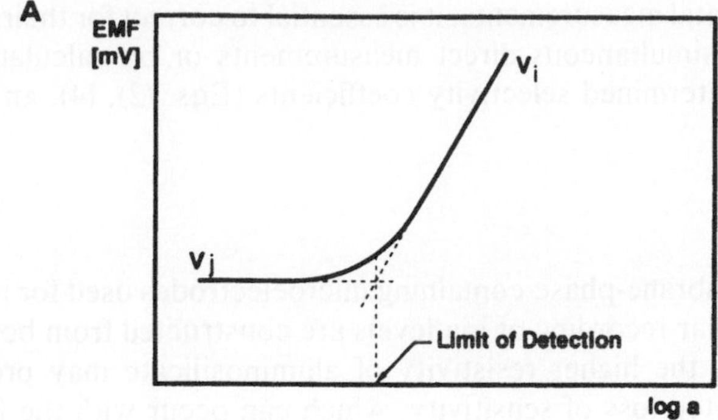


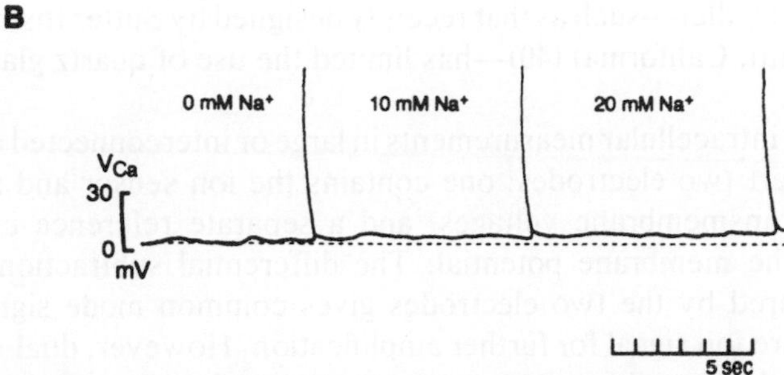


FIG. 3 Selectivity tests. (A) Fixed interference method (FIM): a graph of the data plots the voltage response (electromotive force; EMF) of the microelectrode to solutions of varying activities (log *a*) of the primary ion *i* in the presence of a fixed background of activity for the secondary ion *j*. The point of intersection of extrapolation of the linear sectors $V_i$ (with high sensitivity to the primary ion) and $V_j$ (where there is no sensitivity) provides a value for the limit of detection. [Redrawn after Simon *et al.* (39), with permission.] (B) Fixed primary method (FPM): responses to varying activities of secondary ion *j* in the presence of a fixed value for the primary ion *i*; example shows responses of a $Ca^{2+}$-selective microelectrode to changes in $[Na^+]$ in Ringer, in the absence of $Ca^{2+}$. [Redrawn from Krnjević *et al.* (21), with permission.]

is the highest expected activity for the interfering ion *j*, and $p_{ij}$ is the highest tolerable error in the activity of $a_i$ caused by $a_j$—expressed as a percentage. In practical terms this means that when the calculated value of selectivity equals that measured, for even the worst case the error in $a_i$ due to $a_j$ is less than $p_{ij}$.

When important changes in interfering ions are known to occur during experimental measurements, it is essential to correct for their influence either by using simultaneous direct measurements or by calculations using previously determined selectivity coefficients [Eqs. (2), (4), and (5); Table I].

## *Glass Capillaries*

Most membrane-phase-containing microelectrodes used for intracellular and extracellular recording of ion levels are constructed from borosilicate glass. However, the higher resistivity of aluminosilicate may prevent electrical shunt-related loss of sensitivity, which can occur with the fine tips needed for penetration and recording inside small cells (40). Although quartz glass also has useful characteristics (hardness, resistivity) that might make electrodes especially suitable for intracellular recording, the requirement of a special puller—such as that recently designed by Sutter Instrument Company (Novato, California) (40)—has limited the use of quartz glass to few laboratories.

For intracellular measurements in large or interconnected cells it is possible to insert two electrodes; one contains the ion sensor and records both ion and transmembrane voltages, and a separate reference electrode records only the membrane potential. The differential subtraction of the voltages measured by the two electrodes gives common mode signal rejection and the pure ion signal for further amplification. However, dual-channel or multichannel electrodes are more commonly used and are essential for measurements in small cells. Two types of glass tubing are used to construct these electrodes: (1) separate thick-walled capillaries, which are fused or glued (or otherwise attached) and rotated 180° in the first stage of heating and pulling, or (2) thick septum theta tubing. For both sensitivity and accuracy of measurements the presence of an adequate interface between the channels is important to maintain proper separation and electrical insulation (40). Figure 4 (A and B) shows examples of tips of micropipettes pulled from thick septum theta glass (R & D Scientific Glass Co., Spencerville, Maryland) (in Fig. 4B, broken back to $\approx 3\ \mu m$). Figure 4C shows the tip of a fused dual-capillary electrode; Fig. 4D illustrates the intact septal separation of a fine-tipped electrode made from thick septum theta glass (Glass Company of America, supplied through World Precision Instruments, Inc., Sarasota, Florida). It is important not only to have separation of the channels at the tip but at the top end of the electrode as well. This is achieved either by using capillaries of different lengths or by breaking back the reference barrel with forceps or by drilling (40, 41). A coat of insulating material (e.g., Antispread M 2/200; Etsyntha Chemie, Horb-Ahldorf, Germany) can be usefully added, and attention must be paid to careful sealing with wax.

Although, for extracellular recording of ion levels and field or focal potentials in tissue, dual-channel electrodes that are pulled or broken back to tips $\geq 1\ \mu m$ show excellent sensitivity and stability, the smaller tip size needed for intracellular recording can introduce technical difficulties, requiring solution and compromise and affecting success and yield. Figure 5 illustrates the different pathways of electrical resistance at the electrode tip that contribute to and can shunt and attenuate the voltage signal of the membrane sensor. Significant interference does not occur when the shunt resistances through the glass wall ($R_{g1}$ and $R_{g2}$) and along the glass surface ($R_s$) are large in comparison to the resistance of the membrane phase ($R_e$). $R_e$ must be $>0.01 R_s$ (or $R_g$) to avoid interference (42). Slope values become increasingly sub-Nernstian as $R_s/R_e$ and $R_g/R_e$ ratios decrease, whereas $K_{ij}$ [Eq. (2)] becomes larger as $R_g/R_e$ diminishes: neither sensitivity nor selectivity is constant. Approaches to solve shunt problems include lowering of $R_e$ by beveling or breaking back the tip, optimization of silanization (43), stabilization of the membrane by incorporation of PVC (17), and use of aluminosilicate glass (44, 45).

*Silanization*

Proper silanization to make the ion sensor channel hydrophobic so that it may accept and retain the membrane phase is particularly critical for intracellular dual microelectrodes. It is essential that the extreme tip of the channel be properly silanized or both sensitivity and selectivity may be affected. It is also important that there be no silanization of the reference channel, because this would allow uptake of the sensor and neutralize electrode sensitivity after differential subtraction of the recorded voltage signals. For measurements from large cells and with extracellular recording, imperfections of tip silanization can often be eliminated when the tip is broken back to $\geq 1\ \mu m$.

In preparation for the silanization and to promote easier filling of both channels with electrolytes, the glass capillaries should be carefully cleaned by immersion in acid—for example, in phosphoric acid for 30 min to 1 hr, followed by multiple rinses with distilled or deionized water (ensuring that large quantities of water are forced through each capillary), drainage, acetone rinse (1 hr), and drying in an oven at 200°C (20). Although the cleaned glass may be covered and stored (for example, in a desiccator) for some time, in practice electrodes should be silanized immediately after they are pulled. Ideally they should be dried again, as still hydrated inner surfaces of the glass are exposed as a result of the pulling.

For intracellular electrodes, silanization methods are recommended that utilize baking or heating of the glass following or during exposure to the silane vapor. One variation is the introduction of the liquid silane into the

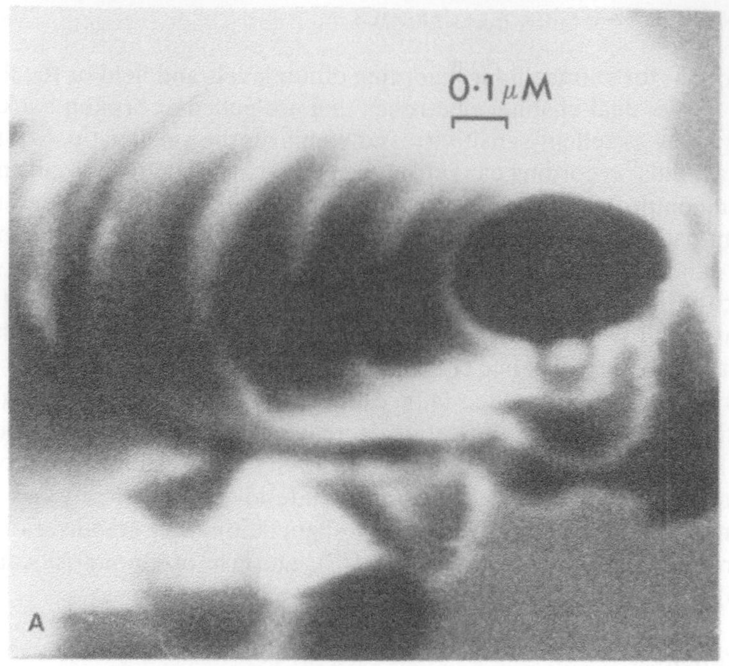
0·1μM
A

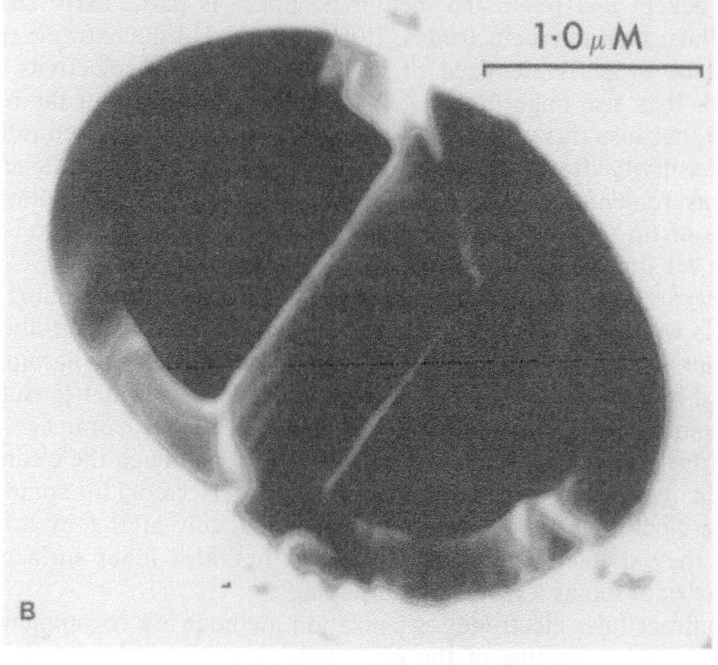
1·0μM
B

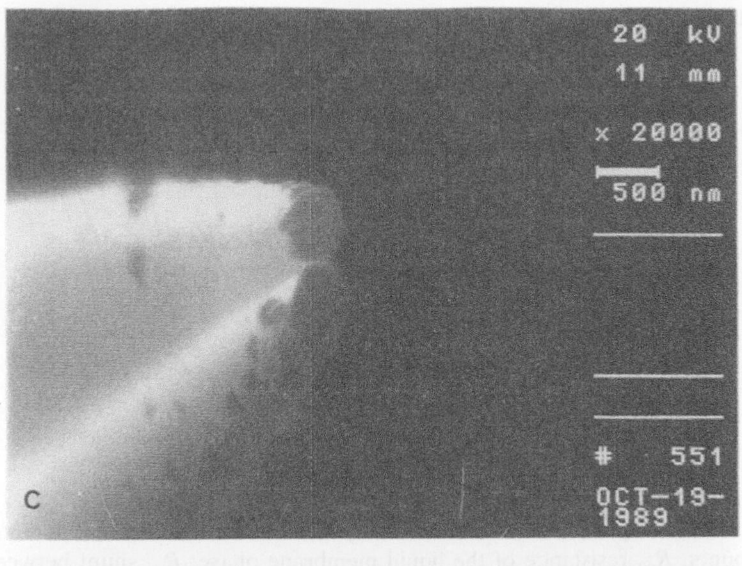


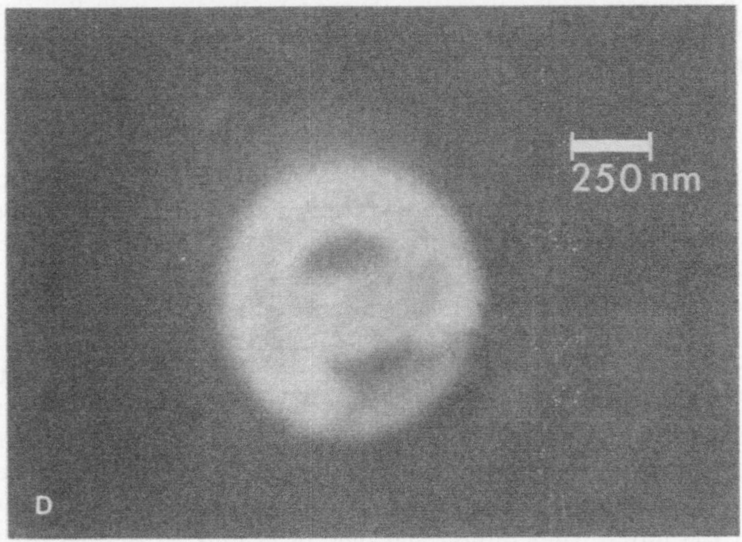


FIG. 4 Scanning electron micrographs of tips of microelectrodes, prepared with gold-plate/back-scatter technique. (A and B) Thick septum theta glass pipettes pulled from glass (supplied by R & D Scientific Glass Co., Spencerville, Maryland); (A) the septum is probably present but could not be demonstrated; (B) the tip of an electrode was broken back to show the septum. (C) The tip of of a fused dual-capillary pipette. (D) The tip of a pipette pulled from thick septum theta glass (Glass Company of America, supplied by World Precision Instruments, Inc., Sarasota, Florida).

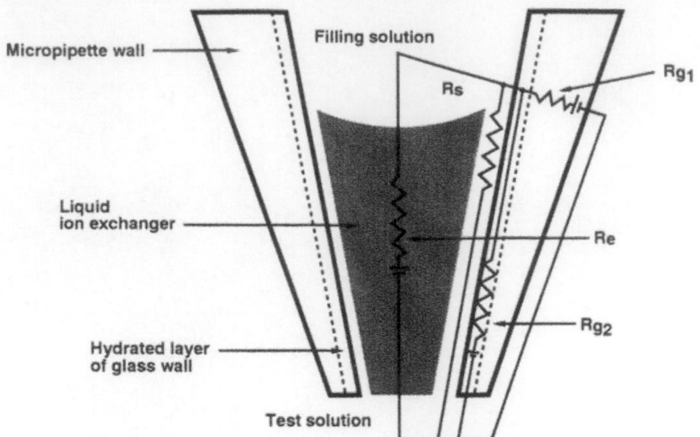


FIG. 5 Diagram of liquid ion exchanger microelectrode tip and potential electrical shunts: $R_e$, resistance of the liquid membrane phase; $R_s$, shunt between the liquid ion exchanger and the glass; $R_{g1}$, shunt across the glass wall; $R_{g2}$, shunt along the inner hydrated layer of the glass wall. $R_s$, $R_{g1}$, and $R_{g2}$ need to be at least 100× greater than $R_e$. [Redrawn from Hinke (42), with permission.]

ion sensor channel by backfill injection and, following its vaporization or the evaporation of the diluent solvent, exposure of the electrodes to heat transfer from a metal conducting surface on top of a hot plate (≥450°C) (46) and/or heating in an oven or with a hair dryer. An alternative technique, developed by Coles and Tsacopoulos (47), uses $N_2$ gas as a carrier to introduce silane vapor into the ion sensor channel, while applying localized heat (≤300°C) to the electrode tip. Figure 6 illustrates a modification of this method, in which $N_2$ flow with silane is delivered by Teflon tubing into the tip region of the ion-measuring channel, a second shorter tube carries outflow to a fume hood, plain $N_2$ gas flows through the reference channel, and heat from a microforge element is applied to the tip. This method allows variation of gas flows, temperature, and time of heating as well as isolation of the silane fumes. The use of a single inflow to the channel being silanized [as in the original method (46)] necessitates that the procedure be carried out in a fume cupboard, dry box, or closed container. A number of silane compounds are available (e.g., Fluka 41716 or 39853; Fluka Corporation, Ronkonkama, New York/Fluka Chemie AG, CH9470 Buchs, Switzerland) and their important properties are well described by J. A. Coles in Chapter 18.

Although silanized electrodes can be stored for 1 to 2 weeks before filling, intracellular electrodes are usually used immediately or shortly after they are filled, because storage in solutions may hydrate their tips, alter sharpness

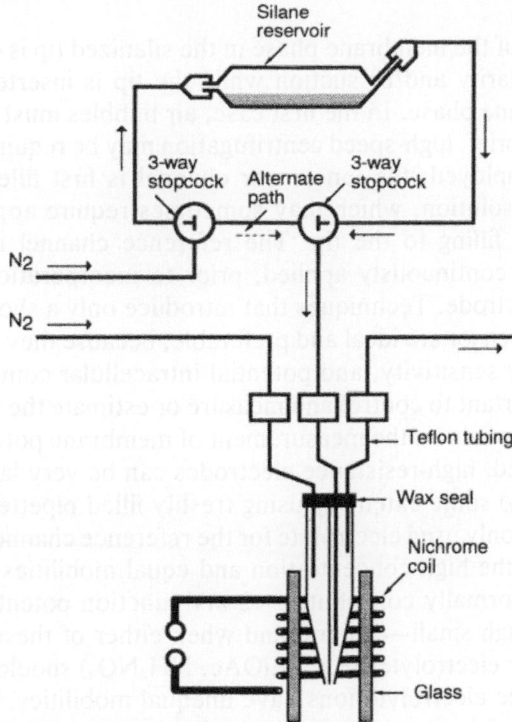


FIG. 6 Silanization of a microelectrode. Nitrogen flow through the ion sensor channel can be diverted and exposed to silane vapor; the shorter Teflon tube distributes gas to a water trap in a fume hood, and allows monitoring of gas flow. A higher rate of $N_2$ flow is used for the reference channel. A nichrome coil wrapped around the glass tubing is heated to ≈240°C. Duration of flow and heating is varied as required to ensure silanization and absence of tip blockage; the total time of $N_2$ flow before, during, and after pickup of silane is usually 3, 8, and 3 min, respectively, for large-diameter (OD > 2 mm) theta glass capillary pipettes. [Modification of original method of Coles and Tsacopoulos (47), as suggested by Jonathan A. Coles.]

and sensitivity, and increase tip potentials (48). Beveling of the tip is often used before or after filling to increase sharpness and lower resistance, provided it does not introduce the problem of differential intracellular localization of channels on penetration. To control final tip size, resistance of the reference channel can be monitored during brief contact with an abrasive surface (as in systems supplied by World Precision Instruments, Inc., Sarasota, Florida, and Sutter Instrument Company, Novato, California). One highly recommended method of dry beveling has been described by Kaila and Voipio (49).

*Filling*

Placement of the membrane phase in the silanized tip is either by back-filling or by capillarity and/or suction while the tip is inserted into a solution of the membrane phase. In the first case, air bubbles must be avoided or eliminated, and brief, high-speed centrifugation may be required (22). When front-filling is employed, the ion sensor channel is first filled with the required electrolyte solution, which may sometimes require application of pressure to promote filling to the tip. The reference channel is then filled, or air pressure is continuously applied, prior to incorporation of the membrane into the electrode. Techniques that introduce only a short column (<50–100 $\mu$m) of ion sensor are ideal and preferable, because they decrease resistance, temperature sensitivity, and potential intracellular contamination (50).

It is important to control and measure or estimate the influence of junction and tip potentials on the measurement of membrane potential. Tip potentials of fine-tipped, high-resistance electrodes can be very large (48); this can be controlled to some extent by using freshly filled pipettes and beveling. The most commonly used electrolyte for the reference channel is 3 *M* KCl, which, because of the high concentration and equal mobilities of $K^+$ and $Cl^-$ (see Table II), normally contributes <2 mV junction potential. However, leakage—although small—occurs, and when either of these ions is being measured, other electrolytes (e.g., LiOAc, $NH_4NO_3$) should be used (51, 52). If the reference electrolyte ions have unequal mobilities, the measured membrane potentials should be compared to those recorded using electrodes made with 3 *M* KCl (53). An alternative is the use of a reference liquid ion exchanger (RLIE), which has equal selectivity for $Na^+$ and $K^+$ and therefore maintains constant potential with constant ion activity and does not contrib-

TABLE II Ionic Mobilities[a]

| Ion | $(\nu_+)_0 \times 10^8$ m sec$^{-1}$/(V m$^{-1}$) | Ion | $(\nu_-)_0 \times 10^8$ m sec$^{-1}$/(V m$^{-1}$) |
|---|---|---|---|
| $H^+$ | 36.3 | $OH^-$ | 20.5 |
| $K^+$ | 7.61 | $SO_4^{2-}$ | 8.27 |
| $NH_4^+$ | 7.60 | $Br^-$ | 8.09[b] |
| $Ba^+$ | 6.60 | $I^-$ | 7.95 |
| $Ca^{2+}$ | 6.16 | $Cl^-$ | 7.91 |
| $Mg^{2+}$ | 5.50[b] | $NO_3^-$ | 7.41 |
| $Na^+$ | 5.19 | $HCOO^-$ | 5.65[b] |
| $Li^+$ | 4.01 | $CH_3COO^-$ | 4.23 |

[a] Mobilities expressed as ionic velocities [$(\nu_+)_0$ or $(\nu_-)_0$] at infinite dilution in meters per second per unit field strength.

[b] Values from Saunders (51); all others from Barrow (52).

ute any ion leakage (54), but presents the disadvantage of a considerable increase in resistance.

*Electronic Testing*

Although the resistance of the reference channel of the electrode can be monitored with standard circuitry, the very high impedance of the ion sensor channel (in the order of $10^{11}$ Ω) demands special circuitry—e.g., with division of a known voltage applied between the bath and ground, between the electrode and a closely matched high-impedance resistor in parallel (as drawn in Fig. 7):

$$R_e/R_s = V_1/V_2 - 1, \tag{8}$$

where $R_e$ is the unknown electrode resistance, $R_s$ is the standard resistance, $V_1$ is the voltage applied, and $V_2$ is the voltage recorded.

Intracellular electrodes should also be tested for their voltage rejection

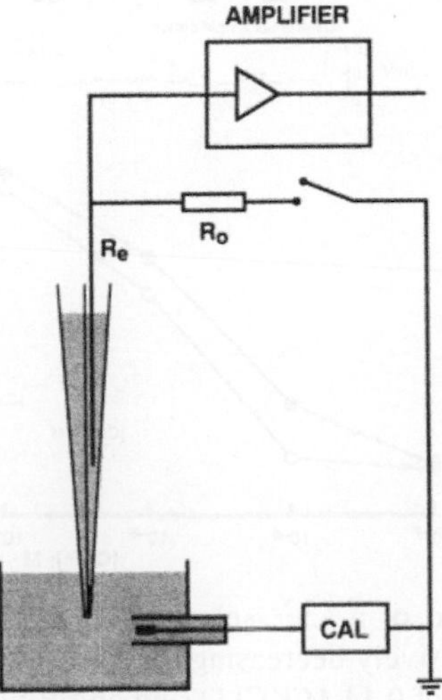


FIG. 7 Diagram of circuit for testing the resistance of the liquid ion exchanger channel of microelectrode. A DC pulse is applied from a square-wave calibrator (CAL) between the bath electrode and ground. The voltage of the ion sensor channel is divided by placing a matching resistor ($R_0$) of high impedance (≈$10^{11}$ Ω) in parallel with that of the electrode ($R_e$). The reference electrode should not be connected to the differential amplifier.

capacity (as in the example in the upper left insert of Fig. 8, obtained using a circuit similar to that of the diagram of Fig. 9). A negative capacitance circuit in series with the reference channel can be used to compensate for impedance and time constant differences between the ion sensor and reference channel and so eliminate transient artifacts in the ion signal (such as is evident in Figs. 2A and 8). A high-input-impedance differential amplifier that provides both resistance monitoring and negative capacitance control is the AXOPROBE model supplied by Axon Instruments, Inc. (Foster City, California).

## *Calibration of Electrodes*

Ion-selective electrodes measure activity rather than concentration:

$$a = cf, \tag{9}$$

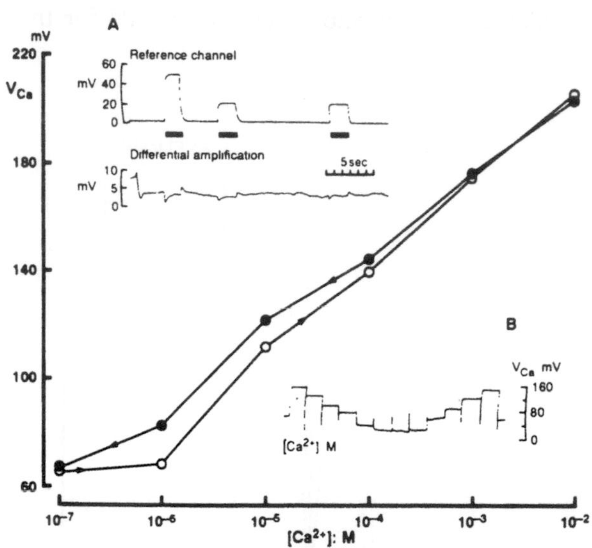


FIG. 8 Calibration of a $Ca^{2+}$-selective microelectrode. Graph shows voltage responses to progressively decreasing (●) and increasing (○) $[Ca^{2+}]$ in $Ca^{-}$- and $H^{+}$-buffered solutions of 0.14 *M* KCl [Tsien and Rink (17)]. Inset B shows pen recorder traces of electrode. Inset A shows responses of another electrode to 50- and 20-mV pulses applied between the bath and ground (see circuit in Fig. 9) to test the common mode rejection capacity; traces show the response of 3 *M* KCl reference channel above and that recorded from both channels below. Note transient changes at the start and end of pulse application due to the different time constants of the channels. [Redrawn from Krnjević *et al.* (21), with permission.]

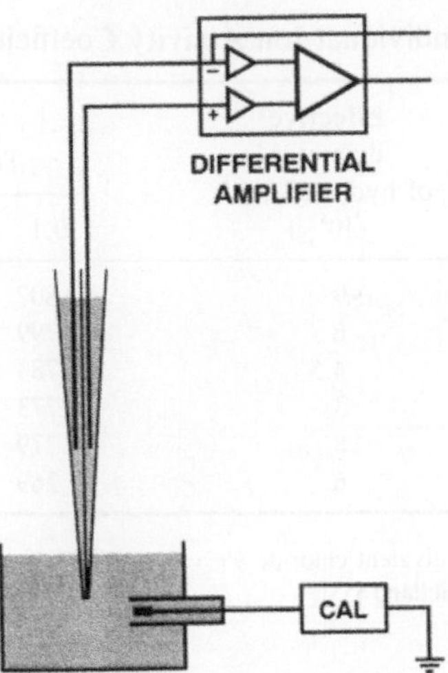


FIG. 9 Diagram of circuit for testing the common mode signal rejection capacity of an ion-sensitive microelectrode. Square-wave pulses of sufficient duration and slow frequency are applied from a DC calibrator (CAL) between the bath electrode and ground; responses of the two channels recorded separately and after subtraction can be compared (see inset A, Fig. 8, for an example of single-ended recording from a reference channel and differential subtraction of voltage responses of both channels).

where $a$ is the thermodynamic activity of an ion, $c$ is the concentration, and $f$ is the activity coefficient. Therefore, calibrations before and after measurements should ideally be made in solutions approximating the intracellular milieu and using known or measured activity coefficients, which are dependent on a number of factors, including temperature, ionic strength, and composition. With calculation of the ionic strength of the calibration solution, standard tables of values of mean activity coefficients as a function of ionic strength for single-electrolyte solutions (55) can be used to calculate activity [see Table III; note that the symbol $\gamma$ is also used (see Chapter 18 by Coles)]. After calibrations using activity, calculations of free ion concentrations must subsequently incorporate the value of the activity coefficient, which should be stated. Calibrations using concentration and an assumption of $f = 1$ are sometimes used for convenience and depending on the application may be quite acceptable (see Chapter 18 by Coles).

Calibrations should be made in a measuring range including the anticipated experimental conditions, in solutions containing a sufficient number of different concentrations of the ion to be measured (which will depend on the

TABLE III Individual Ion Activity Coefficients[a]

| Ion | Effective diameter of hydrated ion[b] ($10^8\,d_i$) | Total ion concentration ($m$) | | | |
|---|---|---|---|---|---|
| | | 0.1 | 0.2 | 0.5 | 1.0 |
| $H^+$ | 9 | 0.807 | 0.788 | 0.812 | 0.940 |
| $Li^+$ | 6 | 0.799 | 0.775 | 0.786 | 0.882 |
| $Na^+$ | 4.5 | 0.783 | 0.744 | 0.701 | 0.697 |
| $K^+$ | 3 | 0.773 | 0.722 | 0.659 | 0.623 |
| $Mg^{2+}$ | 8 | 0.279 | 0.239 | 0.234 | 0.344 |
| $Ca^{2+}$ | 6 | 0.269 | 0.224 | 0.204 | 0.263 |

[a] Univalent and divalent chloride $\gamma \pm$ values at 25°C from Bates *et al.* (55).
[b] Values from Kielland (55).

linearity and limit of detection) and the known background of other ions in the internal milieu. Total ionic strength or activity should be maintained constant by substitution of other ions that produce no or minimal interference. Calibrations should test for hysteresis (see example in Fig. 8), ideally at the temperature of measurement. However, if electrode slope is shown to be temperature independent, corrections can easily be made for voltage offset using the value obtained in the superfusant solution before and after measurement in the tissue. Figure 10 shows a useful calibration system that allows the electrode to remain stationary during changes of solution that are rapidly achieved by programmed or manual selection using solenoid valves and allow estimation of the time constant of the electrode (57, 58).

Response times of electrodes are not easily measured accurately and always include a component arising from the amplifier. A simple technique [suggested by Coles (58)] can be used to obtain an estimate by positioning—as closely as possible to the tip of the electrode in a narrow, shallow, and slanted bath—flow from the points of two small needles, leading from tubing and reservoirs for two different calibration solutions and controlled by pneumatic valve switches.

A number of computerized programs allow on-line data storage or off-line plotting of calibration curves, calculation of electrode slopes and selectivity coefficients, and conversion of chart recordings in millimeters and millivolts to activity or concentration; these programs have been described and are available from individual experimenters (57, 59, 60).

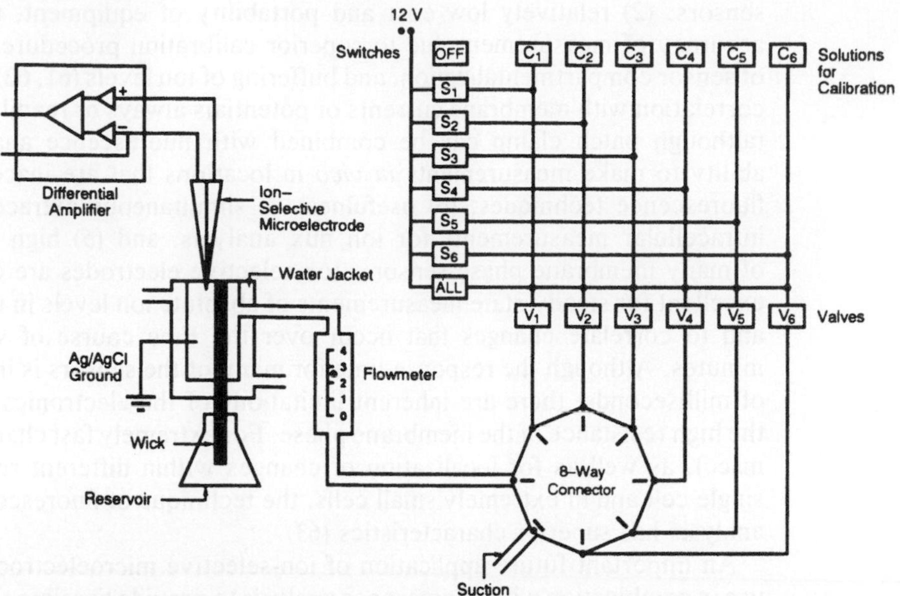


FIG. 10 Diagram of a system for delivery of solutions for calibration of a microelectrode. Switches (S), powered by a 12-V DC rechargable battery (GC 1290; Electrosonic Inc., Willowdale, Ontario), operate open/close positions of solenoid valves (V) Model # 3-188-900; General Valve Corporation, Fairfield, New York) to allow flow through an eight-way valve (# N-1103; Omnifit Ltd., Atlantic Beach, New Jersey), which can be monitored with a flowmeter (Gilmont # F 2100-11; Cole-Parmer Instrument Co., Chicago, Illinois). [Redrawn from Barolet *et al.* (57), with permission.]

## Advantages and Disadvantages of Ion-Selective Microelectrode Measurements

Although direct measurement of intracellular ion levels with ion-sensitive microelectrodes is an invasive technique, fine electrode tips normally permit good membrane seal and minimal leakage of electrode contents. When desired, rigorous control of intracellular ion composition can be maintained by using the insensitive reference liquid ion exchanger (54), instead of an electrolyte in the reference channel. Although leakage of the membrane phase may theoretically or potentially contribute functional differences, as suggested by Ammann (11), there has to date been little evidence of interference, at least for membrane excitability and ion channel function.

Significant advantages for the use of ion-selective microelectrode measurements are (*1*) a wide measuring range and high selectivity of membrane phase sensors, (*2*) relatively low cost and portability of equipment, (*3*) greater accuracy of measurement due to superior calibration procedures, absence of sensor compartmentalization, and buffering of ion levels (61, 62), (*4*) direct correlation with membrane currents or potentials always or readily available (although patch clamp can be combined with fluorescence analysis), (*5*) ability to make measurements *in vivo* in locations that are inaccessible to fluorescence techniques, (*6*) usefulness of simultaneous extracellular and intracellular measurements for ion flux analysis, and (*6*) high selectivity of many membrane phase sensors. Ion-selective electrodes are considered excellent for steady-state measurements of absolute ion levels in the cytosol and to correlate changes that occur over the time course of seconds or minutes. Although the response time for many of the sensors is in the order of milliseconds, there are inherent limitations of the electronics, including the high resistance of the membrane phase. For extremely fast changes ($\leq$100 msec), as well as for localization of changes within different regions of a single cell and in extremely small cells, the technique of fluorescence image analysis has superior characteristics (63).

An important future application of ion-selective microelectrodes is their use in combination with fluorescence analysis to provide baseline calibrations for the fluorophore technique. This should be easily achievable using specially constructed combinations of patch and ion-selective microelectrodes. An added advantage of such an approach is the ability of the latter to respond to changes in the cytosol above the saturation level of the fluorophore and to indicate errors of measurement due to ion complexation.

## Acknowledgments

Appreciation is extended to Jonathan Wu, Edna Leech, and David Conner for faithful assistance in the institution and modification of techniques of fabrication of ion-selective microelectrodes in our laboratory over the years. Special thanks are due to Andrei Rosen for preparation of the illustrations.

## References

1. P. C. Caldwell, *J. Physiol. (London)* **126,** 169 (1954).
2. J. A. M. Hinke, *Nature (London)* **184,** 1257 (1959).
3. A. A. Lev and E. P. Bushinsky, *Tsitologiya* **3,** 614 (1961).
4. P. G. Kostyuk and Z. A. Sorokina, *in* "Membrane Transport and Metabolism" (A. Kleinzeller and A. Kotyk, eds.), p. 193. Academic Press, New York, 1961.
5. A. A. Lev, *Nature (London)* **201,** 1132 (1964).

6. R. C. Thomas, *J. Physiol.* (*London*) **210,** 82P (1970).
7. R. C. Thomas, "Ion-Sensitive Intracellular Microelectrodes." Academic Press, New York, 1978.
8. F. N. Orme, *in* "Glass Microelectrodes" (M. Lavallee, O. F. Schanne, and N. C. Hébert, eds.), p. 376. Wiley, New York, 1969.
9. J. L. Walker, *Anal. Chem.* **43,** 89A (1971).
10. M. Kessler, K. Hajek, and W. Simon, *in* "Ion and Enzyme Electrodes in Biology and Medicine" (M. Kessler, L. C. Clark, Jr., D. W. Lübbers, I. A. Silver, and W. Simon, eds.), p. 562. Univ. Park Press, Baltimore, Maryland, 1976.
11. D. Ammann, "Ion-Selective Micro-Electrodes," p. 346. Springer-Verlag, Berlin, 1986.
12. A. M. Brown, J. L. Walker, Jr., and R. B. Sutton, *J. Gen. Physiol.* **56,** 559 (1970).
13. M. Deschenes and P. Feltz, *Brain Res.* **118,** 494 (1976).
14. P. Grafe, J. Rimpel, M. M. Reddy, and G. ten Bruggencate, *Neuroscience* (*Oxford*) **7,** 3213 (1982).
15. C. P. Bührle and U. Sonnhof, *Pfluegers Arch.* **396,** 144 (1983).
16. K. Ballanyi, P. Grafe, M. M. Reddy, and G. ten Bruggencate, *Neuroscience* (*Oxford*) **12,** 917 (1984).
17. R. Y. Tsien and T. J. Rink, *Biochim. Biophys. Acta* **599,** 623 (1980).
18. F. J. Alvarez-Leefmans, T. J. Rink, and R. Y. Tsien, *J. Physiol.* (*London*) **315,** 531 (1981).
19. G. R. J. Christoffersen and L. Simonsen, *Comp. Biochem. Physiol. C: Comp. Pharmacol. Toxicol.* **76C,** 351 (1983).
20. M. E. Morris, K. Krnjević, and J. F. MacDonald, *Neuroscience* (*Oxford*) **14,** 563 (1985).
21. K. Krnjević, M. E. Morris, and N. Ropert, *Brain Res.* **374,** 1 (1986).
22. I. A. Silver and M. Erecińska, *J. Gen. Physiol.* **95,** 837 (1990).
23. I. A. Silver and M. Erecińska, *J. Cereb. Blood Flow Metab.* **12,** 759 (1992).
24. T. Bührer, H. Peter, and W. Simon, *Anal. Sci.* **4,** 547 (1988).
25. "Selectophore Ionophores for Ion-Selective Electrodes and Optodes." Fluka Chemie AG, Buchs, Switzerland, 1991.
26. U. Schaller, U. E. Spichiger, and W. Simon, *Pfluegers Arch.* **423,** 338 (1993).
27. B. P. Nicolsky, *Zh. Fiz. Khim.* **10,** 495 (1937).
28. G. Eisenman, "Glass Electrodes for Hydrogen and Other Cations." Dekker, New York, 1967.
29. R. A. Steiner, M. Oehme, D. Ammann, and W. Simon, *Anal. Chem.* **51,** 351 (1979).
30. D. Ammann, P. Chao, and W. Simon, *Neurosci. Lett.* **74,** 221 (1987).
31. P. Chao, D. Ammann, U. Oesch, W. Simon, and F. Lang, *Pfluegers Arch.* **411,** 216 (1988).
32. M. F. Wilson, E. Haikala, and P. Kivalo, *Anal. Chim. Acta* **74,** 395 (1975).
33. R. C. Thomas, W. Simon, and M. Oehme, *Nature* (*London*) **258,** 754 (1975).
34. M. Oehme, Thesis, ETH Zürich, No. 5953 (1977).
35. M. Oehme, M. Kessler, and W. Simon, *Chimia* **30,** 204 (1976).
36. F. Lanter, R. A. Steiner, D. Ammann, and W. Simon, *Anal. Chim. Acta* **135,** 52 (1982).

37. U. Schefer, D. Ammann, E. Pretsch, U. Oesch, and W. Simon, *Anal. Chem.* **58,** 2282 (1986).
38. IUPAC Recommendations for Nomenclature of Ion-selective Electrodes, *Pure Appl. Chem.* **48,** 127 (1976).
39. W. Simon, D. Ammann, M. Oehme, and W. E. Morf, *Ann. N.Y. Acad. Sci.* **307,** 52 (1978).
40. K. T. Brown and D. G. Flaming, "Advanced Micropipette Techniques for Cell Physiology." IBRO Handbook Series, Methods in the Neurosciences, Vol. 9, 1986.
41. K. Krnjević, M. E. Morris, and R. J. Reiffenstein, *Can. J. Physiol. Pharmacol.* **58,** 579 (1982).
42. J. A. M. Hinke, *Can. J. Physiol. Pharmacol.* **65,** 873 (1987).
43. F. Deyhimi and J. A. Coles, *Helv. Chim. Acta* **65,** 1752 (1982).
44. M. Désilets and C. M. Baumgarten, *Am. J. Physiol.* **251,** C197 (1986).
45. H. Yamaguchi, *Can. J. Physiol. Pharmacol.* **65,** 1006 (1987).
46. M. J. Borrelli, W. G. Carlini, W. G. Dewey, and B. R. Ransom, *J. Neurosci. Methods* **15,** 141 (1985).
47. J. A. Coles and M. Tsacopoulos, *J. Physiol. (London)* **270,** 12P (1977).
48. R. H. Adrian, *J. Physiol. (London)* **133,** 631 (1956).
49. K. Kaila and J. Voipio, *J. Physiol. (London)* **369,** 8P (1985).
50. R. D. Vaughan-Jones and K. Kaila, *Pfluegers Arch.* **406,** 641 (1986).
51. L. Saunders, "Principles of Physical Chemistry for Biology and Pharmacy," Second Ed., p. 420. Oxford Univ. Press, Oxford, 1971.
52. G. M. Barrow, "Physical Chemistry," Fifth Ed., p. 859. McGraw-Hill, New York, 1988.
53. R. D. Vaughan-Jones and C. C. Aickin, *in* "Microelectrode Techniques. The Plymouth Workshop Handbook" (N. B. Standen, P. T. A. Gray, and M. J. Whitaker, eds.), p. 256. The Company of Biologists Limited, Cambridge, 1987.
54. R. C. Thomas and C. J. Cohen, *Pfluegers Arch.* **390,** 96 (1981).
55. R. G. Bates, B. R. Staples, and R. A. Robinson, *Anal. Chem.* **42,** 867 (1970).
56. J. Kielland, *J. Am. Chem. Soc.* **59,** 1675 (1937).
57. A. W. Barolet, R. Andrews, and M. E. Morris, *J. Neurosci. Methods* **30,** 263 (1989).
58. J. A. Coles, *in* "Practical Electrophysiological Methods" (H. Kettenmann and R. Grantyn, eds.), p. 449. Wiley-Liss, New York, 1992.
59. J. P. C. Boerrigter and A. Lehmenkühler, *Pfluegers Arch. Suppl.* **400,** R56 (1984).
60. S. Levy, L. Tillem, and D. L. Tillotson, *J. Neurosci. Methods* **15,** 253 (1985).
61. M. W. Roe, J. J. Lemasters, and B. Herman, *Cell Calcium* **11,** 63 (1990).
62. D. A. Williams and F. S. Fay, *Cell Calcium* **11,** 75 (1990).
63. E. D. W. Moore, P. L. Becker, K. E. Fogarty, D. A. Williams, and F. S. Fay, *Cell Calcium* **11,** 157 (1990).

# [3] Use of NMR Spectroscopy to Measure Intracellular Ion Concentrations

Stephen J. Karlik

## Introduction

Nuclear magnetic resonance (NMR) spectroscopy emerged as a tool in the field of analytical chemistry in the 1950s. Since 1980 there has been an explosion in the application of the basic NMR technique to biochemical and clinical fields through the development of magnetic resonance imaging (MRI) technology. The purpose of this chapter is to place previous developments and applications in perspective with new advances that expand the role of the NMR technique. Several excellent review articles have appeared within the past 5 years, and the reader is referred to these articles for discussion of general applications of NMR spectroscopy (1), NMR spectroscopy in physiology (2), noninvasive biochemistry from NMR studies (3), NMR studies of cellular metabolism (4), and NMR spectroscopy of cellular systems (5). These reviews are summarized in this chapter; additional new material that has appeared since their publication is also discussed, with a specific emphasis on studies of metal ions and small metabolites. The format will consist of a series of questions and answers on background material. The reader may find this helpful before proceeding to the chapters on specific ions to be found elsewhere in this volume. I begin with definitions, and then move on through physical principles and a description of the equipment. Some applications will be addressed that are different from, but related to, those found elsewhere in this volume. Finally, there will be discussion of quantification in the NMR system, and NMR spectroscopy of the central nervous system at the level of cells, brain slices, and whole organ will be introduced. This chapter provides a general framework on which the reader can build specific information concerning individual intracellular ions.

## What Are NMR, MRS, and MRI?

The usefulness of nuclear magnetic resonance spectroscopy in the field of chemistry has produced a rapid development since its discovery in 1946. NMR has revolutionized the identification and characterization of a variety of molecules. NMR spectroscopy involves radiofrequency-induced transitions

*Methods in Neurosciences, Volume 27*

between energy states of magnetic nuclei in magnetic fields. NMR spectroscopy is possible because about half the known nuclei possess spin or angular momentum. When these atomic nuclei are placed in a powerful and uniform magnetic field, they tend to assume an orientation parallel to the field. The energy difference of the transition of these nuclei lined up with the external field (lower energy state) or against it (higher energy state) is very small. Thus, NMR is a relatively insensitive technique.

Magnetic resonance spectroscopy (MRS) is simply the name given to medical uses of NMR spectroscopy, and for the purposes of our discussion, NMR spectroscopy and MRS will be used interchangeably. The use of MRS as a clinical and analytical tool for human diagnosis has gained slow acceptance, although there is a vast potential for clinical chemical analysis. For example, it is possible to measure a patient's creatine phosphate, adenosine triphosphate (ATP), and pH values in various portions of the heart after an infarct. This valuable clinical information can be obtained noninvasively only by MRS. However, MRS has its own problems. The combination of high costs, low sensitivity, and difficulties in localization and quantification precludes its routine clinical use. To date, there is no identified clinical problem for which MRS has proved to be a fundamental diagnostic tool. New applications for MRS continue to appear, and it may be in the neurological and psychiatric disorders that routine use of this technology will appear (6, 7).

Magnetic resonance imaging is the use of the NMR in combination with large-bore magnets and magnetic field gradients to produce images of internal body structures. This technology has been widely used in the clinical setting and the applications and uses of this technology continue to grow. Because the components of an MRI scanner include a large-bore (typically 1 m), uniform high magnetic field (1.5 T) and have sophisticated computer, gradient, and radiofrequency (rf) capabilities, it is possible to adapt many of the scanners to perform MRS. The composite time required to reconfigure the MR scanner and perform an MRS examination presents problems of time management and cost recovery in a clinical setting. Although meter-bore 4-T magnets have appeared, which can decrease spectroscopy acquisition times, siting difficulties and cost will limit their distribution. Shielded 80-cm-bore 3-T systems have become commercially available and may be a significant step forward, increasing the *in vivo* use of NMR. Standard "mid-field strength" clinical systems (1.5 to 2 T) have sufficient magnetic field strength such that human spectroscopy is attainable on large, but not unreasonable, volumes.

## What Are the Physical Principles of NMR Spectroscopy?

Approximately half the known atomic nuclei possess spin angular momentum, and hence have a magnetic moment. When these nuclei are placed in a powerful magnetic field, the field aligns their individual magnetic moments with its direction against the disordering tendencies of thermal processes. This yields a sample magnetic moment. However, the nuclei do not align perfectly parallel to the field but are inclined to the field and rotate or precess about the field direction. A spinning top that is wobbling has been used to illustrate this phenomenon. Increasing the field strength does not increase the alignment of the nuclei; it makes the nuclei rotate faster. These ordered nuclei undergo a transition to a higher energy level by applying a radiofrequency pulse at right angles to the uniform, static magnetic field. The nuclei will absorb the rf energy when the frequency of the rotating component of the rf field is the same as the precession frequency.

The higher the magnetic field, the higher the precession frequency for any individual nuclei. There is a characteristic frequency, called the Larmor frequency, for each nucleus in a given magnetic field. Some of these characteristic frequencies are listed in Table I for several nuclei relevant to biological systems. These are relatively easily available to the researcher on conventional multinuclear NMR spectrometers. It is possible to obtain data from many nuclei in one magnet because each nucleus has a specific Larmor frequency at that static field. Protons ($^{1}H$) are extremely important because they enable proton NMR spectroscopy analysis of a number of relevant metabolites and MRI imaging of water molecules. Most of the biological NMR spectroscopy studies are limited to an examination of $^{31}P$, $^{1}H$, $^{13}C$, $^{2}H$, $^{23}Na$, and $^{19}F$ (4).

The limitation of the application of *in vivo* NMR spectroscopy to the above list of nuclei is the result of a combination of factors related to detectability.

TABLE I NMR Properties of Some Biologically Relevant Nuclei

| Isotope | Natural abundance (%) | Relative sensitivity (constant field) | Spin | NMR frequency (MHz) (1.0 T) |
|---|---|---|---|---|
| $^{1}H$ | 99.98 | 1.000 | $\frac{1}{2}$ | 42.58 |
| $^{2}H$ | $1.56 \times 10^{-2}$ | $9.65 \times 10^{-3}$ | 1 | 6.54 |
| $^{7}Li$ | 92.57 | 0.294 | $\frac{3}{2}$ | 16.55 |
| $^{13}C$ | 1.108 | $1.59 \times 10^{-2}$ | $\frac{1}{2}$ | 10.71 |
| $^{15}N$ | 0.365 | $1.04 \times 10^{-3}$ | $\frac{1}{2}$ | 4.315 |
| $^{17}O$ | $3.7 \times 10^{-2}$ | $2.9 \times 10^{-2}$ | $\frac{5}{2}$ | 5.772 |
| $^{19}F$ | 100 | 0.833 | $\frac{1}{2}$ | 40.06 |
| $^{23}Na$ | 100 | $9.25 \times 10^{-2}$ | $\frac{3}{2}$ | 11.26 |
| $^{25}Mg$ | 10.05 | $2.68 \times 10^{-3}$ | $\frac{5}{2}$ | 2.606 |
| $^{31}P$ | 100 | $6.63 \times 10^{-2}$ | $\frac{1}{2}$ | 17.236 |
| $^{35}Cl$ | 75.4 | $4.70 \times 10^{-3}$ | $\frac{3}{2}$ | 4.172 |
| $^{39}K$ | 93.1 | $5.08 \times 10^{-4}$ | $\frac{3}{2}$ | 1.987 |
| $^{43}Ca$ | 0.13 | $6.40 \times 10^{-2}$ | $\frac{7}{2}$ | 2.865 |

First, there is the field strength, which gives a higher signal with a higher magnetic field. Second, the size of the sample must be adequate. Obviously a smaller sample size, for example, 1 million cells versus 10 $cm^3$ of cerebral grey matter, will have a very different quantity of NMR-available nuclei. The sample can be made larger, but this raises the significant problem of heterogeneity. Third, all nuclei do not have the same relative sensitivity or natural abundance. Table I lists the sensitivity (relative to that of protons) of several relevant nuclei for biological systems. Although one can compensate for the low natural abundance by enriching samples or tissues with exogenous nuclei ($^{13}C$, for example), the nuclear sensitivity is a fixed value. The researcher can increase the signal-to-noise ratio by signal averaging, i.e., acquiring spectra over a long period of time. This is trivial for characterization of chemical systems that are stable. However, this is problematic when animals or humans are investigated. Fourth, the metabolites or ions must be mobile to give an NMR signal. Small metabolites or ions that are tightly bound to macromolecules can be undetected in NMR spectra. Furthermore, large macromolecules such as deoxyribonucleic acid similarly are undetectable. Although magic-angle sample spinning, an advanced technique, allows the NMR spectroscopist to obtain high-resolution spectra from solid samples, this is not appropriate for *in vivo* examinations. The investigator must also be aware of the fact that the NMR signals obtained are from all of the material within the rf coil unless a localization technique identifies a specific volume. The available localization techniques are difficult and the merits of each type are widely debated.

An NMR study defines a complete data acquisition in which the solution, cells, or tissue are irradiated by one or more radiofrequency pulses and a signal has been received for processing. In the simplest case of pulsed NMR, an rf pulse is applied to the sample, which rotates the net magnetization in the sample 90° with respect to the surrounding main magnetic field. The rf transmitter is turned off and the same coil is used to receive a signal emitted back from the perturbed nuclei as they relax to equilibrium. The characteristics of this energy dissipation provide considerable information about the perturbed system. If more than one chemical species is present, then a Fourier transform (mathematical tool) of the resultant signal will give a spectrum with peaks that can be readily identified. The NMR spectrum has quantitative information and a description of the environments of the different nuclei based on their chemical shift position along the ordinate compared with an external or internal standard. The abscissa is unitless and all measurements should have a standard for comparison. By modifying the pulse program or sequence, the relaxation times of the sample [spin lattice ($T_1$) or spin–spin ($T_2$) relaxation times] can also be determined. The $T_1$ relaxation time is a measure of the time required for the net nuclear magnetic moment to return to its original orientation by losing energy to its surroundings (lattice). The $T_2$ relaxation time is the characterization of the process whereby excited nuclei lose energy to other nuclei. It is also called the transverse relaxation time. The differences in these two relaxation times, and the total amount of water present in a tissue (proton density), enable clinical MRI to produce detailed pictures of different tissue structures from only water proton signals.

Magnets for NMR and MRI are divided into two basic categories: wide bore, large enough for human or animal studies, and narrow bore, suitable for tissue or cell work. The magnet bore sizes are about 1 m for human studies, 15–40 cm for animal studies, and between 50 and 150 mm for cells and tissues. The magnetic fields vary as widely as the bore sizes. Clinical systems with bore sizes suitable for human use range from ultralow magnetic fields (<0.15 T) to 4.0 T. Magnetic fields of the conventional chemistry and biochemistry systems, with narrow bores, range from 4.7 to 23.5 T (200–1000 MHz for protons).

## Are There Significant Limitations?

For any test system, a requirement must first be satisfied: can the subject, animal, tissue preparation, or cellular preparation fit into the available magnet? The second critical consideration is whether the object (cells, slices, or animals) can be maintained at physiological conditions during the prolonged spectroscopy sessions that are usually necessary to obtain adequate results. This can be especially critical for examining those nuclei that have low natural abundance or sensitivity and, consequently, a long acquisition time, because a large number of spectra must be averaged. The researcher must also consider if there are any manipulations to be done while the test system is within the magnet. In an animal experiment, for example, the protocol may involve altering the inspired carbon dioxide concentration to investigate hypercapnic sequelae in brain metabolism. The monitoring/maintenance devices must adjust the physiological parameters in the bore of the magnet at some distance from the equipment. Normal monitoring devices can be incompatible with a high magnetic field associated with the NMR experiment (8). Specialized equipment and precautions must be taken even when adequate physiological monitoring and maintenance conditions have been established outside the magnet bore (9).

## What About Safety?

To perform physiologically relevant studies on cells, brain slices, or intact animals, it is essential to provide sufficient monitoring and support facilities.

These may range from simple temperature regulation and control of inspired oxygen and carbon dioxide tension to more advanced animal respiratory or cardiovascular support. In certain circumstances, it may also be advantageous to look at brain electrical activity. Most devices available for the use in the conventional laboratory are unsuited for use in the magnetic resonance suite. The high magnetic fringe field associated with unshielded magnets can produce unwanted side effects in the equipment. There are really two safety considerations. The first is to protect the animal or patient and the second is to protect the monitoring equipment from the deleterious effects of the magnetic field. Extensive monitoring of patients has been achieved even at 1.5 T (9), and similar arrangements are possible for cells or animals. Another safety consideration is also due to the magnet fringe field and relates to the attraction of ferromagnetic materials to the magnet. Certain safety precautions must be established when metallic objects such as gas tanks, laboratory carts, or surgical tools are brought within the magnetic field.

## What Are the Principle Application Areas?

In the clinical setting, MRI will be used for anatomical studies for structural and functional assessment of patients. New areas of MRI applications include blood flow, angiography, and cardiac dynamics. An important developing area for MRI is tissue perfusion (or functional imaging; discussed in the next section). In the chemical arena, the technology finds more applications every year and conventional NMR spectroscopy continues to be a widely expanding field with many new applications (1). In the biochemical arena, new areas of examination include two-dimensional and three-dimensional sequences that allow the investigator to perform molecular structure determinations. In addition, there is a widening expansion of medical applications to studies of blood products and cerebrospinal fluid, for example.

## What Is Functional Imaging?

Neuroscientists are particularly enthusiastic about the potential of functional brain imaging using advanced MRI techniques. Functional MRI is defined as anatomic maps of physiological and metabolic parameters. These images can change during normal and pathological processes. Functional imaging techniques, e.g., echo planar imaging (EPI) and $T_2$*-weighted imaging, visualize rapidly changing physiological events through changes in blood flow and volume. Physiological imaging is based on changes in the water proton signal and has a temporal resolution in EPI of $<100$ msec. Maps of tissue oxygenation, relative cerebral blood volume, and apparent diffusion coefficient have been reported (10–13). The temporal resolution of EPI using exogenous MRI contrast and endogenous deoxyhemoglobin contrast has allowed visualization of changes in blood flow and blood volume during cortical activation. Although some of the pioneering work was carried out on 4-T scanners, recent results have clearly shown that the EPI at 1.0 T (12) and the gradient echo technique at 1.5 T are also possible (14, 15). However, the conventional gradient echo acquisitions take longer than echo planar imaging, so that the temporal resolution is lower. Thus, functional imaging advances MRI from purely an anatomic role to one of determining the physiological changes in the active brain. Although magnetic resonance spectroscopy has provided interesting biochemical information, MRI and MRS have not merged into a single clinical examination. It is possible to produce these functional maps in much less time than the localized spectroscopy of very low-concentration metabolites such as lactate (<5 m*M*), because the brain is 80% water (concentration of 88 *M*). Functional imaging is a new modality to display changes in metabolic state with increased temporal and spatial resolution compared to that of MRS. However, at this time, functional imaging cannot detect changes in intracellular ion levels.

## How Do We Quantify in the NMR System?

NMR spectroscopy is a quantitative technique. The signals received back in a free induction decay represent the total magnetization of the number of nuclei in the observed volume. In a simple chemical system with all mobile nuclei, the ratios between the various peaks will be a quantitative measure of the number of atoms of each molecular environment. In combination with an intensity and chemical shift standard, it is possible to determine directly the concentrations of the various components in that mixture. The situation gets more complex when examining biological systems in which some nuclei are present in nonmobile or severely restricted conditions. These can be considered NMR invisible, and the investigator is limited to a consideration of the detectable mobile nuclei even if the nonmobile species are relevant. Consequently, the extraction of a tissue to determine the concentration of a specific metabolite may yield an answer different from that gained in an NMR investigation.

The NMR measurement is of low sensitivity even under the best conditions. Therefore, two criteria must be considered when we propose to study an ion with NMR spectroscopy: there must be sufficient ion concentration and mobility. The detection limit is approximately 500 $\mu M$ for $^{31}P$ spectroscopy and 100 $\mu M$ for protons. Table I gives the change in sensitivity of various nuclei with protons being normalized to one.

Without special techniques, NMR cannot look at individual compartments. For example, even the simplest case of a preparation of cells in an NMR tube has an extracellular fluid medium perfusate, in addition to the intracellular environment. Tissue slices or whole organs contain a broader range of compartments, including intravascular, extravascular, interstitial, and intracellular, all of which can contain mobile nuclei observable by NMR. Without additional specialized techniques, it is impossible to differentiate the contributions to the signals obtained from each compartment. Two ways of identifying different tissue compartments are the use of chemical shift reagents (16) and diffusion-weighted spectroscopy (17). A chemical shift reagent alters the molecular environment of neighboring nuclei and consequently alters their position in an NMR spectrum. This would clearly identify a particular population. In diffusion-weighted spectroscopy, the identification of the populations depends on different diffusion characteristics of different sample compartments.

Quantification and assignment of NMR spectroscopy peaks have become routine in solution studies. Similar measurements in the medical field have many problems. The reader is referred to an examination of $^{1}$H and $^{31}$P NMR spectroscopy results from plasma, brain, and heart (18), illustrating the problems associated with the resolution of spectra into quantifiable peaks for determination metabolite levels. The use of large voxels to acquire data and the controversy surrounding the localization techniques lead to an unfortunate confusion about the effectiveness of MRS. Investigators who choose to examine *in vivo* data from patients or animals may wish to follow the suggested guidelines contained therein.

It can be difficult to make sense of complex spectra arising from NMR spectroscopy in the cell, tissue slice, or whole animal. Line-fitting techniques and peak assignments can be variable and external markers and references should be used to assist in these determinations. An excellent overview of the advantages and limitations of measuring metabolite concentrations by NMR has been published (19). These suggestions are directly relevant and applicable to determining intracellular ion levels in the CNS by NMR spectroscopy.

## What NMR Spectroscopy Has Been Performed in the Central Nervous System?

Clinical applications of MRS have been concentrated in three basic areas: brain, cardiac, and musculoskeletal. A lack of movement and a variety of relevant pathophysiological conditions in the brain have led to an extensive use of *in vivo* $^{31}$P NMR studies of animals and patients. The clinical importance of cerebral disease has focused NMR spectroscopic methods on ischemia, seizures, and hypoxia. $^{31}$P NMR studies during ischemia showed that concentrations of ATP and phosphocreatine (PCr) decreased, and inorganic phosphate ($P_i$) concentration increased simultaneously in a variety of different animal models (20). With controlled ischemia, the following sequence was observed: the pH decreases, the concentration of $P_i$ increases, the concentration of PCr decreases slowly, and then the concentration of ATP decreases (21). In hypoxia, changes in $P_i$ and PCr concentrations were observed at an inspired oxygen of 6%, but pH and ATP concentration were constant even at 4% oxygen in the inspired air (22). Extensive studies have been performed concerning phosphate metabolites in seizures induced by bicuculline (23). In cerebral insults, $^{31}$P metabolite levels show changes in energy metabolism similar to corresponding alterations in EEG activity. Many initial studies in humans were performed on neonates because the earlier magnets had smaller bores (24). The variability in reported values for phosphorous metabolites is one of many problem areas in the application of *in vivo* spectroscopy (18). Most of the quantitative measurements by MRS have used the *relative* concentrations of the different metabolites. To obtain absolute concentrations, the NMR experiment must be calibrated with standards, as referred to previously (19). Although water has been suggested as an internal standard, the $H_2O$ concentration can change prior to and during acquisition and may also be partially NMR invisible. Neither the critical problem of quantification nor the quality control of the data has been addressed satisfactorily (25).

## How Do You Localize the Spectrum?

A significant problem that arises when dealing with a heterogeneous large volume, such as the brain, is that the researcher wishes to localize a particular anatomic structure. This problem is addressed in MRS by using a series of field gradients that isolate the voxel, or volume of interest. There are a variety of strategies to produce localization, which go under a variety of acronyms, including ISIS (26), VSE (27), SPACE (28), DIGGER (29), and SPARS (30). These localization techniques were developed using three rapidly switched orthogonal gradients. Another means to obtain *in vivo* spectroscopic data is the technique of chemical shift imaging (CSI). Whereas the non-CSI techniques isolated only a certain voxel of the brain to obtain an NMR spectrum, it is possible using CSI to obtain $^{31}$P and proton spectra from many voxels, usually describing one slice in MRI, in one long acquisition. It is wise to remember that the cell types are not homogeneous in these voxels and they all may contribute to the observed results.

## What Ions Can Potentially Be Measured in the Brain by NMR?

Determination of the proton concentration ($[H^+]$), or pH, was one of the first measurements to be made, because it is proportional to the chemical shift position of the $P_i$ peak in $^{31}P$ NMR spectra of the brain (18, 31, 32). Phosphatidylethanolamine (33) and 2-deoxyglucose-6-phosphate (34) resonances from $^{31}P$ spectra have been used to determine intracellular pH ($pH_i$). A variety of exogenous reporter molecules have also been suggested for the measurement of $pH_i$ (5). Multinuclear ($^1H$, $^{13}C$, and $^{31}P$) NMR spectroscopy has been used to demonstrate a rebound alkalosis in infarcted rat brain (35). $^{31}P$ NMR spectroscopy has also been used to measure brain intracellular $Mg^{2+}$ concentrations indirectly through the use of changes in the chemical shift position of $\beta$-ATP (36). Because $^{23}Na$ is a spin $\frac{3}{2}$ nucleus, with good sensitivity (see Table I), direct observation is possible. However, intra- and extracellular $Na^+$ ions resonate at the same chemical shift and thus show only one peak in a spectrum. Although chemical shift reagents can help differentiate the NMR signals from these two environments (37), they have not been used *in vivo* due to toxicity problems. They may, however, find application in $^{23}Na$ NMR studies of brain cell or slice (38) preparations. Brain intracellular $Ca^{2+}$ has been measured indirectly using $^{19}F$ NMR of fluorine-containing, $Ca^{2+}$-sensitive chelators (39). It has been necessary to substitute $^{87}Rb$ for $K^+$ in a few non-CNS studies because the $^{39}K$ nucleus has a receptivity of $4.73 \times 10^{-4}$ compared to protons (40, 41). Recently, however, $^{133}Cs^+$ (another $K^+$ analog) has been fed to rats and the excised brain spectrum reported (42). *In vivo* $^7Li$ NMR brain spectra of treated rats has also been possible (43). Although not yet used for brain studies, $^{14}N$ NMR spectroscopy with a dysprosium-containing chemical shift reagent (44) differentiated $NH_4^+$ ions, which may allow cell or brain slice studies, similar to that of $^{23}Na$. The $Ca^{2+}$-fluorinated chelator, (5-FBAPTA) 1,2-bis(*o*-aminophenoxy)ethane-*N,N,N',N'*-tetraacetic acid (39), has a different $^{19}F$ chemical shift when bound to $Zn^{2+}$ and in consequence may have potential to measure intracellular $Zn^{2+}$ (45). Although the abundance of $^{35}Cl$ is relatively high (Table I), the receptivity is only $4.70 \times 10^{-3}$ of that of protons. In one study using erythrocytes, $Cl^-$ was NMR invisible. It is possible that a chemical shift reagent, such as $Co^{2+}$, although toxic, could separate intracellular and extracellular $Cl^-$ compartments (4).

## Conclusion

The process of detection and characterization of ions in the brain is in its infancy and a wide range of investigations is possible. The specific chapters in this volume that address NMR measurements detail studies of pH (31), $Ca^{2+}$ (46), and $Mg^{2+}$ (47). It will be a challenge to incorporate the other techniques, e.g., fluorescent probes or microelectrodes, into protocols that allow simultaneous measurements. The great value of using NMR techniques is the noninvasiveness of the procedure.

## Acknowledgments

Stephen J. Karlik received a Career Scientist Award from the Sterling-Winthrop Imaging Research Institute. The helpful comments and suggestions of T. L. Lee and S. J. Kent are gratefully acknowledged.

## References

1. L. W. Jelinski, *Anal. Chem.* **62,** 212R (1990).
2. B. M. Hitzig, J. W. Pritchard, H. L. Kantor, W. R. Ellington, J. S. Ingwall, C. T. Burt, S. I. Helman, and J. Kowtcher, *FASEB J.* **1,** 22 (1987).
3. G. K. Radda, *FASEB J.* **6,** 3032 (1992).
4. P. Lundberg, E. Harmsen, C. Ho, and H. J. Vogel, *Anal. Biochem.* **191,** 193 (1990).
5. B. S. Szwergold, *Annu. Rev. Physiol.* **54,** 775 (1992).
6. C. D. Smith, L. G. Gallenstein, W. J. Layton, R. J. Kryscio, and W. R. Markesbery, *Neurobiol. Aging* **14,** 775 (1992).
7. J. W. Pettegrew, M. K. Keshavan, K. Panchalingham, S. Strychor, D. Kaplan, M. Tretta, and M. Allen, *Arch. Gen. Psychiatry* **48,** 563 (1991).
8. E. Kanal and F. G. Shellock, *Radiology* **185,** 623 (1992).
9. S. J. Karlik, T. Heatherley, F. Pavan, J. Stein, F. Lebron, B. Rutt, L. Carey, R. Wexler, and A. Gelb, *Magn. Reson. Med.* **7,** 210 (1988).
10. S. G. Kim, J. Ashe, K. Hendrich, J. M. Ellermann, H. Merkle, K. Ugurbil, and A. P. Georgopoulous, *Science* **261,** 615 (1993).
11. P. A. Bandetti, A. Jesmanowicz, E. C. Wong, and J. S. Hyde, *Magn. Res. Med.* **30,** 161 (1993).
12. M. K. Stehling, F. Schmitt, and R. Ladebeck, *J. Magn. Reson. Imaging* **3,** 471 (1993).
13. V. S. Vexler, A. J. S. deCrespigny, M. Wendland, R. Kuwatsuru, A. Muhler, R. C. Brasch, and M. E. Moseley, *J. Magn. Reson. Imaging* **3,** 483 (1993).
14. L. S. Schad, V. Trost, M. V. Knopp, E. Muller, and W. J. Lorenz, *Magn. Reson. Imaging* **11,** 461 (1993).
15. R. T. Constable, G. McCarthy, T. Allison, A. W. Anderson, and J. C. Gore, *Magn. Reson. Imaging* **11,** 451 (1993).
16. G. A. Elgavish, *Invest. Radiol.* **24,** 1028 (1989).

17. P. C. Van Zijl, C. T. Moonen, P. Faustino, J. Pekar, O. Kaplan, and J. S. Cohen, *Proc. Natl. Acad. Sci. U.S.A.* **88,** 3228 (1991).
18. P. A. Bottomly, *J. Magn. Reson. Imaging* **2,** 1 (1992).
19. P. S. Tofts and S. Wray, *NMR Biomed.* **1,** 1 (1988).
20. R. D. Oberhaensli, D. J. Taylor, B. Rajagopalan, and G. K. Radda, *Lancet* **2,** 931 (1987).
21. N. M. Bolas, B. Rajagopalan, F. Mitsumori, and G. K. Radda, *Stroke* **19,** 608 (1988).
22. L. Gymlai, B. Chance, L. Ligeti, G. McDonald, and J. Cone, *Am. J. Physiol.* **254C,** 699 (1988).
23. O. A. C. Petroff, J. W. Prichard, T. Ogino, M. J. Avison, J. R. Alger, and R. G. Shulman, *Ann. Neurol.* **20,** 1985 (1986).
24. D. P. Younkin, M. Delivoria-Papadopoulos, J. Maris, E. Donlon, R. Clancy, and B. Chance, *Ann. Neurol.* **20,** 513 (1988).
25. W. Vine, *Arch. Pathol. Lab. Med.* **114,** 453 (1990).
26. R. J. Ordidge, A. Connelly, and J. A. B. Lohman, *J. Magn. Reson.* **66,** 283 (1986).
27. W. P. Aue, T. A. Muller, T. A. Cross, and J. Seelig, *J. Magn. Reson.* **56,** 350 (1984).
28. D. M. Doddrell, W. M. Brooks, J. Bulsing, J. Field, M. Irving, and H. Baddeley, *J. Magn. Reson.* **68,** 367 (1986).
29. G. Galloway, W. M. Brooks, J. Bulsing, I. Brereton, J. Field, M. Irving, H. Baddeley, and D. M. Doddrell, *J. Magn. Reson.* **73,** 360 (1987).
30. P. R. Luyten, C. M. Anderson, and J. Hollander, *Magn. Reson. Med.* **4,** 431 (1987).
31. O. A. C. Petroff and J. W. Prichard, Chapter 11, this volume.
32. D. Malhotra and J. I. Shapiro, *Concepts Magn. Reson.* **5,** 123 (1993).
33. J. T. Corbett, A. R. Laptook, A. Hassan, and R. L. Nunnally, *Magn. Reson. Med.* **6,** 99 (1988).
34. K. J. Brooks, R. Porteous, and H. S. Bachelard, *J. Neurochem.* **52,** 604 (1989).
35. T. Nakada, K. Houkan, K. Hida, and I. L. Kwee, *Magn. Reson. Med.***18,** 9 (1991).
36. J. S. Taylor, D. B. Vigneron, J. Murphy-Boesch, S. J. Nelson, H. B. Kessler, L. Coia, W. Curran, and T. R. Brown, *Proc. Natl. Acad. Sci. U.S.A.* **88,** 6810 (1991).
37. N. Askenasy, A. Vivi, M. Tassini, and G. Navon, *Magn. Reson. Med.* **28,** 249 (1992).
38. R. A. Kauppinen, H. Kokko, and S. R. Williams, *J. Neurochem.* **58,** 967 (1992).
39. R. S. Bader-Gofter, O. Ben-Yoseph, S. J. Dolin, P. G. Morris, G. A. Smith, and H. S. Bachelard, *J. Neurochem.* **55,** 878 (1990).
40. J. L. Allis, Z. H. Endre, and G. K. Radda, *Renal Physiol. Biochem.* **12,** 171 (1989).
41. P. D. Syme, R. M. Dixon, J. L. Allis, J. K. Aronson, D. G. Grahame-Smith, and G. K. Radda, *Clin. Sci.* **78,** 303 (1990).
42. B. P. Shehan, R. M. Welland, W. R. Adams, and D. J. Craik, *Magn. Reson. Med.* **30,** 573 (1993).

43. S. Ramaprasad, J. E. Newton, D. Cardwell, A. H. Fowler, and R. A. Komoroski, *Magn. Reson. Med.* **24,** 308 (1992).
44. D. Burstein, H. I. Litt, and E. T. Fossel, *Magn. Reson. Med.* **9,** 66 (1989).
45. G. A. Smith, R. T. Hesketh, J. C. Metcalfe, J. Feeney, and P. G. Morris, *Proc. Natl. Acad. Sci. U.S.A.* **80,** 7178 (1983).
46. G. A. Smith and H. S. Bachelard, Chapter 8, this volume.
47. J. A. Helpern and H. R. Halvorson, Chapter 16, this volume.

# [4] Use of Ionophores for Manipulating Intracellular Ion Concentrations

G. Andrew Woolley, Douglas R. Pfeiffer, and Charles M. Deber

## Introduction

Since their discovery in the 1960s, ionophores have been used as agents for manipulating cellular ion concentrations (Ovchinnikov *et al.*, 1972; McLaughlin and Eisenberg, 1975; Pressman, 1976; Reed, 1979). The term *ionophore* reflects the ability of this class of substances to "bear ions" across lipid membranes. Lipid membranes normally present a formidable barrier to ion movement, because transferring an ion from aqueous solution to the hydrophobic environment of the membrane is costly in energetic terms. Although they are relatively simple, low molecular mass compounds (500–2500 Da), ionophores are able to circumvent this barrier effectively and permit the passage of even highly charged cations such as $Ca^{2+}$. They do this by forming specific amphipathic structures that envelop an ion in polar groups while presenting a hydrophobic exterior to the surrounding membrane.

Most useful ionophores are isolated from natural sources or are derivatives of naturally occurring molecules. The number of wholly synthetic ionophores is growing, however. Many of these are based on the crown ethers and cryptands discovered by Cram, Lehn, Pedersen, and co-workers (for which they were awarded the Nobel Prize in 1987) (Fyles, 1990; Vogtle, 1991). Peptide-based synthetic ionophores and mimics of natural systems have also been explored (Deber *et al.*, 1980; Fyles, 1990; Akerfeldt *et al.*, 1993; Montal *et al.*, 1993). Eventually such efforts in molecular design will provide compounds with improved selectivities or with novel properties tailored for specific applications. Aside from their usefulness to cell biologists, many ionophores are important commercially. Monensin, for example, is widely used in the poultry industry to combat coccidiosis, a protozoan infection of chickens (Ruff, 1982). Valinomycin is a key component of potassium-selective electrodes, which are used widely in clinical and industrial settings (Ammann *et al.*, 1987). These applications play a significant role in guiding research into ionophores and their mechanisms.

To use ionophores effectively as agents for manipulating cellular ion concentrations, one must have a practical understanding of their properties and


*Methods in Neurosciences, Volume 27*

modes of action. We shall briefly review the major mechanistic classes of ionophore and then outline practical considerations for their use. Salient features of the most commonly used ionophores will then be reviewed.

## Mechanisms

Ionophores may be divided broadly into two groups: channel-formers and carriers. Carriers bind an ion at one interface, shuttle the ion across the hydrocarbon region of the membrane, and deposit it at the other interface. Channel-formers allow ions to flow across the membrane by creating a hydrophilic pathway through the hydrophobic interior (Fig. 1).

### *Carriers*

Carrier ionophores form discrete complexes with the transported ions and shuttle them back and forth across the membrane. Carrier-catalyzed ion transport proceeds at rates several orders of magnitude slower than channel-catalyzed transport, but is generally more selective. Carrier-mediated transport is also more sensitive to the physical state of the membrane and is drastically reduced in gel-state membranes. Carrier-mediated transport may be electrogenic or electroneutral, depending on whether a net movement of charge accompanies transport.

#### *Electrogenic Carriers*

Ionophores that are neutral and contain no ionizable groups will, in general, carry out electrogenic transport. This process is outlined in Fig. 2A. The binding of a cation to the neutral ionophore results in a charged complex that is nevertheless membrane soluble because the ionophore envelops the charge with polar groups and presents a hydrophobic surface to the lipid. The dissociation of the cation at the other interface results in the transfer of net charge across the membrane. In the presence of an ion concentration gradient, this transport causes an electrical potential to be established. For cation transport to be maintained, cotransport of an anion or countertransport of a cation (e.g., proton) must occur.

Electrogenic ionophores will alter ion concentrations so that permeant ions move toward electrochemical equilibrium. This relationship is described by the Nernst–Planck equation:

$$\Delta\Psi = \frac{RT}{zF} \ln \frac{[M^{z+}]_{out}}{[M^{z+}]_{in}},$$

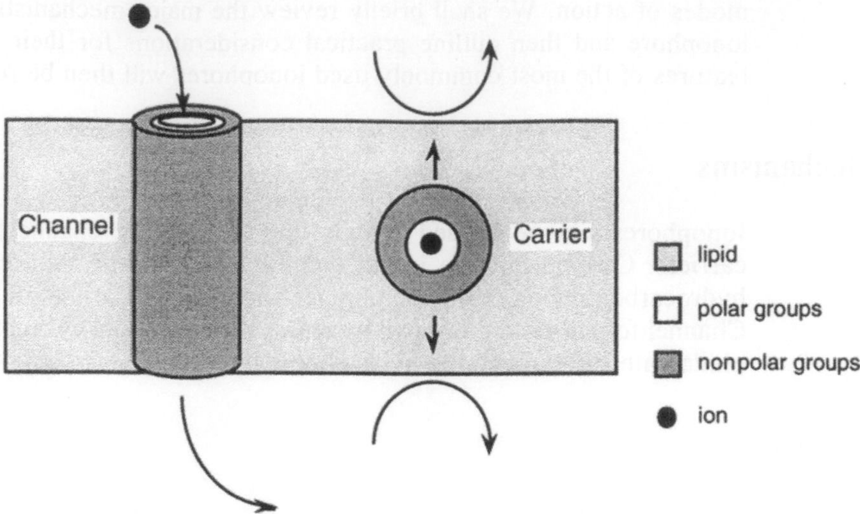


FIG. 1 Carrier versus channel-mediated ion transport.

where $\Delta\Psi$ is the membrane potential, $R$ is the gas constant, $T$ is the absolute temperature, $F$ is the Faraday constant, and $z$ is the charge on the permeant ion. If more than one species is permeant, then the Goldman–Hodgkin–Katz (GHK) equation is used to describe the final equilibrium condition (for monovalent ions):

$$\Delta\Psi = \frac{RT}{zF} \ln \frac{[M^+]_{out} + S_{mn}[N^+]_{out} + \cdots}{[M^+]_{in} + S_{mn}[N^+]_{in} + \cdots},$$

where $S_{mn}$ is the selectivity ratio for $N^+$ vs. $M^+$. [For a detailed discussion of the origins of these equations, the assumptions they embody, and their extension to more complex cases, see Schultz (1980).]

*Electroneutral Carriers*

A variety of ionophores carry out transport that is electrically silent. These compounds are also called carboxylic acid ionophores because their action depends on the protonation and deprotonation of a carboxylic acid group within the molecule. The transport mechanism of these ionophores is outlined in Fig. 2B. The deprotonated form of the molecule is negatively charged and can bind a cation at one membrane interface. The net-neutral complex then diffuses across the membrane to the other interface and the cation dissociates.

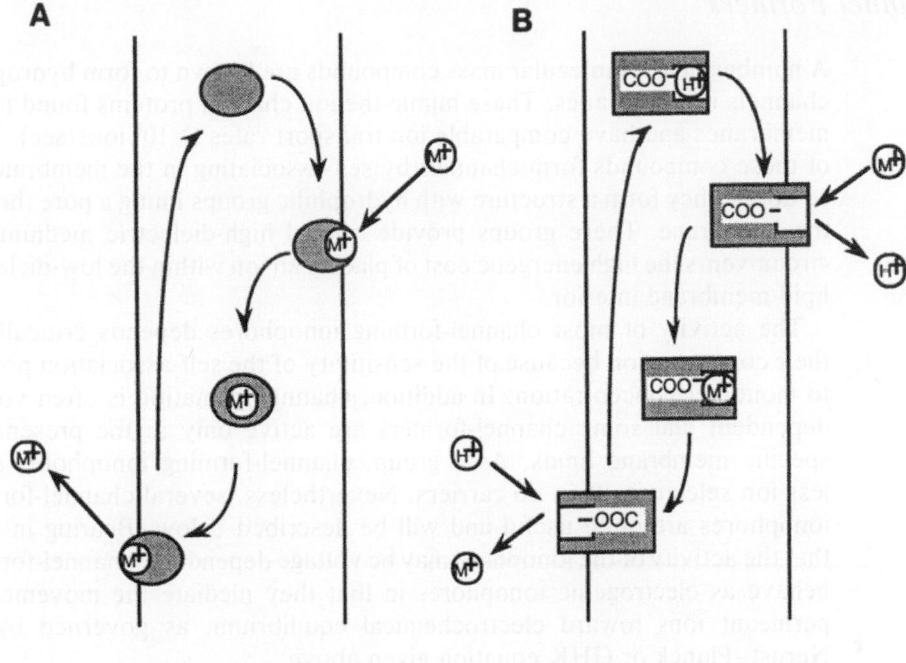


FIG. 2 Electrogenic versus electroneutral ionic flux.

At this point the ionophore is protonated and returns across the membrane in a neutral form. This mechanism results in an exchange of cations for protons, with no net movement of charge. As a result the transport is unaffected by membrane potential and the final ion equilibrium is given by

$$\frac{[H^+]_{out}}{[H^+]_{in}} = \frac{[M^+]_{out}}{[M^+]_{in}} = \frac{[N^{z+}]_{out}^{1/z}}{[N^{z+}]_{in}^{1/z}} = \ldots$$

Note that in this expression the final distribution of ions does not depend on the ion selectivity of the ionophore. If the ionophore can transport a given cation at all, then the distribution of that cation will eventually be affected as described by the equation. If the ionophore transports one ion more efficiently, however, then the distribution of that ion will be affected more rapidly.

### *Channel Formers*

A number of low molecular mass compounds are known to form hydrophilic channels in membranes. These mimic the ion channel proteins found in cell membranes and have comparable ion transport rates ($>10^6$ ions/sec). Most of these compounds form channels by self-associating in the membrane. In so doing, they form a structure with hydrophilic groups lining a pore through the membrane. These groups provide a local high-dielectric medium that circumvents the high energetic cost of placing an ion within the low-dielectric lipid membrane interior.

The activity of most channel-forming ionophores depends critically on their concentration because of the sensitivity of the self-association process to monomer concentration. In addition, channel formation is often voltage dependent and some channel-formers are active only in the presence of specific membrane lipids. As a group, channel-forming ionophores show less ion selectivity than do carriers. Nevertheless, several channel-forming ionophores are quite useful and will be described below. Bearing in mind that the activity of the ionophore may be voltage dependent, channel-formers behave as electrogenic ionophores in that they mediate the movement of permeant ions toward electrochemical equilibrium, as governed by the Nernst–Planck or GHK equation given above.

## Practical Considerations for Ionophore Use

### *Selectivity Is Not Absolute*

Ionophore activity is the net result of a complex series of reaction steps. Ion selectivity reflects the balance of forces and reaction rates that apply at each of these steps. For instance, the rate at which solvating water molecules are replaced by ionophore polar groups will usually differ for different ions. The inherent stability of the ion/ionophore complex, both in interfacial and hydrocarbon environments, will also be a function of ion type. These parameters of ion selectivity can be environment dependent. Variables such as membrane type, temperature, ionophore concentration, membrane potential, and the ionic composition of solutions will influence selectivity. It is important not to accept selectivities as absolute or results can be misinterpreted. For instance, monensin, commonly used as a sodium ionophore, will transport potassium almost as easily in specific instances (Rabaste *et al.*, 1992). Also, X-537A (lasalocid), often thought of as a calcium ionophore, will transport a wide variety of cations with almost equal efficiency (Pressman and Fahim, 1982). Wherever possible, a test of selectivity in the membrane of interest should be made. The presence of a variety of very specific indicators (described elsewhere in this volume) has made this a fairly straightforward task in many cases.

## *Use the Lowest Concentrations Possible: Beware of Side Effects*

As amphiphilic compounds, ionophores have the potential to interact with a variety of membrane-bound systems and other hydrophobic or amphipathic environments within the cell. A range of side effects may result, and several have been documented (Grosman and Nielsen, 1990; Mollenhauer *et al.*, 1990; Takemura *et al.*, 1992). The simplest solution to this problem is to employ the ionophore at the lowest concentration that still produces the desired effect on ion transport. As concentrations increase, the likelihood of other interactions is greater and most ionophores will exhibit nonspecific detergent-like effects on membranes when present at high concentrations (Pressman, 1976; Reed, 1979). High ionophore concentrations can also affect the transport process. For instance, the relative amounts of 1 : 1 vs. 1 : 2 ion/ionophore complexes depend on the ion/ionophore concentration ratio and will affect transport directly (Stiles *et al.*, 1991; Alva *et al.*, 1992).

Ionophore concentrations that are appropriate for a particular set of experimental conditions may need to be reevaluated when conditions are changed. The cell or membrane concentration utilized is especially important in this regard because ionophores generally partition strongly to the lipid phase. An ionophore concentration that is optimal at a relatively high membrane level may then provoke side effects if the amount of membrane in the system is reduced.

## *Test for Solvent Effects: Use Reproducible Protocols*

Ionophores are usually stored in powder form at −20°C or below, to limit hydrolysis and oxidation, and fresh solutions are made just before use. Most ionophores are not very soluble in water and as a result are added to cells from stock solutions in organic solvents. Typical solvents are methanol, ethanol, and dimethyl sulfoxide. The addition of organic solvents to cell membranes can have undesirable effects in certain cases (Reed, 1979). The amount of solvent should be minimized and tested alone (without ionophore) for possible effects. Because of low water solubility, the investigator should be alert to possible carryover of ionophores from incubation to incubation. Rinsing glassware with the solvent employed to prepare stock solutions is recommended. The procedure for adding the ionophore to the cells can also be the source of some difficulty. Some ionophores are slow to incorporate into membranes and can precipitate in solution, especially if added in a concentrated form. Although there is no general solution to this problem for every cell type, it is useful to carry out the addition in a reproducible manner (i.e., the same volumes, concentrations, stirring, and rate of addition).

With these principles in mind, we now summarize the activity of specific ionophores that are commonly used in cell biological studies for manipulating ion concentrations. For a more extensive compilation of the known compounds, the reader is referred to Taylor *et al.* (1982).

## Carrier Ionophores

### *Potassium Selective*

#### *Valinomycin*

The compound valinomycin was one of the first ionophores to be discovered (Pressman *et al.*, 1967) and is one of the most reliable. The structure is a macrocyclic ring containing both amide and ester linkages [it is thus designated a depsipeptide (Ovchinnikov *et al.*, 1972)].

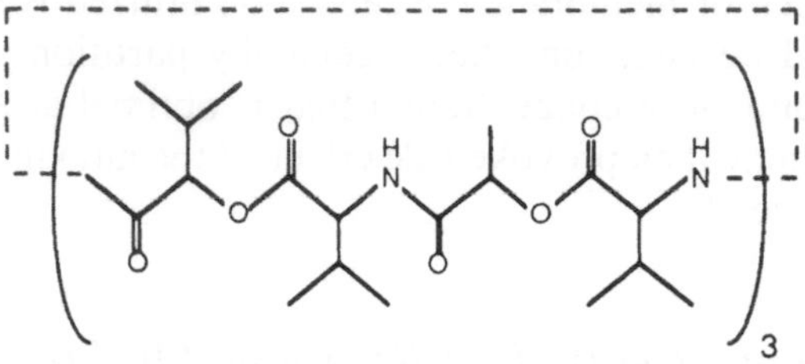

Valinomycin is an electrogenic ionophore that has a selectivity for potassium over sodium of more than 10,000 to 1 (Pressman, 1976). It will also transport $Cs^+$ and $Rb^+$ about as efficiently as $K^+$, but will not transport $Li^+$. Valinomycin is soluble in a variety of organic solvents and inserts quite readily into membranes. It is effective at concentrations as low as $10^{-7}$ to $10^{-9}$ *M*. The rate of ion transport by valinomycin is rather sensitive to the lipid composition of the membrane and can vary over several orders of magnitude (Reed, 1979). Valinomycin may be reliably used to set membrane potential in the presence of a known $K^+$ gradient. It will also mediate selective $K^+$ movement on a large scale if charge balance is provided by another mechanism [e.g., a protonophore (see below)].

*Nigericin (X-464, Polyetherin A)*

Nigericin is an electroneutral ionophore that exchanges $K^+$ for $H^+$. It has also been referred to as X-464 and polyetherin A*. The selectivity sequence of nigericin is $K^+ > Rb^+ > Na^+ > Cs^+ > Li^+$ and the selectivity for $K^+$ over $Na^+$ is about 50 : 1 (Pressman, 1968). Typically, concentrations less than $10^{-6}$ *M* are used to effect $K^+/H^+$ exchange.

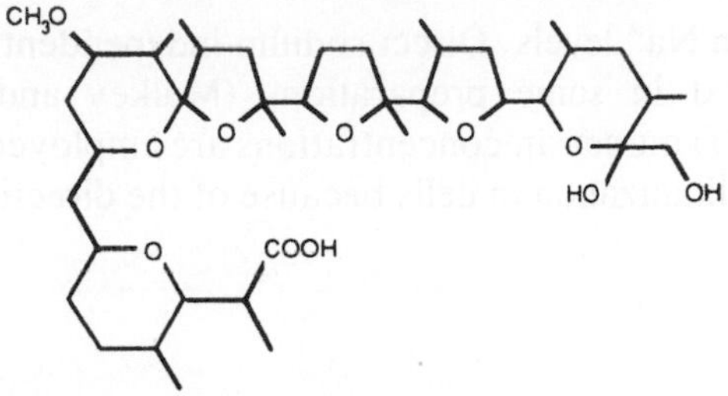

At concentrations above $10^{-6}$ *M*, nigericin can catalyze a pH- and concentration-dependent electrogenic transport of potassium that is believed to involve nigericin dimers (Alva *et al.*, 1992). Nigericin normally produces a transient intracellular acidification because of the direction of the normal potassium gradient in cells.

## Sodium Selective

*Monensin (A3823)*

Monensin is an electroneutral ionophore similar to nigericin except that it is selective for sodium. The selectivity sequence of monensin is $Na^+ > K^+ > Rb^+ > Li^+ > Cs^+$ and the selectivity for sodium over potassium is about 10 : 1. This selectivity is not high and recent studies have emphasized the importance of checking the selectivity in the system of interest (Rabaste *et al.*, 1992). Optimal selectivities appear to be obtained with low ($10^{-7}$ *M*) ionophore concentrations (Takemura *et al.*, 1992).

Monensin has a wide variety of cellular effects that may or may not be related to its ionophore activity. Many of these have been tabulated by Mollenhauer *et al.* (1990). Monensin will often affect intracellular calcium levels (Pressman and Fahim, 1982). This can happen indirectly through activity of the cellular $Na^+/Ca^{2+}$ transporter following monensin-mediated alter-

*The nomenclature of ionophores can be rather confusing and the same compound can go by several names. Compounds designated by Hofmann–LaRoche begin with X- or Ro-; Eli Lilly uses A-; U- refers to the Upjohn Co.

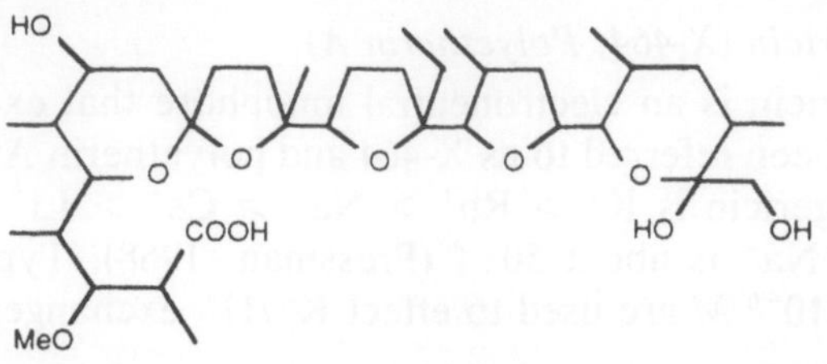

ation in $Na^+$ levels. Direct sodium-independent $Ca^{2+}$ transport has also been reported in some preparations (Mulkey and Zucker, 1992) when high ($10^{-5}$ *M*) monensin concentrations are employed. Monensin produces a transient alkanization in cells because of the direction of the normal sodium gradient.

## *Calcium Selective*

### *A23187*

The importance of $Ca^{2+}$ as an intracellular messenger accounts for much of the interest in $Ca^{2+}$ ionophores. The compound designated A23187 and its brominated analog 4-Br-A23187 are widely used (bromination is on the benzoxazole ring at the position adjacent to the $CH_3NH$ substituent). A23187 has also been referred to as calcimycin. An extensive tabulation of the biochemical and biological effects of A23187 in a variety of systems has been published (Reed, 1982).

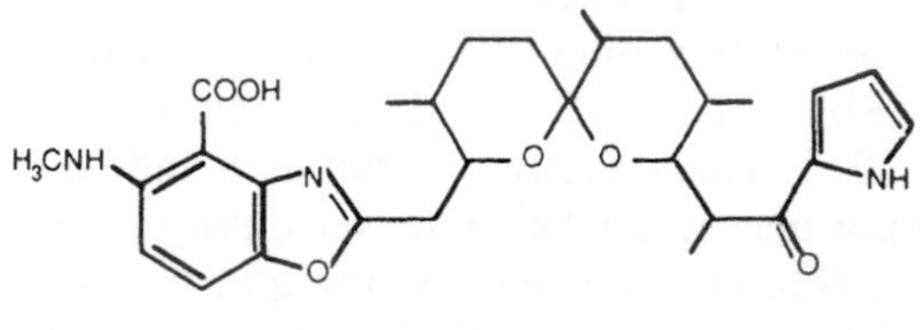

A23187 operates primarily as an electroneutral $Ca^{2+}/2H^+$ exchanger (Pohl *et al.*, 1990; Erdahl *et al.*, 1994). Other divalent cations ($Mn^{2+}$, $Mg^{2+}$) are transported, but monovalent cations (i.e., $K^+$ and $Na^+$) are not transported effectively when divalent ions are present (Reed, 1982). Note that two protons are exchanged for each calcium ion. Two ionophore molecules bind to one calcium ion to form a neutral complex. One-to-one complexes can also form but these are charged and do not cross the lipid membrane. A23187 may form channels under certain conditions, although it is not yet clear to what extent this activity contributes to overall transport (Balasubramanian *et al.*, 1992). The ionophore is typically added in concentrations between $10^{-7}$ and $10^{-5}$ *M*. Variations in protocol have been reported to affect the efficacy of A23187 (Grosman and Nielsen, 1990). The brominated version of A23187 is nonfluorescent (in contrast to the parent compound), which makes it useful where interference with a fluorometric assay might be a problem (Deber *et al.*, 1985). 4-Br-A23187 is also somewhat more selective for $Ca^{2+}$ over $Mg^{2+}$ compared to the parent compound (Debono *et al.*, 1981).

*Ionomycin*

Ionomycin is another electroneutral carrier for calcium. It has an equal or greater transport efficiency than A23187 for calcium, depending on the concentration considered (Erdahl *et al.*, 1994). In this case a neutral 1 : 1

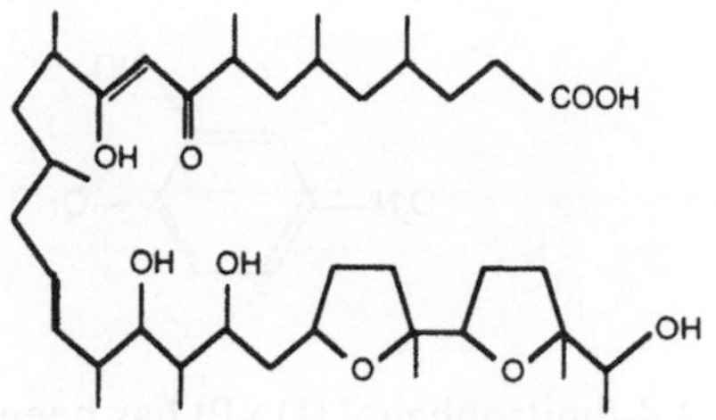

COOH
OH O
OH OH
OH
O O

complex forms. The selectivity of ionomycin is similar to that of A23187 (Liu and Hermann, 1978; Stiles *et al.*, 1991).

*Synthetic Cyclic Peptides*

The potential of cyclic peptides to complex ions via a template of amide carbonyl groups has been explored in several studies. Although the transport properties of these compounds have not yet been optimized, the activities of cyclo[Glu-Sar-Gly-(*N*-decyl)Gly]$_2$ (DECYL-2) and cyclo[Glu(OBz)-Sar-Gly-(*N*-cyclohexyl)Gly]$_2$ (CYCLEX-2E) are notable. The former appears to carry out electroneutral calcium/proton exchange, e.g., in dog and human lymphocytes (Deber and Hsu, 1986), whereas the latter can act as an electrogenic calcium ionophore in phospholipid membranes if calcium is present in the solution (Deber *et al.*, 1978, 1980). The overall mechanism by which CYCLEX-2E mediates transmembrane calcium transport must differ in detail from the mode of calcium transport by ionophore A23187. Because CYCLEX-2E (dibenzyl ester) lacks exchangeable protons, its mode of cation transport may be more formally analogous to the prototypic $K^+$ ionophore valinomycin (vide infra). Thus, both substances generate positively charged species on complexation with cations, which require a compensating process of charge balance to complete a transport cycle. In Pressman cells, transport by CYCLEX-2E requires picrate (trinitrophenol) as a lipophilic anion to solubilize cation complexes in organic phases, as well as to assure the overall charge balance (Deber *et al.*, 1980). Yet, CYCLEX-2E-mediated $Ca^{2+}$ efflux from phospholipid vesicles was noted to be essentially independent of directly accompanying anions (such as chloride, acetate, succinate, or sulfate); however, hydroxide was not ruled out as a functional anion for the CYCLEX-2E–$Ca^{2+}$ complex.

## *Proton Ionophores*

### *Dinitrophenol*

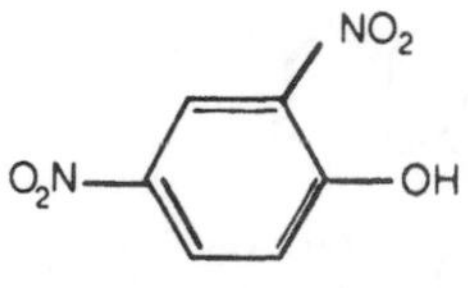

$NO_2$

$O_2N$— —OH

The compound 2,4-dinitrophenol (DNP) has been used for many years as an uncoupler of oxidative phosphorylation. Because the negative charge of the phenolate anion is delocalized, the deprotonated as well as the protonated forms of DNP can cross lipid membranes. DNP thus acts to conduct protons and dissipate pH gradients. Relatively high concentrations ($10^{-5}$ to $10^{-4}$ *M*) are required and the solutions are colored, which may be a problem if spectrophotometric measurements are performed.

### *Carbonylcyanide*-p-*trifluoromethoxyphenylhydrazone*

F-C(F)(F)-O— —N(H)—N=C(CN)(CN)

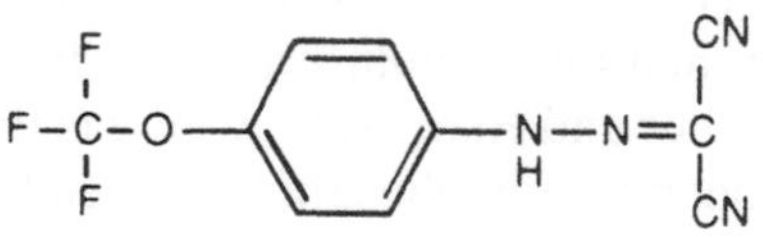

Like dinitrophenol, carbonylcyanide-*p*-trifluoromethoxyphenylhydrazone (FCCP) can cross membranes in protonated and deprotonated form and so conducts protons across membranes. Together with carbonylcyanide-*m*-chlorophenylhydrazone (CCCP), FCCP is a commonly used uncoupler. These compounds are effective at $10^{-7}$–$10^{-8}$ *M* concentrations.

## Other Carriers

*Nonactin (Macrotetralide Actins)*

The compound nonactin is a member of a family that includes dinactin, trinactin, and tetranactin—the macrotetralide actins. These are electrogenic carriers with properties similar to those of valinomycin, although with lower $K^+/Na^+$ selectivities.

*Enniatins*

Enniatins are a family of macrocyclic carriers that are electrogenic like the actins and valinomycin. They have selectivity ratios for $K^+$ over $Na^+$ of less than 10:1 (Ovchinnikov *et al.*, 1972; Pressman, 1976).

*Lasalocid (X-537A; RO 2-2985)*

Lasalocid is used commercially as a feed additive and coccidiostat. It has been employed as a calcium ionophore but is of limited use in this regard because it will also transport a wide variety of other cations, including biogenic amines (Pressman and Fahim, 1982).

*Synthetic Polyethers and Cryptands*

A wide variety of synthetic ion-complexing polyethers (crown ethers) and cryptands have been synthesized (Fyles, 1990; Vogtle, 1991). These can complex specific ions in solution and form stable ionic complexes. Selectivity is related to the size of the central cavity formed by the molecule. For instance, dicyclohexyl-14-crown-4 has a $Na^+/K^+$ selectivity of about 10:1, whereas dibenzo-21-crown-7 has a 5:1 selectivity for $Cs^+$ over $K^+$ (Pedersen, 1968; Pressman, 1976). The transport activity of many of these compounds is complex, however (Eisenman *et al.*, 1972; Fyles, 1990). As this field continues to develop, compounds that will be of specific use in cell biology are likely to be identified.

*ETH 129, ETH 1001*

Several compounds developed at the Swiss Federal Institute of Technology (ETH) show promise as calcium ionophores. In particular, ETH 129 (*N,N,N',N'*-tetracyclohexyl-3-oxapentane-diamide) has been reported to be an effective electrogenic calcium ionophore in natural and artificial membranes (Prestipino *et al.*, 1993).

*Anion Carriers*

Useful naturally occurring anion ionophores are virtually unknown. Trialkyltin compounds have found application in anion-selective electrodes (Wuthier *et al.*, 1984) and other synthetic anion-complexing compounds have been described (Fyles, 1990; Dietrich, 1993). Although several of these form selective, stable complexes, they are usually highly charged and thus are not very efficient as transporters in membranes.

## Channel-Forming Ionophores

### *Gramicidin D (Gramicidins A, B, and C)*

The commercially available substance, gramicidin D, is a mixture of gramicidins A (88%), B (7%), and C (5%), which all have similar properties (Woolley and Wallace, 1992). Each is a 15-amino acid linear peptide with alternating L and D amino acids. The linear gramicidins dimerize in membranes to form

HCO-L-Val-Gly-L-Ala-D-Leu-L-Ala-D-Val-L-Val-D-Val-
L-Trp-D-Leu-L-Trp-D-Leu-L-Trp-D-Leu-L-Trp-$NHCH_2CH_2OH$

channels that are highly selective for monovalent cations. Divalent cations and anions are effectively excluded from the channel. Molecules larger than about 4.5 Å in diameter are also excluded. The peptide is soluble in polar organic solvents and must undergo conformational reorganization to assume its membrane-active form (Woolley and Wallace, 1992). This process can be slow, and together with the possibility of precipitation of the peptide in the aqueous phase, can present problems in getting it into a membrane in certain instances. However, the peptide is effective at very low concentrations ($<10^{-9}$ *M*) because its channel mechanism permits much higher transport rates than do carrier mechanisms per active unit. [Gramicidin S (a cyclic decapeptide) is structurally and functionally unrelated.]

### *Nystatin*

Nystatin is a polyene antibiotic, so named for the series of conjugated double bonds it contains. Nystatin forms channels in membranes by self-associating in a complex with cholesterol or other sterols. Thus, the presence of sterols is a prerequisite for nystatin activity. Because several nystatin molecules are needed to form one channel, the activity is highly concentration dependent [fourth to twelfth power, depending on the membrane (McLaughlin and Eisenberg, 1975)]. The pores formed have a radius between 4 and 8 Å and will permit the passage of monovalent cations and anions. Multivalent ions appear to be excluded. The cation/anion selectivity is sensitive to membrane properties and depends on whether the antibiotic is added to one or both sides of the membrane (McLaughlin and Eisenberg, 1975; Bolard, 1986). The activity of nystatin is also temperature sensitive, decreasing at higher temperatures. Stock solutions [commonly in dimethyl sulfoxide (DMSO)] should be made fresh because polyenes are quite sensitive to degradation. Nystatin is often used in electrophysiology to create "perforated patches" (Horn, 1991). By permitting the passage of small ions, it allows electrical access to the interior of a cell without causing large-scale disruption or leakage of cell contents. It thus permits electrical measurements to be made relatively noninvasively.

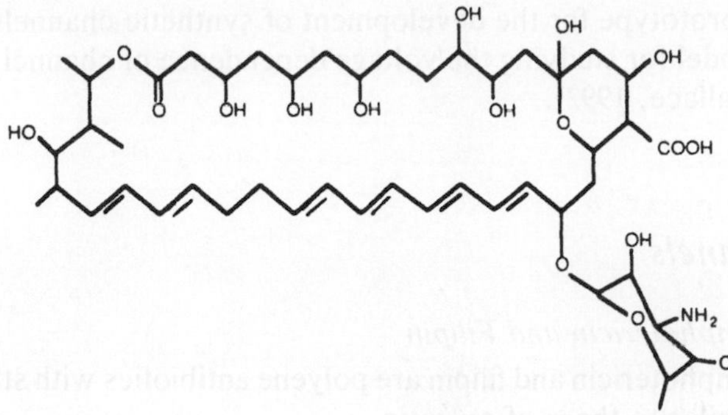

### *Alamethicin*

The peptide alamethicin forms relatively nonselective channels in membranes (Woolley and Wallace, 1992). Both anions and cations (monovalent and

Ac-Aib-Pro-Aib-Ala-Aib-Ala-Gln-Aib-Val-Aib-Gly-Leu-
Aib-Pro-Val-Aib-Aib-Gln-Gln-Pheol

Aib = aminoisobutyric acid
Pheol = phenylalaninol

divalent) can pass through alamethicin channels. Like nystatin, the activity of alamethicin is concentration dependent. Channel formation by alamethicin is voltage dependent; if added to one side of a membrane, it is most active when the opposite side is made negative. Alamethicin has found some use as a permeabilizing agent (Ritov *et al.*, 1993), although rather large molecules can pass through alamethicin channels and the voltage dependence of activity can complicate electrical measurements. Alamethicin has proved useful as a prototype for the development of synthetic channels (see below) and as a model for studying the voltage dependence of channel activity (Woolley and Wallace, 1992).

## *Other Channels*

### *Amphotericin and Filipin*

Amphotericin and filipin are polyene antibiotics with structures and activities similar to those of nystatin.

### *Synthetic Channels*

There has been considerable progress in creating synthetic channels (Fyles *et al.*, 1990a,b; Woolley and Wallace, 1992; Akerfeldt *et al.*, 1993; Montal *et al.*, 1993). Peptidic channels appear to form in the same manner as the peptide alamethicin by self-associating in membranes. Peptides with sequences corresponding to putative pore-lining segments of known ion channel proteins have in some cases been found to form channels with activity similar to that of native channels (Montal *et al.*, 1993). The properties of these channels are better defined when an aggregate of peptides (four or five monomers) is held together covalently (Akerfeldt *et al.*, 1993; Montal *et al.*, 1993). Nonpeptide synthetic channels are also under investigation (Fyles *et al.*, 1990a,b; Pregel *et al.*, 1992). If selectivity can be controlled, these compounds offer considerable promise as tools for use in cell biology because they would be highly active at very low concentrations.

## Summary

The preceding inventory of ionophores and their activities illustrates their versatility and applications in a variety of systems. Although there are caveats associated with their use, ionophores remain an extraordinarily powerful tool for cell biologists. An understanding of the limitations of ionophores is perhaps an inevitable accompaniment to the excitement generated by discoveries of their remarkable activity. Developments in the design and synthesis of new ionophores will continue. In the meantime, with due attention to the practical considerations involved in their use, ionophores can be used effectively in many situations to manipulate ion concentrations and answer significant biological questions.

## References

Akerfeldt, K. S., Lear, J. D., Wasserman, Z. R., Chung, L. A., and DeGrado, W. F. (1993). *Acc. Chem. Res.* **26,** 191.

Alva, R., Lugo-R, J. A., Arzt, E., Cerbon, J., Rivera, B. E., Toro, M., and Estrada-O. S. (1992). *J. Bioenerg. Biomembr.* **24,** 125.

Ammann, D., Oesch, U., Buhrer, T., and Simon, W. (1987). *Can. J. Physiol. Pharmacol.* **65,** 879.

Balasubramanian, S. V., Sikdar, S. K., and Easwaran, K. R. K. (1992). *Biochem. Biophys. Res. Commun.* **189,** 1038.

Bolard, J. (1986). *Biochim. Biophys. Acta* **64,** 257.

Deber, C. M., and Hsu, L. C. (1986). *Biochem. Biophys. Res. Commun.* **134,** 731.

Deber, C. M., Adawadkar, P. D., and Tom-Kun, J. (1978). *Biochem. Biophys. Res. Commun.* **81,** 1357.

Deber, C. M., Young, M. E. M., and Tom-Kun, J. (1980). *Biochemistry* **19,** 6194.

Deber, C. M., Tom-Kun, J., Mack, E., and Grinstein, S. (1985). *Anal. Biochem.* **146,** 349.

Debono, M., Molloy, R. M., Dorman, D. E., Paschal, J. W., Babcock, D. F., Deber, C. M., and Pfeiffer, D. R. (1981). *Biochemistry* **20,** 6865.

Dietrich, B. (1993). *Pure Appl. Chem.* **65,** 1457.

Eisenman, G., Szabo, G., McLaughlin, S. G. A., and Ciani, S. M. (1972). *in* "Molecular Mechanisms of Antibiotic Action on Protein Biosynthesis and Membranes" (E. Munoz, F. Garcia-Ferrandiz, and D. Vazquez, eds.), p. 545. Elsevier, Amsterdam.

Erdahl, W. L., Chapman, C. J., Taylor, R. W., and Pfeiffer, D. R. (1994). *Biophys. J.* **66,** 1678.

Fyles, T. M. (1990). *in* "Bioorganic Chemistry Frontiers" (H. Dugas, ed.), p. 71. Springer-Verlag, Berlin.

Fyles, T. M., James, T. D., and Kaye, K. C. (1990a). *Can. J. Chem.* **68,** 976.

Fyles, T. M., Kaye, K. C., James, T. D., and Smiley, D. W. M. (1990b). *Tetrahedron Lett.* **31,** 1233.

Grosman, N., and Nielsen, J. K. (1990). *Agents Actions* **30,** 131.

Horn, R. (1991). *Biophys. J.* **60,** 329.

Liu, C. M., and Hermann, T. E. (1978). *J. Biol. Chem.* **253,** 5892.

McLaughlin, S., and Eisenberg, M. (1975). *Annu. Rev. Biochem.* **43,** 335.

Mollenhauer, H. H., Morre, D. J., and Rowe, L. D. (1990). *Biochim. Biophys. Acta* **1031,** 225.

Montal, M. O., Iwamoto, T., Tomich, J. M., and Montal, M. (1993). *FEBS Lett.* **320,** 261.

Mulkey, R. M., and Zucker, R. S. (1992). *Neurosci. Lett.* **143,** 115.

Ovchinnikov, Y. A., Ivanov, V. T., and Shkrob, A. M. (1972). *in* "Molecular Mechanisms of Antibiotic Action on Protein Biosynthesis and Membranes" (E. Munoz, F. Garcia-Ferrandiz, and D. Vazquez, eds.), p. 459. Elsevier, Amsterdam.

Pedersen, C. J. (1968). *Fed. Proc.* **27,** 1305.

Pohl, P., Antonenko, Y. N., and Yaguzhinsky, L. S. (1990). *Biochim. Biophys. Acta* **1027,** 295.

Pregel, M. J., Jullien, L., and Lehn, J. M. (1992). *Angew. Chem., Int. Ed. Engl.* **31,** 1637.

Pressman, B. C. (1968). *Fed. Proc.* **27,** 1283.

Pressman, B. C. (1976). *Annu. Rev. Biochem.* **45,** 501.

Pressman, B. C., and Fahim, M. (1982). *Annu. Rev. Pharmacol. Toxicol.* **22,** 465.

Pressman, B. C., Harris, E. J., Jagger, W. S., and Johnson, J. H. (1967). *Proc. Natl. Acad. Sci. U.S.A.* **58,** 1949.

Prestipino, G., Falugi, C., Falchetto, R., and Gazzotti, P. (1993). *Anal. Biochem.* **210,** 119.

Rabaste, F., Jeminet, G., Dauphin, G., and Delort, A. M. (1992). *Biochim. Biophys. Acta* **1108,** 177.

Reed, P. W. (1979). *in* "Methods in Enzymology" (S. Fleischer and L. Packer, eds.), Vol. 55, p. 435. Academic Press, New York.

Reed, P. W. (1982). *in* "Polyether Antibiotics" (J. W. Westley, ed.), Vol. 1, p. 185. Dekker, New York.

Ritov, V. B., Murzakhmetova, M. K., Tverdislova, I. L., Menshikova, E. V., Butylin, A. A., Avakian, T. Y., and Yakovenko, L. V. (1993). *Biochim. Biophys. Acta* **1148,** 257.

Ruff, M. D. (1982). *in* "Polyether Antibiotics" (J. W. Westley, ed.), Vol. 1, p. 303. Dekker, New York.

Schultz, S. G. (1980). "Basic Principles of Membrane Transport." Cambridge Univ. Press, New York.

Stiles, M. K., Craig, M. E., Gunnell, S. L. N., Pfeiffer, D. R., and Taylor, R. W. (1991). *J. Biol. Chem.* **266,** 8336.

Takemura, H., Li, Z., and Ohshika, H. (1992). *Biochem. Pharmacol.* **44,** 1395.

Taylor, R. W., Kauffman, R. F., and Pfeiffer, D. R. (1982). *in* "Polyether Antibiotics" (J. W. Westley, ed.), Vol. 1, p. 103. Dekker, New York.

Vogtle, F. (1991). "Supramolecular Chemistry." Wiley, Chichester.

Woolley, G. A., and Wallace, B. A. (1992). *J. Membr. Biol.* **129,** 109.

Wuthier, U., Pham, H. V., Zund, R., Welti, D., Funck, R. J. J., Bezegh, A., Ammann, D., Pretsch, E., and Simon, W. (1984). *Anal. Chem.* **56,** 535.

# [5] Fluorescence Imaging of Cytosolic Calcium: An Introduction to Basic Experimental Principles

David A. Williams

Many excellent articles have reviewed the general field of $Ca^{2+}$ imaging, and special issues of the journal *Cell Calcium** dedicated to this area are of particular note and provide valuable reference material. The discussion herein is limited to the imaging of cellular ionized calcium in living cells. The majority of studies in this area have utilized fluorescence techniques, thus this chapter is predominantly a description of some of the techniques most commonly used in the author's laboratory for imaging of fluorescent $Ca^{2+}$-sensitive indicators in cells and tissues.

The object of this type of study is to relate a fluorescence image or intensity to some feature of cell structure or function. If these images are collected over a period of time, the changes within them give some indication of cell behavior and changing function. However, it is important to ensure that the tools for the measurement (i.e., the fluorophores and the imaging technology) do not affect the measurement that is being made (see Fig. 1). This chapter will indicate some of the problems that must be avoided and will provide some suggestions and, where applicable, protocols to avoid or minimize these potential problems.

Figure 2 shows a flow chart that indicates the main steps and substeps carried out in undertaking the imaging of cellular $Ca^{2+}$. The initial steps deal with the choices of tissue preparations and the types of microscopic techniques and fluorophores that should be used. These decisions are generally of a "once-off" nature, but are crucial in the overall success of the measurements.

## Isolation of Tissues

Although an infinite number of tissues are suitable for imaging studies, these generally fall into four main classes: (1) freshly isolated cells, (2) cultured cells, (3) strips of cells (small tissue samples), and (4) intact organs or muscles.

*Imaging of Cell Calcium (Feb./Mar., 1990); Oscillations in Cell Calcium (Feb./Mar., 1991); $Ca^{2+}$ Signalling: Waves and Gradients (Nov., 1993).

*Methods in Neurosciences, Volume 27*

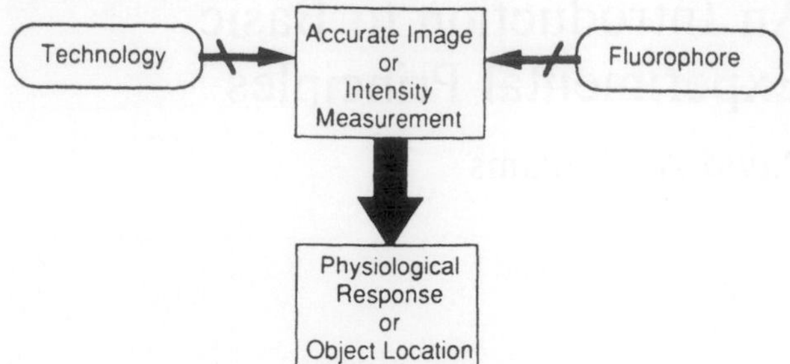


FIG. 1 Schematic representation of the aims of a $Ca^{2+}$ imaging experiment. The link between image and physiological response should not be affected deleteriously by the technology or fluorophore used in the imaging procedures.

The choice of preparation will mainly determine the type of microscopy that must be employed in the investigation. Laser-scanning confocal microscopy (LSCM; see Refs. 1 and 2 for review) can be used with all of these tissue types, whereas conventional, wide-field fluorescence microscopy is most accurately used with cells (options 1 and 2), because of significant contributions of out-of-focus data to the two-dimensional images of the larger tissue samples. The majority of studies from this laboratory have employed freshly isolated cells, thus descriptions will be based on this type of preparation unless otherwise indicated.

## Sample Loading

The loading of cell preparations with fluorescent $Ca^{2+}$-sensitive indicators is largely a matter of exposing the tissue to esterified derivatives of the desired fluorophore, and after a period of time, removing nonsequestered dye. However, there are many effectors of the success of this loading process, (see Fig. 3). The ultimate aim of this process is to incorporate enough indicator to allow sufficient signal to record, while minimizing buffering of $[Ca^{2+}]$ (see later) and avoiding an inner filter effect, both of which may result from higher than necessary intracellular concentrations. Arbitrarily, we aim to maintain the intracellular level of fluorophore at less than, or around, 100 $\mu M$. The inner filter effect is largely dependent on both the internal fluorophore concentration and the size (or thickness) of a cell and represents the distortion from linearity of the relationship between fluorescence output and absolute dye concentration. We have previously quoted values of 12,000 $\mu M \cdot \mu m$ as an upper level for the desirable range for the product of cell density and fluorophore concentration.

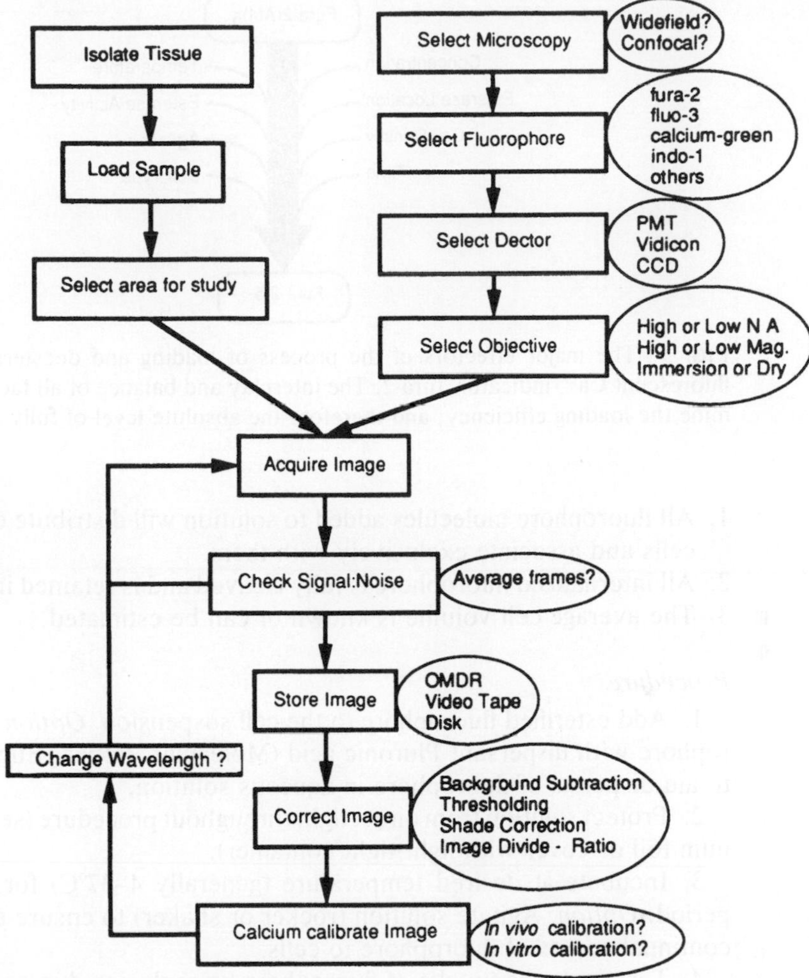


FIG. 2 Flow chart of the main procedural steps and decision processes of a $Ca^{2+}$ imaging experiment. PMT, Photomultiplier tube; CCD, charge-coupled device; NA, numerical aperture; OMDR, optical memory disk recorder.

## *Loading Procedure*

### *Initial Calculations*

Determine cell density or total cell number for isolated or cultured cells. For a chosen concentration of esterified fluorophore, calculate the average concentration that will be achieved by each cell given the following assumptions:

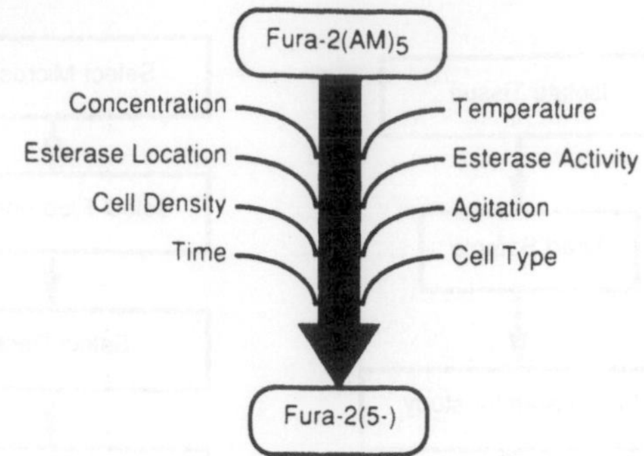


FIG. 3 The major effectors of the process of loading and deesterification of the fluorescent $Ca^{2+}$ indicator, fura-2. The interplay and balance of all factors will determine the loading efficiency, and therefore the absolute level of fully cleaved fura-2.

1. All fluorophore molecules added to solution will distribute evenly among cells and associate exclusively with them.
2. All internalized fluorophore is fully cleaved and is retained in cell cytosol.
3. The average cell volume is known or can be estimated.

*Procedure*

1. Add esterified fluorophore to the cell suspension. *Option:* Premix fluorophore with dispersant Pluronic acid (Molecular Probes, Eugene, Oregon) to aid dispersal of fluorophore in aqueous solution.

2. Protect solution from direct light throughout procedure (seal with aluminum foil or cover with light-tight container).

3. Incubate at desired temperature (generally 4–37°C) for defined time period. *Option:* Agitate solution (rocker or shaker) to ensure enhanced and continual access of fluorophore to cells.

4. Terminate the uptake of fluorophore by replacing dye-containing solution with fresh cell buffer. This is relatively easy for cells in culture or for tissue and organ samples. Cells in suspension, particularly noncontractile cells, can be concentrated through gentle centrifugation (200–500 rpm) before removal and replacement of cell medium. If centrifugation is not desirable, loading can effectively be terminated by a large dilution of the loading medium with fresh buffer.

5. Allow time (20 min is generally sufficient) for the complete intracellular cleavage of all masking acetoxymethyl ester groups from internalized fluorophore.

A number of features of this process can be modified to influence the success of the loading procedure (summarized in Fig. 3), of which the two main effectors are time (access time of cells to fluorophore) and temperature (affecting the enzymatic removal of ester groups). Indeed, it is essential to adjust these parameters when initiating experiments with a new or different cell type. A single recipe does not suffice for all cell types because the amount, activity, and intracellular location of nonspecific cell esterases vary immensely between cell types. One key strategy that should be employed when optimizing this process involves performing the loading procedure with a large range of initial concentrations (e.g., 1–25 $\mu M$) of the fluorophore, with otherwise identical loading parameters. This should result in a large range of intracellular concentrations of the fluorophore, and at the highest levels, basal $Ca^{2+}$ levels and agonist-induced $Ca^{2+}$ responses may be markedly affected by $Ca^{2+}$ buffering (3). The ideal initial loading concentration should result in the highest possible intracellular concentration (thereby providing the highest signal level), but the loading concentration should not affect the variables under investigation.

## Select Cell Area for Study

The absolute density of seeding of cells onto the base of the experimental chamber is not crucial, apart from ensuring that individual cells can be clearly distinguished and do not overlap each other. There are many effective designs for experimental chambers and these are largely dictated by the other (nonimaging) requirements of the experimental protocol. The only essential feature of a vessel for imaging studies is that the interface between the microscope objective and the cell of interest does not interfere with image acquisition or contribute to background fluorescence as many plastics do. A thin (size 1; 0.017 mm) coverglass is often incorporated into the base of even the most complicated experimental chambers. Another viable alternative is to use water immersion objectives, but this generally restricts the experiment to an upright fluorescence microscope. The field of view may contain a single cell or a group of cells because the fluorescence of each cell can be separately analyzed following image collection.

## Choice of Fluorescent $Ca^{2+}$-Sensitive Probe

A large number of $Ca^{2+}$-sensitive fluorophores have been used for imaging studies in living cells and tissues, and the list is growing constantly. To a large extent the properties of the chosen probe dictate the nature of the experiment. However, it is absolutely essential that something is known about the biological response under investigation, and, in particular, the expected magnitude and speed of $Ca^{2+}$ response will influence both the choice of fluorophore and the equipment required for recording the response.

The $Ca^{2+}$-sensitive fluorophores fall into several classes based on the type of spectral response to $Ca^{2+}$ binding. One group of probes exhibits a change in the quantum yield or fluorescence efficiency (i.e., change in brightness) with $Ca^{2+}$ binding. Included in this group are some of the most commonly used probes, such as fluo-3 and calcium green, as well as rhod-2, calcium orange, and calcium crimson, for which there are few published reports of use.

The second class includes fura-2, quin-2, fura-red, and mag-fura-2 (furaptra). These molecules exhibit large decreases in the wavelengths required for peak absorbance (excitation) with $Ca^{2+}$ binding, concomitant with a change in fluorescence quantum yield. The final group, of which the most notable member is indo-1, exhibits shifts in both the absorption and emission characteristics. The latter two groups of fluorophores offer the distinct advantage of allowing highly quantitative investigation of $Ca^{2+}$. Each of these indicators displays a unique excitation and/or emission spectrum for a given concentration of ionized $Ca^{2+}$, and the presence of these spectrally distinct forms results in the presence of a distinct isosbestic point in the spectra (point where fluorescence is independent of ion concentration). These spectra can be uniquely and accurately described mathematically by the ratio ($R$) of two points on any spectral curve, providing calibration curves relating $[Ca^{2+}]$ to the fluorescence ratio.

The fluorescent probe chosen should have a dissociation constant ($K_d$) that falls within the range of $[Ca^{2+}]_i$ that is expected during the physiological response. These constants, as well as other important characteristics of most of the commercially available fluorophores, are detailed in the catalogs of the major vendors (Molecular Probes, Texfluor Labs, Calbiochem). However, it is absolutely essential for each individual user of these fluorophores to establish these characteristics with their own experimental system and cells.

Other features of the fluorophore–cell combination that should be measured are the rates of leakage of fluorophore from cells and photobleaching of fluorophore within cells. Both of these processes can result in significant reductions in the absolute fluorescence intensities of cells, although for probes that can be used in a dual-wavelength (ratiometric) mode, determinations of $Ca^{2+}$ levels are largely independent of these fluorescence reductions. However, the loss of internalized fluorophore to the external medium is a problem because of high (dye-saturating) levels of $Ca^{2+}$ in this compartment. Although it is possible in most imaging systems, and particularly confocal systems, to minimize the fluorescence contribution of this compartment in image analysis, background fluorescence levels should always be minimized to reduce the propagation of errors in the $Ca^{2+}$ determination (4).

## Conventional or Confocal Microscopy?

There is little doubt that images with the highest spatial resolution will result from the use of laser-scanning confocal microscopy with highly fluorescent, $Ca^{2+}$-sensitive fluorophores. However, most of the commonly used probes for $Ca^{2+}$ measurements are UV excitable and are not readily utilized with the majority of confocal microscopes, which are largely equipped with argon ion lasers. As such, the majority of imaging studies of intracellular calcium still employ standard (wide-field) fluorescence microscopy. The high cost of confocal microscopes is also not a trivial factor in the present distribution of techniques.

It is easy to assume that good video images come simply from using the detection device with the best specifications for the task. However, even imaging devices with the best quantum efficiencies are limited if only low numbers of photons reach the imaging surface. It is therefore absolutely essential to optimize transmission characteristics of the optical system used in association with the imaging device.

## Optimizing the Collection of the Fluorescent Signal of Interest

Figure 4 shows a schematic diagram of a conventional epifluorescence system used for the imaging of intracellular $Ca^{2+}$. To obtain the best performance from such a system, a number of features can be optimized for the cell–fluorophore combination used in an experiment.

The following features should be considered in optimizing excitation of the $Ca^{2+}$ fluorophore:

1. The incident (excitation) light should be clearly separated from the cell–fluorophore signal by careful choice of a highly selective dichroic mirror (DM). Where the Stoke's shift of the fluorophore is small (e.g., fluorescein-like fluorophores), a compromise is required between efficiency of excitation and emission collection.

2. The contributions of cell autofluorescence and other background signals (including glass and container, immersion medium, or mounting medium) should be minimized. Excite the fluorophore at or close to its excitation wavelength maximum by employing narrow-bandpass filters (a 10-nm band-

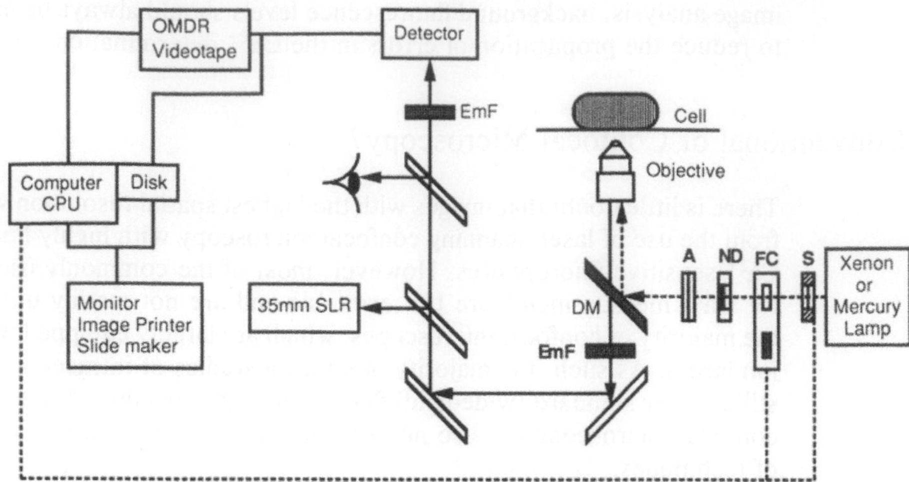


FIG. 4 Schematic diagram of a common arrangement of the elements of a $Ca^{2+}$ imaging system. S, Shutter; FC, filter changer for excitation filters; ND, neutral-density filters; A, variable aperture; DM, dichroic mirror; EmF, emission filter (note: two common locations for this element are shown); SLR, single-lens reflex camera; CPU, central processing unit; OMDR, optical memory disk recorder. Arrowed lines indicate light paths, solid lines represent cable connections, and dashed lines illustrate the optional computer control of some components of the system.

pass is common) or use a monochromatic light source (lamp and monochromator or a suitable laser). Limit the field of view in the specimen to the cell or cells of interest by constricting the illuminated field with a variable aperture (A) in the excitation light path.

3. Ensure that the fluorescence light source provides ample intensity at the desired excitation wavelength. A xenon lamp is most commonly used with the UV-excitable $Ca^{2+}$ probes because of a relatively flat intensity spectrum between 300 and 400 nm. A stabilized power supply will also help by preventing random fluctuations in excitation intensity during the acquisition of data.

4. Ensure that the excitation wavelengths are not absorbed by any element of the optical system of the microscope, particularly important for use of UV-excitable fluorophores. Most commercially available microscopes are now designed for efficient transmission of UV wavelengths, or at least have an optional fluorescence condenser that is. Suspect optical elements, particularly neutral-density (ND) lenses, heat-absorbing filters, and objective lenses, should be checked by obtaining specification sheets from the manufacturer,

or by removing the element and directly measuring the absorbance/transmission characteristics in a standard spectrophotometer.

5. Select objectives very carefully. Ideally, objectives for high-resolution $Ca^{2+}$ imaging should have the greatest combination of the following features: (*i*) high transmission of both excitation and fluorescence wavelengths (see Ref. 4), (*ii*) large working distance, (*iii*) oil or water immersion, and (*iv*) high numerical aperture. Magnification is often dictated by the absolute size of the specimen and the area of interest. In addition, the useful imaging area of many objectives is limited to the central portion of the field of view, presenting the potential of distortion of image content near image edges.

6. Exposure of the sample to light must be minimized to limit photobleaching of the fluorophore and photodamage of the cells. Use a light-tight shutter in the excitation path and expose the fluorophore and cells only during actual image collection. The intensity of incident light exposure can be reduced with neutral-density filters whose wavelength transmission characteristics are known.

Although care is essential to make the incident light highly specific to the excitation characteristics of the fluorophore, it may be even more important to maximize the efficiency of the capture of emitted fluorescence because of the need to minimize exposure duration and intensity. This optimization includes two key features:

1. Match the emission filters (EmF) to the emission spectrum of the fluorophore, while avoiding the inclusion of any competing wavelengths (i.e., autofluorescence emissions or those of other fluorophores if multidye experiments are performed). There are many manufacturers of good quality, high-transmission filters, and companies such as Molecular Probes (Eugene, Oregon) now specifically supply optical filter combinations tailored for a range of fluorescent probes.

2. Choose a detector that is sensitive to the wavelengths of the fluorescence emission (see the following discussion on choice of detector).

## Choice of Detector

Two main types of detectors have been successfully used to image intracellular $Ca^{2+}$: silicon-intensified target (SIT) and charge-coupled device (CCD) cameras. In addition, photomultiplier tubes (PMTs) are almost exclusively employed with point-scanning confocal microscopes. All of these devices have different characteristics, and although the ultimate choice of detector is primarily determined by the temporal and spatial resolution desired, the performance specification advantages of cooled or intensified CCDs have made these devices the detector of choice in $Ca^{2+}$-imaging studies. These advantages include higher quantum efficiency, higher dynamic range, lack of geometric distortion, and a linear relationship between input and output intensities (4).

## Noise in Fluorescent Images

Images of fluorescently labeled cells are notoriously noisy because of the relatively low numbers of photons that are collected at each pixel in a single video frame. Noise in the signal basically comes from the instrumentation (instrumentation noise) and from the stochastic nature of the interaction of photons with the detection device, a process described by Poisson statistics (4). The average intensity of cell areas of fluorescence images should be statistically significantly above that of background image areas. The distribution of error sources within images has been described in extensive detail (4). However, as a working guide to an acceptable signal-to-noise ratio, we aim, with loading conditions and detector settings (gain and black level), to achieve mean signal levels that are two standard deviations above the mean intensity level of background areas (i.e., think about and treat digital images as you would any other large collection of numbers).

## Calibration of Fluorescent Dyes

The response of the $Ca^{2+}$ probe in use must be accurately known over the full range of ion concentrations at which fluorescence can be measured, a range determined by the $K_d$ of the fluorophore. This requires that techniques are employed that precisely buffer the ionized $Ca^{2+}$ concentration of a range of solutions to be employed in calibration. The $Ca^{2+}$ buffer ethyleneglycol bis($\beta$-aminoethyl ether)-*N,N,N',N'*-tetraacetic acid (EGTA) is most commonly employed for this process because it is capable of $Ca^{2+}$ control in a similar range to that of the most common fluorophores (pCa 7.5 to 5.5), although 1,2-bis(*o*-aminophenoxy)ethane-*N,N,N',N'*-tetraacetic acid (BAPTA) is also more than suitable for this purpose. Solutions with a range of concentrations can be simply constructed by mixing, in varied proportions, two stock solutions, one containing nominally equimolar $Ca^{2+}$–EGTA and another with only EGTA. Both solutions should contain similar levels of other constituents, such as physiological ions ($Mg^{2+}$, $K^+$, $Na^+$) and pH buffers. Although these solutions are relatively easy to make and mix, measurement of the exact [$Ca^{2+}$] in each solution ultimately determines the accuracy of the calibration process, because this is the variable against which fluorescence is ultimately plotted. Potentiometric titration is one way to establish these values (5). Alternatively, stock solution kits can be purchased from Molecular Probes (Eugene, Oregon), and it is perhaps best if newcomers to these methods use such standardized kits, until familiarity with the techniques increase.

*In vitro* calibration of the fluorophore is performed by adding small, identical aliquots of fluorophore to $Ca^{2+}$-buffered solutions, preferably contained in the vessel to be used in experiments. Solutions should be well-mixed to ensure even distribution of the fluorophore in solution. Fluorescence images should be collected for all solutions, including a blank (fluorophore-free solution), at all wavelength combinations to be used in experiments. When ratiometric probes such as fura-2 are used, individual images should be background subtracted (subtract the average intensity of the blank solution or subtract images on a pixel-by-pixel basis), before calculation of a ratio image (pixel-by-pixel image division). The average intensity (ratio) values of these ratio images is then plotted against the solution $[Ca^{2+}]$ and represents a standard curve, which may be directly used to determine experimental $[Ca^{2+}]$. This information also allows calculation of the $K_d$ and the limiting fluorescence ratio values (i.e., $R_{max}$ and $R_{min}$) for the experimental system. These values are related to $[Ca^{2+}]$ by the often quoted equation (Ref. 6),

$$[Ca^{2+}] = [(R - R_{min})/(R_{max} - R)]K_d\beta.$$

*In vivo* calibration is a much more difficult process but it is the preferred method, because it fully reveals the intracellular relationship between the fluorophore and $Ca^{2+}$. This process has requirements similar to those for *in vitro* calibration, with the additional need to be able to clamp the intracellular ion concentration to a value identical to that of the extracellular medium. In this respect mixtures of ionophore are used to overcome the inherent pumping capacities of the cellular membranes. Most commonly, we have used either ionomycin or Br-A23187 (concentrations between 1 and 25 $\mu M$), to aid in equilibration of intracellular $Ca^{2+}$ levels, with the latter preferred because of the relative low pH dependence of the $Ca^{2+}$-transporting capacity of this molecule. Despite the use of ionophores, in some cell types the extracellular $Ca^{2+}$ level may still significantly exceed that of the cytosol. In these cells it is essential to overcome the activity of pump processes that will antagonize this equilibration process, and we have commonly added other ionophores (such as monensin, nigericin, and gramacidin) to an "ionophore cocktail" to minimize the driving forces for other ions (such as $K^+$, $Na^+$, and $H^+$).

## References

1. C. J. R. Sheppard, *in* "Fluorescent and Luminescent Probes for Biological Activity: A Practical Guide to Technology for Quantitative Real-Time Analysis," p. 229. Academic Press, London, 1993.
2. D. A. Williams, S. H. Cody, and P. N. Dubbin, *in* "Fluorescent and Luminescent Probes for Biological Activity: A Practical Guide to Technology for Quantitative Real-Time Analysis," p. 320. Academic Press, London, 1993.
3. D. A. Williams and F. S. Fay, *Am. J. Physiol.* **250,** C779 (1986).
4. E. D. W. Moore, P. L. Becker, K. E. Fogarty, D. A. Williams, and F. S. Fay, *Cell Calcium* **11,** 157 (1990).
5. D. A. Williams and F. S. Fay, *Cell Calcium* **11,** 75 (1990).
6. G. Grynkiewicz, M. Poenie, and R. Y. Tsien, *J. Biol. Chem.* **260,** 3440 (1985).

# [6] Fluorescence Measurements of Cytosolic Calcium: Combined Photometry with Electrophysiology

W. T. Mason, J. Dempster, R. Zorec, John Hoyland, and P. M. Lledo

## Introduction

The ability to interface photomultiplier tubes, fast video cameras, and computer technology to the conventional microscope has made it possible not only to make qualitative observations, but also to derive quantitative data from single cells, at speeds of up to 30 video frames per second if we use video cameras or confocal laser-scanning technology, or many hundreds of samples per second if we use photon-counting technology with photomultipliers (Mason *et al.*, 1993).

Every cell has ionic charge gradients generated by ions such as $Ca^{2+}$, $Na^+$, and $K^+$. Gradients in intracellular hydrogen ion concentration also are important to many biological processes. Ionic concentrations inside of cells change quickly and dramatically and underlie a wide range of cellular processes, including development, growth, secretion, and reproduction, so it is important to observe and understand them. As we have been able to apply image processing and low light level image capture from signals emitted by optical probes, it has also become clear that large standing ionic gradients occur within cells, and may persist for many seconds during stimulus or suppression of cellular activity. The source of ionic changes may also occur in widely different parts of the same cell, and even with small cells of only 10 $\mu$m or so, these ionic pools may be detected.

In this chapter, we discuss the technology required to use fast photometric measurements of optical probes and the additional power that can be gained by combining such measurements with electrophysiology at the single-cell level. The main focus will be on how photon acquisition and analysis are making it possible to study ionic gradients in living cells by using optical probes. Some specific applications and technology from our laboratory using this technology will be used to illustrate the potential for such techniques.

*Methods in Neurosciences, Volume 27*

## Real-Time Measurement of Ion-Sensitive Fluorescent Dyes

Fluorescence ratiometry is at the heart of dynamic quantitative photometric measurements. Table I shows the range of dyes that may be used for such measurements, in both dual-excitation mode and dual-emission mode (and see Haugland, 1994).

If fluorescent signals are obtained as a pair at 340 and 380 nm (with fura-2, for instance) or at 405 and 490 nm (with indo-1, for instance) and the number of emitted photons or light intensity is ratioed on a point-by-point basis with respect to time, the resulting "ratio measurement" is proportional to $Ca^{2+}$ concentration and reduces the chance of possible artifacts due to uneven loading or partitioning of dye within the cell, or varying cell thickness and dye concentration (Grynkiewicz *et al.*, 1985).

Indo-1 (Fig. 1) is generally used for photometric measurements—it has a slightly faster time response than fura-2 in terms of dissociation time constant (typically estimated at 5–20 msec as opposed to 30–40 msec), and can be used with static optical beam splitters to separate the emitted light and focus it onto two photomultiplier tubes as a continuous signal. Ratio measurements of the emitted photon channels can also be employed. This approach has

TABLE I Dual-Emission versus Dual-Excitation Fluorescent Fluorochromes

| Dye | Ionic specificity | Wavelength (nm)[a] | |
|---|---|---|---|
| | | Excitation | Emission |
| **Dual-excitation dyes** | | | |
| Fura-2 | Calcium | 340 and 380 | 500–520 |
| BCECF | Hydrogen (pH) | 440 and 490 | 510–550 |
| SNAFL-1 | Hydrogen (pH) | 470–530 and 550 | 600 |
| SNARF-6 | Hydrogen (pH) | 490–530 and 560 | 600–610 |
| SBFI | Sodium | 340 and 380 | 500–520 |
| PBFI | Potassium | 340 and 380 | 500–520 |
| Mag-fura-2 | Magnesium | 340 and 380 | 500–520 |
| **Dual-emission dyes** | | | |
| Indo-1 | Calcium | 405 and 490 | 340–460 |
| DCH | Hydrogen (pH) | 435 and 520 | 405 |
| SNAFL-1 | Hydrogen (pH) | 540–550 and 630 | 510–540 |
| SNARF-2 | Hydrogen (pH) | 550 and 640 | 490–530 |
| FCRYP-2 | Sodium | 405 and 480 | 340–350 |
| Mag-indo-1 | Magnesium | 405 and 480 | 340–350 |

[a] These wavelengths are approximate. It is usual to employ interference filters for separation of discrete wavelengths with 7–15 nm half-bandwidth (i.e., the wavelength spread at half-maximum transmission for the filter center frequency).

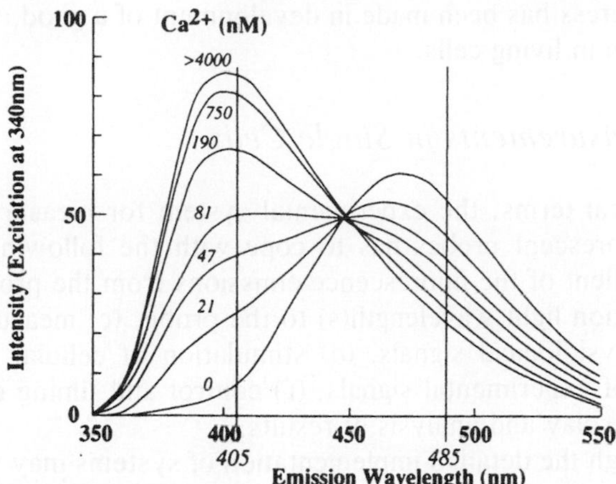


FIG. 1 Fluorescence emission spectra of indo-1 measured in saline in free solution. It should be noted that optical probes loaded into cells often show different emission spectra than those measured under free solution conditions.

the advantage that no movement need take place in order to change filter position, and so measurements can be fast and vibration free.

Another potential combination of optical probes that enables measurements in the visible region is the combination of fluo-3 and fura-red. Both are excited near the fluorescein excitation of about 490 nm. The former increases fluorescence output at 520 nm with increasing $Ca^{2+}$ and the latter decreases fluorescence output at 560 nm with increasing $Ca^{2+}$.

Another dye, 2′,7′-bis(carboxyethyl)-5,6-carboxyfluorescein (BCECF), for example, measures intracellular pH as an optical signal, using similar dual-wavelength imaging technology. This dye is excited at 440 and 490 nm and is measured at about 510–520 nm. The pH increases will increase the fluorescence of dye excited at 490, but have little effect on 440-nm fluorescence. This dye is being successfully combined with fura-2 to provide simultaneous imaging of $Ca^{2+}$ and pH. For this type of work, a four-position filter wheel is used to excite, in turn, the probes at 340, 380, 440, and 490 nm, having loaded both dyes into the single cell simultaneously. Fluorescence is measured with a 515-nm dichroic longpass filter and a 535-nm bandpass filter, and this provides strong signals for both dyes with minimal overlap or interference of the two probes. The dye DCH and other probes of the seminaphthorhodafluor (SNARF) and seminaphthofluorescein (SNAFL) family provide the potential to do either single-wavelength, nonratiometric, or dual-wavelength emission ratiometric experiments.

Optical probes are also available for sodium and chloride, but frustratingly little progress has been made in development of a good, selective probe for potassium in living cells.

## *Photometric Measurements in Single Cells*

In practical terms, the experimental system for measuring cytosolic $Ca^{2+}$ using fluorescent probes has to cope with the following major tasks: (a) measurement of the fluorescence emissions from the probe, (b) application of excitation light wavelength(s) to the probe, (c) measurement of cellular electrophysiological signals, (d) stimulation of cellular responses, (e) recording of experimental signals, (f) control and timing of the experiment, and (g) display and analysis of results.

Although the detailed implementation of systems may vary, certain hardware items are required. Due to the typically low levels of fluorescent light emission, a light-measuring device of high sensitivity is required, such as a photomultiplier tube or a cooled charge-coupled device (CCD) camera. If dual-excitation probes are to be used, a switching system capable of alternating the wavelength of the UV excitation light is required. Recording cellular membrane currents and voltage requires the use of a voltage-clamp device, either a switching voltage clamp, such as the Axoclamp or Axopatch (Axon Instruments, Foster City, California), or a patch voltage clamp, such as the List EPC-7 (List Electronic, Darmstadt, Germany) or the Cambridge Patch Systems SWAMII (Life Science Resources, Great Shelford, Cambridge, England). As will be discussed later, if changes in cell capacity are to be measured, a specialized "lock-in" amplifier may also be needed. If cells are to be stimulated pharmacologically, an ionophoretic or pressure-driven drug application system is required. At least five different signal channels of experimental data are produced during the experiment, two fluorescent signals and their ratio per probe molecule, membrane current, and voltage. These signals must be stored for later analysis, either on a multichannel instrumentation tape recorder or, more commonly these days, in digital form on computer disk. The computer system and associated software play a crucial role both in the management of these disparate measurement and stimulation devices during the experiment and in the analysis of results afterward. A generalized diagram of a typical combined fluorescence/electrophysiological measurement system is shown in Fig. 2.

## *Measurement of Fluorescent Light—Photometry versus Imaging*

When designing (or choosing) a system for fluorescence measurements from single cells under a microscope, a basic choice has to be made between using an imaging device, such as a high-sensitivity television camera, or a

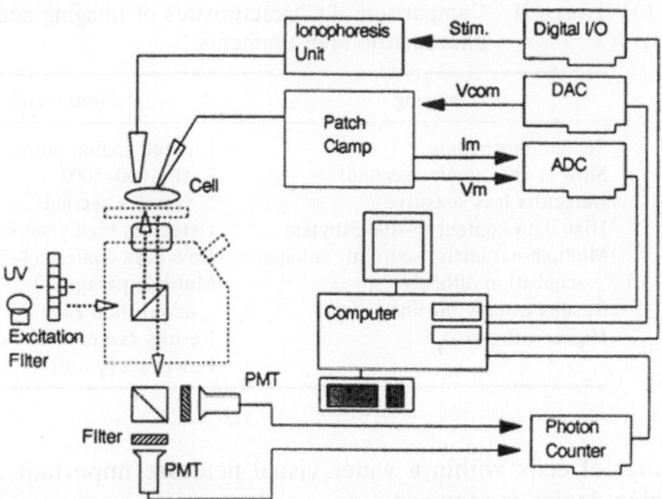


FIG. 2 Diagram of a system for the combined recording of cell fluorescence and membrane current and potential. A cell is epi-illuminated with UV light using a narrow-band interference filter. Fluorescent photon emissions are captured via the microscope camera port and measured with photomultiplier tubes (PMTs). For dual-emission dyes, the emitted light at the characteristic peak emission wavelength for the dye is measured by a pair of PMTs, after being split into two wavelength streams using a combination of a 45° dichroic mirror and bandpass filters. A filter wheel placed into the excitation filter path can be used to provide alternating UV excitation wavelengths for dual-excitation dyes. Cell membrane current and potential are measured via a micropipette attached to the cell and drugs are applied via an ionophoresis unit. Overall timing, application of stimuli, and digital recording of signals are performed by a laboratory computer, equipped with the analog-to-digital converter (ADC), digital-to-analog converted (DAC), and photon counting interface cards and appropriate software such as PhoCal or PhoClamp.

nonimaging photometric transducer, such as a photomultiplier tube (PMT). The pros and cons of imaging versus photometric measurements are presented in Table II. Cameras have the advantage of providing a complete image of the fluorescence within a probe-loaded cell and hence the intracellular distribution of $Ca^{2+}$. The striking false-color images often associated with the technique are derived using cameras. PMTs, on the other hand, provide a signal proportional to the amount of light falling on their light-sensitive surface. They therefore provide a measure of the total fluorescence being emitted from the cell, rather than an image. The choice of camera or PMT depends on the requirements of the experiment. If the spatial distribution of $Ca^{2+}$ within the cell is important, or if the differential responses of a

TABLE II Comparison of Characteristics of Imaging and Photometric Measurements

| Imaging | Photometry |
|---|---|
| Spatial information | Limited spatial information |
| Slow (1–30 samples/second) | Fast (1000–5000 samples/second) |
| Detectors less sensitive | Detectors highly sensitive |
| High data content (5–100 Mbytes) | Low data content (1–2 Mbytes) |
| Multiple-parameter (current, voltage) acquisition difficult | Multiple-parameter acquisition easy |
| Results usually off-line | Results can easily be on-line |
| Higher entry cost | Lower entry cost |

number of cells within a wider visual field are important, then clearly an imaging device is required.

On the other hand, the PMT can be used at lower light levels and can acquire measurements at higher rates than the intensified video camera; the latter is limited to the video frame rate of 25 Hz in Europe or 30 Hz in North America. For example, video frame rate cameras in Europe have a maximum acquisition time of 40 msec, whereas photometric devices can easily accumulate one data point each millisecond or faster, depending on light availability. Thus, the main factor that dominates the quality of low-level light measurements is the quantal nature of light. Light emissions consist of a random stream of photons. In any given time interval, the number of photons captured by a sensor will fluctuate with a standard deviation equal to the square root of the mean photon count. A light sensor (PMT or camera) must accumulate enough photons to reduce these statistical fluctuations to acceptable levels. This, more than anything else, determines the rate at which light measurements can be made. With intracellular probe concentration and UV illumination levels found under typical experimental conditions, an acceptable photon count can be accumulated by a PMT within 5–10 msec. It takes much longer, however, to form an acceptable camera image because each sensing element of the camera is receiving photons from a much smaller area. For instance, a cell might occupy a camera image area consisting of 100 × 100 pixels. If this cell were emitting 1000 photons per 1 msec (an entirely acceptable level for a PMT), each pixel would only accumulate an average of 1 photon each, in a 10-msec exposure.

Camera images also require a great deal more computer processing than does the single measurement from a PMT. With a digital CCD camera, the contents of each pixel must be read out in sequence, digitized, and transferred to the computer. Although there are means with some cameras to digitize only certain parts of image, and to accumulate the contents of blocks of pixels within the camera (a process called on chip binning), nevertheless overheads of 30 ms–1 sec (depending on the type of camera) cannot readily be avoided. The PMT signal, on the other hand, is directly transferred to the data acquisition system with little overhead. [A detailed discussion of cameras, PMTs, and their properties can be found in Mason (1993).]

In general, when fluorescence measurements are being combined with electrophysiological measurements, the above factors favoring the PMT outweigh the advantages of the imaging capabilities of the camera. In any case, it is the total intracellular $Ca^{2+}$ signal that is usually of interest in such experiments. If a camera were to be used, the image would simply have to be reintegrated numerically over the available pixels to obtain an average $Ca^{2+}$ concentration, for example.

## *Photomultiplier Tube Technology*

The PMT is one of the few examples of thermionic tube technology that has not been replaced by semiconductors. The PMT is an evacuated glass tube with a photosensitive surface painted onto one end. Photons striking this surface eject electrons, which are accelerated along the tube by a high voltage (~1000 V) applied to a chain of metal electrodes (the dynode chain). As the fast-moving electrons strike the electrode surfaces, a cascade of further electrons is produced, getting larger with each step. This process of electron amplification results in a measurable pulse of current being produced at the PMT anode whenever a photon strikes the tube surface. The signal from a PMT is thus a series of these randomly occurring short-duration (10–100 nsec) current pulses and it is the rate of occurrence of the pulses that provides an indication of the amount of light falling on the tube. The PMT is a very sensitive light-measuring device capable of detecting single photons of light when combined with an appropriate pulse-counting device. Thus a complete PMT-based light measurement system requires a photomultiplier tube(s), a stable high-voltage (0–2000 V) source, a photon-counting device, and a data acquisition system.

PMTs, their associated power supplies, and counting devices can be obtained from a number of specialist suppliers, e.g., Thorn EMI (Ruislip, Middlesex, England), Oriel (Stratford, Connecticut), Hamamatsu (Hamamatsu City, Japan; Oak Brook, Illinois), and Life Science Resources. It is worth noting that PMTs can vary in terms of sensitivity and other factors, such as their optimal operating voltage. If a pair of PMTs are to be purchased (for use with dual-emission probes), it is worthwhile obtaining a pair with matched operating characteristics with respect to sensitivity and low dark current at a given operating voltage.

## *Photon Counting versus Photocurrent Integration*

Two approaches can be taken to convert the raw stream of photon pulses produced by the PMT into a stable signal indicating light level—photon counting and photocurrent measurement—and the pros and cons are presented in Table III.

In the photon-counting method (Fig. 3A) the current pulses are applied to a high-speed digital counter, with each pulse incrementing the counter by one unit. The counter is allowed to accumulate pulses for a fixed period of time, at the end of which the number of counts is read out, stored, and the counter reset. The number of counts for successive count periods provides a measure of the light level. Photon counting requires a high-speed digital counter capable of responding to short-duration pulses. A typical example of a photon-counting system is employed by the PhoCal system (Life Science Resources, Great Shelford, Cambridge, England); it comprises a combination of a two-channel photon counting interface board and an amplifier-discriminator board. The interface board is suitable for IBM PC-compatible computers providing two high-speed counters, capable of accumulating over 16 million counts each at rates up to 10 MHz. The board is controlled by software running on the PC and can be made to count repeatedly over preset intervals, ranging from 0.5 msec to 20 sec. The current pulses produced by a PMT are too small to be detected directly by the interface board and therefore an

TABLE III Pros and Cons of Photocurrent versus Photon Counting Light Detection

| Photocurrent | Photon counting |
|---|---|
| Low temporal resolution | High temporal resolution |
| Lower dynamic range | High dynamic range |
| Integrated current measurement | Bin counting |
| Time resolution depends on RC filter setting | Time resolution dependent only on bin width setting |
| Integrates many photons, hence less sensitive | Single-photon detection, hence more sensitive |
| Limited at low light levels | Better for low light levels |
| Suited to large cells (20 $\mu$m plus) | Ideal for small cells (5–20 $\mu$m) |
| Lower cost | Higher cost |

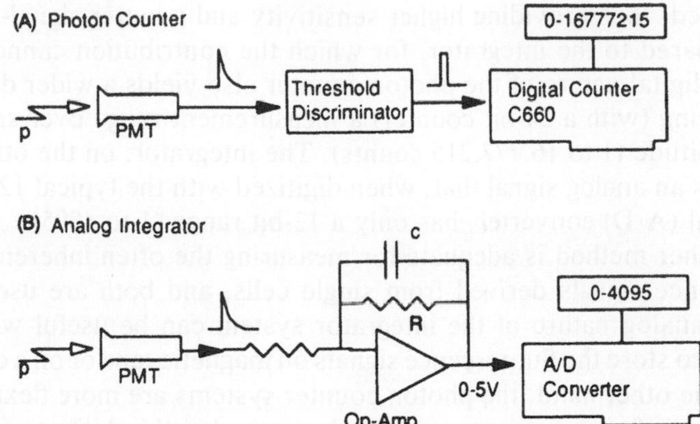


FIG. 3 Photomultiplier tube (PMT) light measurement methods. (A) Photon Counting. Direct counting of PMT output pulses produced by photons striking the tube surface. Pulses are amplified and fed to a discriminator, which produces a digital pulse for pulses exceeding a 1-mV threshold. The digital pulse is used to increment a high-speed digital counter, installed in a computer system. (B) Photocurrent integration. PMT output is fed into an integrator circuit, which produces an analog output voltage proportional to the photon pulse rate.

amplifier/discriminator is used to convert them into standard emitter-coupled logic (ECL) digital pulses to match the counter input requirements. The Stanford Research Instruments SR400 provides similar dual-channel counter functions within a standalone device that can be interfaced to a variety of computer systems via either an RS232 serial link or an IEEE 488 interface bus.

The alternative to photon counting is to integrate the PMT current directly to produce an analog voltage proportional to the average pulse frequency. This is done by feeding the PMT current into an integrator circuit as shown in Fig. 3B. Each photon current pulse adds a small amount of charge to the capacitor $C$, thus increasing the integrator output voltage. The resistor $R$, in parallel with the capacitor, causes the stored charge to "leak" away with a time constant $\tau = RC$. The integrator thus produces an output voltage that is proportional to the balance between the charge added from the photon pulses and that leaking away through $R$. In broad terms, the time constant $\tau$ is equivalent to the counting interval of the photon counter. A similar effect can be achieved by feeding the PMT signal through a lowpass filter.

Overall, the photon-counting method is preferable to photocurrent integration. PMTs produce a random background noise signal known as the "dark current," on which the photon current pulses are superimposed. The threshold discriminator of a photon counter can be set so that this signal is largely ignored, thus providing higher sensitivity and a better signal-to-noise ratio, compared to the integrator, for which the contribution cannot be avoided. The digital nature of the photon counter also yields a wider dynamic range, allowing (with a 24-bit counter) a measurement range over seven orders of magnitude (1 to 16,777,215 counts). The integrator, on the other hand, produces an analog signal that, when digitized with the typical 12-bit analog-to-digital (A/D) converter, has only a 12-bit range (1 to 4095).

Either method is adequate for measuring the often inherently noisy fluorescence signals derived from single cells, and both are used in practice. The analog nature of the integrator system can be useful when there is a wish to store the fluorescence signals on magnetic tape or on a chart recorder. On the other hand, the photon counter systems are more flexible and easier to handle from a computer system. In particular, the ability to set the counting interval precisely from software is very convenient.

## *Excitation Filter Switching*

Systems designed to handle dual-excitation probes must have a means of switching the fluorescence excitation light between the absorbance peaks of the Ca-bound and free forms of the probe. In general, the excitation light wavelength is controlled by means of a rotating wheel containing two or more dichroic bandpass filters [e.g., for fura-2 340 nm (Ca free) and 380 nm (Ca bound)] that are mounted in the UV excitation light path. A 340/380 ratio measurement is obtained by rotating each filter, under computer (or specialized hardware) control, into the light path, taking a light measurement for each, storing the result, then computing the ratio. This must be done as fast as possible to maximize the rate at which fluorescence ratio measurements can be made. The required rate depends on how fast the intracellular $Ca^{2+}$ concentration is varying. In some cases this may be a matter of minutes and filter exchanges times of 1–2 sec can be used. However, some $Ca^{2+}$ concentration changes, such as those associated with action potentials, are very rapid, and in these situations ratio measurement rates of 500 Hz may be required. For such applications, dual-emission measurements are significantly faster because they do not require the filter wheel to be moved.

The complexity of the task of coordinating the rotation of the filter wheel, light measurements, and ratio calculations requires that it be done under the control of a computer or with specialized hardware designed for the purpose. For this reason most filter wheels are sold as part of complete fluorescence ratio measurement packages [such as PhoCal (described below) from Life Science Resources], with appropriate hardware/software to control the wheel. Suppliers may be willing to provide a wheel by itself, but care should be taken to ensure that appropriate documentation and/or software are supplied to allow the wheel to be operated (Sutter Instruments, Ludl Instruments, and Life Science Resources all produce a wheel as an individual item; see below).

Filter wheels can vary quite radically in design and performance. The Life Science Resources filter wheel used in the PhoCal/PhoClamp systems (also sold as a standalone) holds up to 8 filters and blanks controlled by a powerful DC servo motor. It is highly versatile, operating to 1 part in 2048 accuracy and capable of either rapid stepping (~100 msec) or smooth high-speed rotation up to about 120 revolutions per second, thus yielding as many as 480 filter changes per second. This wheel is available with a UV-transmitting liquid light guide for remote operation or with a direct optical coupling to a microscope. The Cairn Research filter wheel can hold up to 10 filters, is controlled by a synchronous motor, and can be made to rotate constantly and smoothly at any rate of between 1 and 500 Hz. However, it is difficult to make this wheel rotate at very slow speeds or to step rapidly between arbitrary filter positions under programmed control. The Lambda-10 wheel (Sutter Instruments, Novato, California), on the other hand, is another 10-filter wheel, but is controlled by a stepper motor. It can be made to jump between any filter position in arbitrary sequence, moving between adjacent filters within 50 msec, but cannot be made to rotate at a constant high speed. If high ratio measurement rates are important, a constantly rotating wheel design, such as the Cairn or Life Science Resources wheel, is preferable because it avoids the need to accelerate and decelerate the wheel constantly. DC servo motors probably provide the best compromise of speed and smoothness overall.

Dual-excitation probes, when combined with filter wheels holding four or more filters, have the advantage that they permit the simultaneous study of more than one molecular species through the use of multiple probes applied to or injected into the cell. However, they have the disadvantage that the mechanical movement of filters inevitably limits the ratio measurement rate and may also cause vibration that can upset the patch electrode seals on the cells. It may be necessary to isolate the light source and filter wheel from the microscope coupling the light, using a flexible light guide.

## *Dual-Emission Probes*

Many of the problems associated with the use of dual-excitation probes (complexity, limited ratio measurement rates, vibration) do not exist for dual-emission probes, where it is the emission peak that is shifted by the binding of $Ca^{2+}$. No filter switching is required because these probes can be excited with a single wavelength. Instead, the emitted light is split into two components using a 45° wavelength-sensitive dichroic mirror and is passed via bandpass filters to two PMTs. Light at the two characteristic peak emission wavelengths for the probe can thus be measured simultaneously, rather than alternately. This readily allows ratio measurement rates as high as 1 kHz, a limit set by the $Ca^{2+}$ binding/unbinding rates of the probe rather than any mechanical consideration.

## *Electrophysiology*

Fluorescence measurements can readily be combined with a wide range of electrophysiological measurements, ranging from cellular action potentials to single-ion channels. The electrophysiological apparatus used in combination with fluorescence studies is quite standard and will not be discussed in detail here (see Standen *et al.*, 1987; Sakman and Neher, 1983 for details).

The aims are generally to observe the effects of variations in an intracellular ion concentration (usually, $Ca^{2+}$, $Na^{+}$, $H^{+}$) on the behavior of an ionic conductance system in the cell membrane, or vice versa. The whole-cell patch-clamp technique is particularly well suited for use in fluorescence studies in that the patch pipette can be used to introduce the fluorescent probe into the cell cytoplasm, as well as to control the cell membrane potential. This has the distinct advantage that it allows well-controlled concentrations of the membrane-impermeant acid form of the probe to be applied directly to the cell interior, rather than (as is normal) applying the permeant ester form to the bathing solution and relying on intracellular esterases to convert it inside the cell to the active form.

## *Data Acquisition*

As can be seen, experiments combining two different measurement techniques can produce quite a large number of separate signal channels. At least five signals are produced with PhoCal or PhoClamp (Life Science Resources; provding two fluorescence channels, ratio, current, voltage, and a marker channel) and more if multiple probes are in use, or for complex electrophysiological measurements such as cell capacity, for which as many as eight are possible. Two approaches can be made to store these data. They can be digitized on-line by the computer system controlling the experiment and stored on computer disk or they can be recorded in analog form on a multichannel tape recorder. There are advantages and disadvantages to each approach, with factors such as the design of the photometric measurement system and the nature of the experimental protocols having a bearing on the choice.

The typical modern instrumentation tape recorders use digital pulse code modulation techniques to record multiple analog channels onto either video tape or digital audio tape (DAT) cassettes. For example, the Biologic DT1802 recorder (Biologic, Claix, France) can store up to eight channels of analog data on a 2-hr DAT cassette. Encoders, which allow a similar number of channels to be stored on video tape, can be obtained from Medical Systems Corp (New York) or Life Science Resources (Great Shelford, Cambridge, England). Tape recorders have the advantage that a complete record of all the data from a long-lasting experiment can be recorded on an inexpensive storage medium. Storing similar amounts of data on computer disk, although possible, is substantially more expensive. Nevertheless, in order to analyze the data it must be eventually played back and digitized by a computer system.

If a tape system is to be used, it may be necessary to ensure that the fluorescence measurement system provides suitable analog outputs of the PMT signals and, ideally, the ratio. This is where an independent hardware-based system such as that of Cairn Research has advantages, because a complete recording system can be created without the need for a computer inasmuch as all relevant PMT and ratio signals are provided in analog form. Systems that are computer based, with many functions implemented as hardware/software combinations, may not always provide such signals. Tape recording can be effective when relatively slow, long-lasting $Ca^{2+}$ signals are under study, where the experimental protocols are relatively simple, without the need to apply complex patterns of stimuli repeatedly.

However, direct storage of data on the computer is certainly possible and has the advantage of avoiding the need for a tape recorder and having the experimental data immediately available in the form required for analysis. The Axon Instruments AxoTape software package, for instance, provides a computer-based emulation of a tape recorder, with the data stored on computer disk. Computer-based fluorescence measurement systems, such as the PhoCal/PhoClamp (Life Science Resources) or Newcastle Photometrics products, usually also provide chart recorder emulation software for storing and analyzing experimental data.

To sum up the difference between these two approaches, tape storage has advantages of versatility and reliability. Data can be played back into any computer system, or onto a simple chart recorder. However, a computer data acquisition analysis system will probably still be required to analyze and display the results. On-line computer-based fluorescence measurement systems are often simpler to use, and all major functions are controlled from one master program. They also can take better advantage of the high-precision results from digital photon counters.

## Voltage-Clamp Studies

There are many types of experiments for which acquisition of experimental data have to be closely synchronized with the application of complex stimuli, often repeatedly applied, and combined later for analysis. In such cases, only a computer-based solution can handle the complexity of the required tasks. The most common example of this type of experiment is the study of transient membrane currents, such as voltage-activated $Ca^{2+}$ currents, which provoke internal $Ca^{2+}$ concentration changes. Voltage-clamp studies such as these are distinctive in that the experimental protocol is split into a series of short, discrete, recording sweeps rather than a single, long, continuous record. Instead of a relatively small number of stimuli, separated by intervals of minutes, hundreds of precise voltage pulse stimuli are applied at 1- to 20-sec intervals. The use of computer-based systems for applying experimental stimuli and simultaneous data acquisition is now the norm in electrophysiological studies. Such systems generally perform the following data acquisition and analysis tasks: generate repeated voltage pulse stimuli of varying heights and/or widths, record membrane current and voltage sweeps, signal averaging, subtraction of "leak" currents, and signal measurement [current/voltage ($I/V$) curves, exponential curve fitting].

The development of such software can be difficult and time consuming and it is now common for most laboratories to make use of commercial (or public domain) electrophysiological analysis software. Several programs are available for the IBM PC computer families, including the Axon Instrument pCLAMP package, the CED Voltage and Patch clamp package (Cambridge Electronic Design, Cambridge, England) (both commercial products), and the Strathclyde Electrophysiology Software (free to academic institutions, supplied by J. Dempster, Dept. of Physiology and Pharmacology, University of Strathclyde, Glasgow, Scotland). Programs such as Axon's Axodata and Pulse (Instrutech Corp.) are available for the Apple Macintosh family. A detailed discussion of electrophysiological analysis software and the acquisition of such data on computer can be found in Dempster (1993).

With some limitations, electrophysiological analysis software can usually be pressed into service to handle fluorescence signals, particularly if dual-emission probes are being used, because the complications of excitation filter switching are avoided. Most packages permit the recording of additional analog data channels, necessary to record the fluorescence signals (which must, of course, be available in analog rather than digital form). Some packages (e.g., CED and pCLAMP) can also calculate the ratio of the two channels. However, not being designed with fluorescence ratio measurements in mind, they often lack more sophisticated features, such as calculating the actual ionic concentration from the ratio. Similarly, they cannot make use of digital photon counters such as the Thorn MI CT1 or the SRI 400.

At present, relatively few software packages combine a full set of features for the acquisition and analysis of both fluorescence and voltage clamp data. Ideally, in addition to the electrophysiological functions discussed above, such a package would record the fluorescence signals from digital photon counters, control the UV excitation filter, apply digital filtering to signals, and compute ratios and $Ca^{2+}$ concentration from the $Ca^{2+}$ binding equation.

One of the few packages that addresses these problems is the PhoClamp system (Life Science Resources), which combines the simultaneous recording of fluorescence signals from dual-emission probes, using a photon counter, with the voltage pulse generation and recording feature necessary for voltage-clamp studies. Unlike most other programs, PhoClamp allows fluorescence and voltage-clamp recording sweeps to be of different durations, allowing them to be better matched to the different time courses of $Ca^{2+}$ currents (0.25–1 sec) and concentration transients (5–20 sec). The program also supports on-line leak current subtraction, a technique commonly used to separate voltage-activated $Ca^{2+}$ currents from background membrane currents carried by other ions. The system is described in more detail below.

As can be seen, the development of a complete fluorescence measurement package requires the careful integration of a variety of different technologies into a functioning system. Issues such as the choice of appropriate PMTs, excitation and emission filters, filter changers, beam splitters, data acquisition hardware, and software can all have a bearing on the effectiveness of the system. Developing such systems in-house can be quite difficult and time consuming unless detailed specifications for the components for each subsystem are available. Many researchers therefore opt to purchase complete fluorescence measurement packages, relying on the supplier to ensure the proper choice of matching components. A number of such "turn-key" systems are available. Most such systems are designed to work with common microscopes such as the Nikon Diaphot or Zeiss Axiovert system, but can vary markedly in general design and performance. Some products, such as the PhoCal/PhoClamp and Newcastle Photometrics systems, are essentially computer based, with much of the functionality of the system dependent on the software. Rates of data acquisition may also vary widely. Others, such as the Cairn Research system, are based on specialist standalone hardware.

Such variations make it prudent, prior to purchase of a system, to observe such packages being applied to experimental work similar to that envisaged.

## PhoCal and PhoClamp: Integrated Systems for Combined Photometry and Electrophysiology

PhoCal and its companion system PhoClamp provide simultaneous analog-to-digital conversion at high frequencies in tandem with photometric detection of optical probe emission, and they are compared as working models in the following section. PhoClamp, for instance, is ideally suited to voltage- and current-clamp experiments because it also controls digital generation of voltage protocols, and, as discussed above, can function on two independent time bases. This makes it possible to accumulate fast electrophysiological signals (up to 35 kHz) while simultaneously recording the somewhat slower responses of $Ca^{2+}$, for example. The technical points that differentiate the PhoCal and PhoClamp systems are given in Table IV, and more technical detail follows below.

TABLE IV PhoCal versus PhoClamp Technical Comparison

| PhoCal | PhoClamp |
|---|---|
| Single time base for physiology and photon counting | Dual time base for physiology and photon counting |
| Slow analog (1000 samples/second) | Fast analog (40,000 samples/second) |
| No internal waveform generator | Internal digital waveform generator |
| No automatic leak subtraction for physiology | Provides automatic leak subtraction for physiology |
| Does not permit superposition of records for display and plotting | Permits superposition of records for display and plotting |
| Triggered sweeps or chart recorder mode | Triggered sweeps or continuous recording mode |
| Dual-emission and multiple-wavelength excitation | Single or dual emission only |
| 1, 2, 3, or 4 wavelengths | 1 or 2 wavelengths only |
| Supports intermittent shuttering for probe bleach reduction | Supports intermittent shuttering for probe bleach reduction |
| Multiple probes if needed | Single probe work generally |
| Photocurrent or photon counting | Photon counting only |
| General-purpose photometric and/or slow electrophysiology | Ideal for dedicated general photometry combined with fast electrophysiology |
| Results usually off-line | On-line results display |
| Limited analysis/fitting with extensive measurement | Statistical analysis and fitting package with extensive measurement |

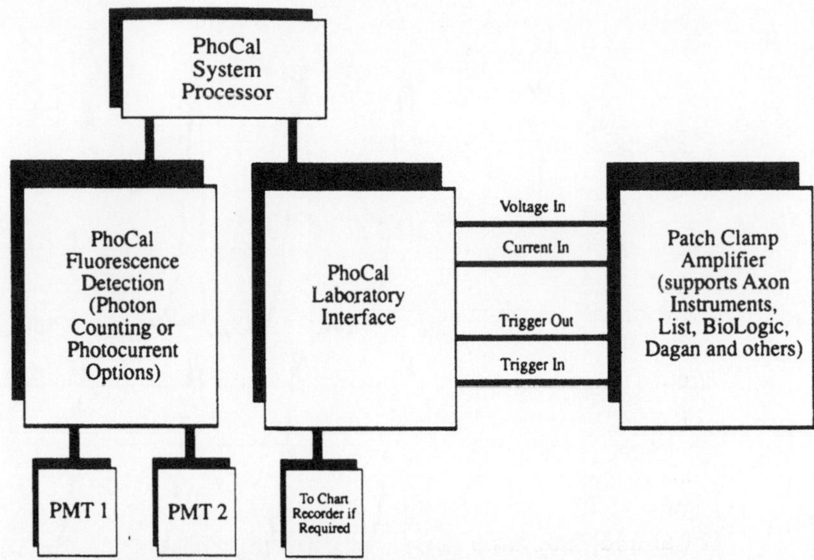


FIG. 4 Schematic input/output diagram of PhoCal system. Characteristics: continuous chart recorder mode or triggered sweep acquisition; single time base collection (sampling rate up to 500 ratios per second + 500 separate analog samples from each of two channels + marker channel); simultaneous recording to hard disk with on-line calibrated ion display; simultaneous monitor display and chart recorder output of voltage proportional to ratio and calculated ion concentration.

## *PhoCal System Characteristics*

The PhoCal system (Life Science Resources) is shown schematically in Fig. 4. Examples of data that can be collected with PhoCal are shown in Figs. 5 and 6. The following characteristics expand on the flow chart in Fig. 4:

1. PhoCal is IBM PC-compatible.
2. PhoCal will detect fluorescent dyes using either dual-wavelength emission ratio analysis (single-wavelength excitation) or vice versa, using dual-wavelength excitation coupled with the Life Science Resources filter wheel. PhoCal is available for dual-emission or dual-excitation work.
3. The PhoCal system can use either a two-channel photon counting system or photocurrent-to-photovoltage measurement.
4. PhoCal is supplied with a high-quality analog-to-digital converter. The software provides the user with the ability to take on two channels of data in addition to the fluorescent measurement. This approach is invaluable for

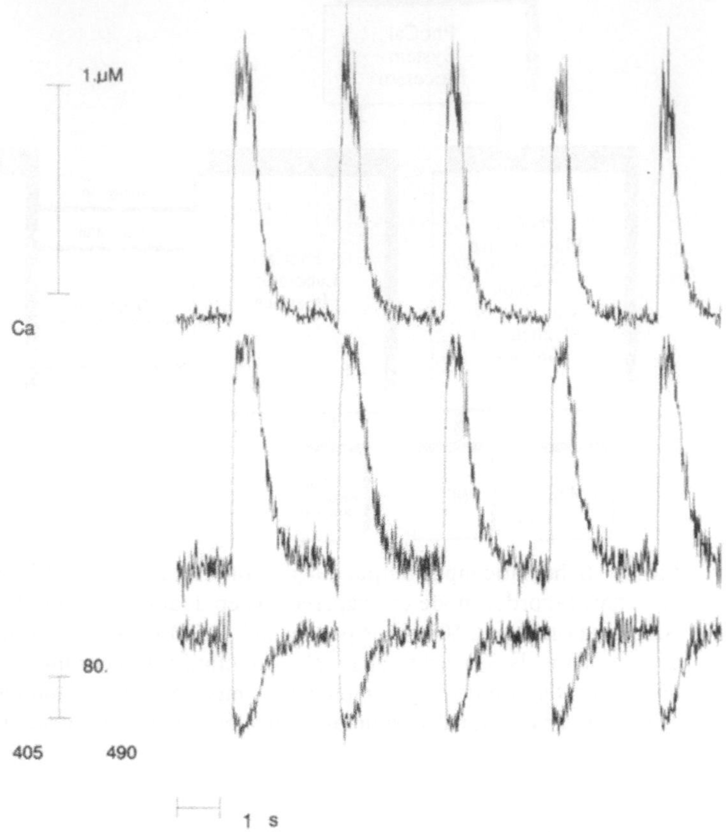


FIG. 5 Dual-wavelength photo current measurements of intracellular $Ca^{2+}$ in an isolated rat cardiac muscle cell, loaded with indo-1. The top trace is the calculated intracellular $Ca^{2+}$ concentration; the bottom two traces show the 405- and 490-nm traces of light emission from indo-1, respectively. Note that one wavelength increases in intensity and the other decreases as $Ca^{2+}$ rises. The muscle cell was stimulated with electrical field stimulation.

enabling other parameters of physiological function to be studied simultaneously. Examples might include patch-clamp or electrophysiological measurement of intracellular voltage and current, simultaneously with a two-channel fluorescent measurement of intracellular $Ca^{2+}$; or use of an extracellular pH electrode (or ion-sensitive electrode) to record extracellular ionic activity in parallel with intracellular ion measurement.

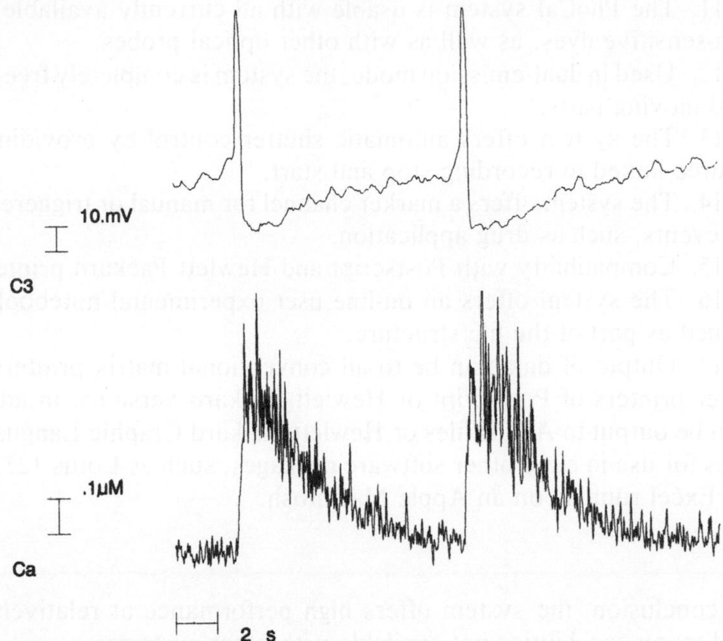


FIG. 6 Photometric experiment using indo-1 and the PhoCal system to measure simultaneously spontaneous action potentials (top record) and $Ca^{2+}$ transients in a spiking GH3 clonal pituitary cell. Two emission wavelengths were recorded using fast photon counting, with about five sample points/millisecond.

5. The PhoCal system has good temporal response. Dual-channel $Ca^{2+}$ measurements can be made as fast as 1500 samples per second per channel.

6. With photon counting, electrophysiological or other measurements can be made up to about 1000 samples per second per channel, as defined by the user.

7. Users have complete control over temporal response of data acquisition.

8. Data files are written directly to disk.

9. The software has routines for data smoothing using a running average method.

10. The system is compatible with most commercially available microscopes. It has been fitted to date to both Nikon and Zeiss microscopes, and is compatible with Leitz and Olympus equipment.

11. The PhoCal system is usable with all currently available fluorescent ion-sensitive dyes, as well as with other optical probes.

12. Used in dual-emission mode, the system is completely free of vibration and moving parts.

13. The system offers automatic shutter control by providing a voltage source linked to recording stop and start.

14. The system offers a marker channel for manual or triggered indication of events, such as drug application.

15. Compatibility with Postscript and Hewlett Packard printer/plotters.

16. The system offers an on-line user experimental notebook that is retained as part of the file structure.

17. Output of data can be to all conventional matrix printers as well as laser printers of Postscript or Hewlett Packard versions; in addition, data can be output to ASCII files or Hewlett Packard Graphic Language (HPGL) files for use in most other software packages, such as Lotus 123, Freelance, or Excel running on an Apple Macintosh.

In conclusion, the system offers high performance at relatively low cost, and many capabilities not available with other systems:

1. Single- or dual-channel photon counting, or photocurrent analysis.
2. Excellent temporal response, including high-performance analog-to-digital conversion for recording of biological-related electrical signals, i.e., membrane voltage and current.
3. Ability to display computed ion concentrations on-line.
4. Ability to record up to two channels of data, in addition to fluorescence channels.
5. On-line display of data, including changes in ion concentration.
6. Ability to export all data as ASCII files to popular spreadsheet packages running on both PC compatibles, Apple Macintoshes, and networks, as standard.
7. Postscript and Hewlett Packard printer/plotter compatibility.
8. Permanent file-structure retention of data with ability to apply new calibration tables at any time.
9. On-line user experimental text entry as a "live" notebook.
10. High-quality Life Science Resources filter wheel.
11. Operation as dual-excitation or dual-emission system, completely interchangeable.

## *PhoClamp System Characteristics*

Examples of data collected from a simultaneous recording of membrane current under voltage-clamp and photometric measurement of $Ca^{2+}$ using indo-1 are shown in Fig. 7. A schematic diagram of the PhoClamp system is shown in Fig. 8. The main points about PhoClamp are as follows:

1. PhoClamp is IBM PC compatible.

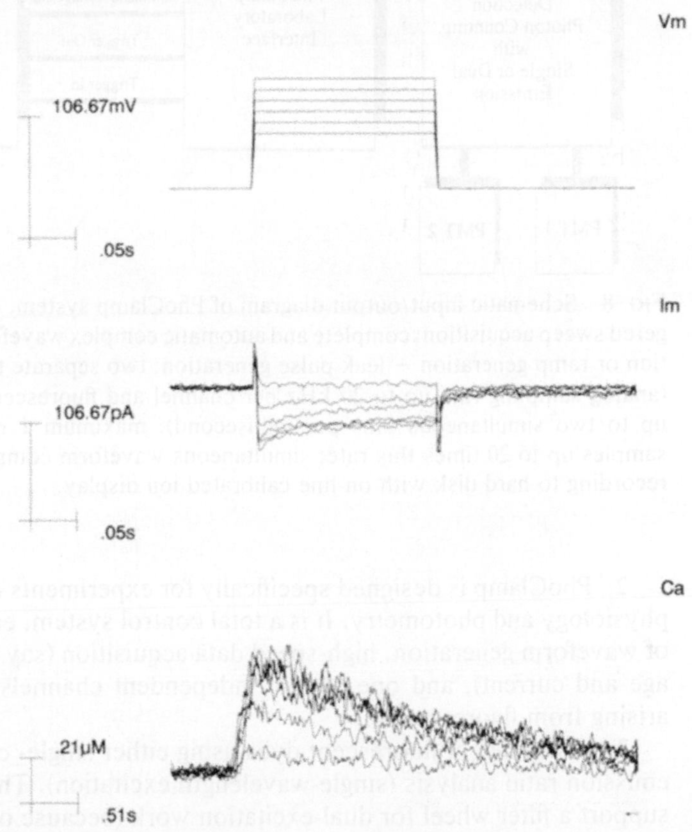


FIG. 7 Measurement of simultaneous intracellular $Ca^{2+}$, together with membrane voltage using PhoClamp, which permits voltage- or current-clamp measurements together with two wavelengths of photon information using photon counting. Here an imposed voltage-clamp step elicits a long-lasting rise in intracellular $Ca^{2+}$, the latter persisting for about 10 times longer than the current transient. Note the intracellular $Ca^{2+}$ trace is at a different time scale than the voltage and current traces. Superimposed traces represent recordings at different voltage-clamp steps.

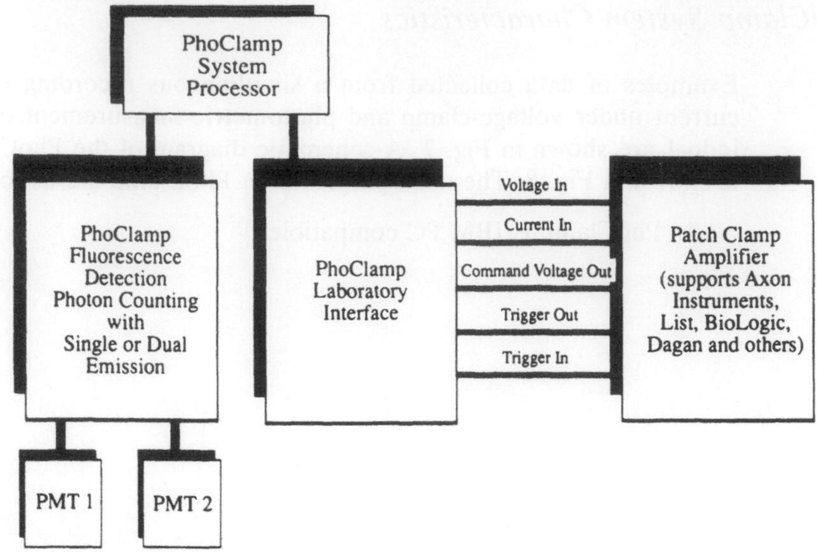


FIG. 8 Schematic input/output diagram of PhoClamp system. Characteristics: triggered sweep acquisition; complete and automatic complex waveform protocol generation or ramp generation + leak pulse generation; two separate time base collections (analog sampling rate up to 20 kHz per channel and fluorescence photon counting up to two simultaneous bins per millisecond); maximum 1 ratio/msec + analog samples up to 20 times this rate; simultaneous waveform command generation and recording to hard disk with on-line calibrated ion display.

2. PhoClamp is designed specifically for experiments combining electrophysiology and photometry. It is a total control system, enabling integration of waveform generation, high-speed data acquisition (say of membrane voltage and current), and one or two independent channels of photon counts arising from fluorescence.

3. It will detect fluorescent dyes using either single- or dual-wavelength emission ratio analysis (single-wavelength excitation). The system does not support a filter wheel for dual-excitation work (because of speed considerations), but ratiometric work with fura-2 is possible using isosbestic point imaging, where the 360-nm wavelength at which fura-2 is insensitive to $Ca^{2+}$ is sampled and used as the numerator or denominator for ratio purposes.

4. Extensive data analysis routines are provided, permitting plots of graphs of data from up to 10 different variables in the *X* and *Y* dimension—for instance, a plot of current versus voltage, or a plot of calibrated $Ca^{2+}$ concentration versus membrane current. Curve-fitting facilities are also provided for one or two exponentials of varying forms.

5. PhoClamp uses two-channel photon counting. Photon counting is particularly useful when dealing with low signals captured on very fast time scales, because it is less noisy.

6. PhoClamp is supplied complete with a high-quality built-in waveform generator under software control of the user. This can be used to generate one or two pulse protocols for electrophysiology, and, additionally, ramps, leak pulses, and membrane holding potentials. These signals are generated simultaneously with data collection and individual protocols can be stored on hard disk for instant recall and execution.

7. PhoClamp is supplied with a fast analog-to-digital and digital-to-analog converter. The software provides the user with the ability to take on two channels of data in addition to the fluorescent measurement. This is a powerful approach, invaluable for enabling simultaneous study of other parameters of physiological function. Examples might include patch-clamp or electrophysiological measurement of intracellular voltage and current, simultaneously with a two-channel fluorescent measurement of intracellular $Ca^{2+}$; or using an extracellular pH electrode (or ion-sensitive electrode) to record extracellular ionic activity in parallel with intracellular ion measurement. PhoClamp also has a wide range of data analysis procedures for multiparametric graphing of data, and time-varying signals—membrane potential, current, fluorescence, or ion concentration—are automatically analyzed and plotted by the system. A number of complex curve-fitting procedures are also provided.

8. The PhoClamp analog-to-digital converter works at about 20 kHz per channel for two channels and the photon counting system operates at bin widths up to 1 msec, or 1 kHz; this means that dual-channel $Ca^{2+}$ measurements can be made as fast as 1000 samples per second in each of two channels, about the maximum response time of the dyes.

9. With photon counting, electrophysiological or other measurements can be made simultaneously up to about 37,000 samples per second, as defined by the user.

10. Users have complete control over temporal response.

11. Data files are written to disk.

12. The software has built-in routines for data smoothing using a running-average method.

13. The system is compatible with most commercially available microscopes. It has been fitted to date to both Nikon and Zeiss microscopes, and is compatible with Leitz and Olympus equipment.

14. The PhoClamp system is usable with all currently available fluorescent ion-sensitive dyes, as well as with other optical probes.

15. The system is completely free of vibration and moving parts.

16. The system offers optional automatic shutter control with a fast Uni-

Blitz shutter (<5 msec open and close) by providing a voltage source linked to recording stop and start. This can be used for synchronization with data acquisition to reduce bleaching from the excitation source. It can also be used for work with photolyzable caged compounds (see Chapter 7).

17. The system offers internal or external triggered sweeps.
18. Compatibility with Postscript and Hewlett Packard printer/plotters.
19. Output of data can be to all conventional matrix printers as well as laser printers of Postscript or Hewlett Packard versions; in addition, data can be output to ASCII files or Hewlett Packard Graphic Language (HPGL) files for use in most other software packages, such as Lotus 123, Freelance, or Excel running on an Apple Macintosh.

In conclusion, the system offers high performance at relatively low cost, and many capabilities not available with other systems:

1. Single- or dual-channel photon counting.
2. High-performance (up to 65 kHz) analog-to-digital conversion for recording of biological-related electrical signals, i.e., membrane voltage and current.
3. Display of computed ion concentration, photon counts, or ratio on-line.
4. Ability to record up to five channels of data, including voltage, current, fluorescence photon counts, ratio, or calibrated ion concentration.
5. On-line display of data, including changes in ion concentration.
6. Ability to export all data as ASCII files to popular spreadsheet packages running on both PC compatibles, Apple Macintoshes, and networks, as standard.
7. Postscript and Hewlett Packard printer/plotter compatibility.
8. Permanent file-structure retention of data with ability to apply new calibration tables at any time.
9. On-line user experimental text entry as a "live" notebook.
10. High-quality 8-position Life Science Resources filter wheel.
11. Operation as single or dual-emission system.

## Loading of Fluorescent Probes for Photometric Experiments

Loading of optical probes into single cells can be either via the microelectrode using the free acid forms of ion-sensitive dyes, where whole-cell patch-clamp or other solute exchange techniques are employed, or simply by using the acetoxymethyl ester version if simple imaging or permeabilized patch techniques (such as with nystatin or other agents) are used.

Routine cell preparation and culturing techniques are employed. The cells intended for study are plated onto thin glass coverslips, usually No. 1.5. This permits focusing from below using an inverted microscope and allows passage of the lower wavelengths of ultraviolet light required to excite many of the available fluorescent probes (<360 nm). Plastic media cannot be used because plastic absorbs ultraviolet excitation light.

Cells may be loaded with ion-sensitive fluorescent probe in one of two ways. The free acid form of the probe may be directly loaded through the cell membrane by micropipette or patch pipette. For fura-2 and indo-1, a concentration of around 50–100 $\mu M$ has been found to be satisfactory. This method, however, requires techniques often unavailable in many laboratories. The great attraction of many of these probes is that they are available in the acetoxymethyl ester form. These nonpolar ester derivatives may be added to the extracellular medium, where they will diffuse across the membrane to be hydrolyzed by nonspecific cytoplasmic esterases, resulting in the membrane impermeable free acid form. Generally, the ester form of the dyes is insensitive to changes in ion concentration. Unfortunately, most plant cells studied so far appear to have low esterase activity, and this has made experiments difficult unless microinjection is used.

For experiments measuring intracellular $Ca^{2+}$ using fura-2 and indo-1 or measuring pH using BCECF or other probes for other ions, the probe can be initially made up as a stock solution of 2 m$M$ in dimethyl sulfoxide (DMSO). The DMSO should always be predried over a molecular sieve to remove any traces of water, which will shorten the life of the probes even when frozen. Cells are normally incubated in a 1–4 $\mu M$ solution of the acetoxymethyl (AM) ester form (fura-2-AM, BCECF-AM, or other) made up in a standard extracellular medium such as 125 m$M$ NaCl, 5 m$M$ KCl, 1.8 m$M$ $CaCl_2$, 2 m$M$ $MgCl_2$, 0.5 m$M$ $NaH_2PO_4$, 5 m$M$ $NaHCO_3$, 10 m$M$ HEPES, 10 m$M$ glucose, and 0.1 m$M$ bovine serum albumin (BSA) (pH 7.2) for 30 min at 37°C. Washing at least three times with fresh medium removes any excess probe, and subsequent microscopic examination reveals a clear signal from the intracellular free acid with minimal compartmentalization within the intracellular organelles. Pluronic acid, a weak detergent (Molecular Probes, Eugene, Oregon), can also help with loading dye into some cells that are resistant to such procedures.

One additional point of considerable importance is that loading solutions should never contain serum as used in culture medium. Inevitably, inclusion of serum—whether horse, fetal calf, etc., will produce poor loading because most of the serum fractions contain strong esterase activity, hydrolyzing the dye before it can be loaded. If serum is required, bovine serum albumin can be substituted safely.

Data collection from cells should start within approximately 2 hr of loading; cells have been seen to lose responsiveness when loaded for extended periods. Maintaining them at room temperature rather than 37°C does, however appear to extend this time. The cause is not fully understood but may be due to the probe partitioning into compartments that do not normally undergo changes in ionic concentration, or simply the normal biological function of the cells may be effected.

Loading methods produce varying results depending very much on the cell type being studied, time in culture, etc. There are no general rules to avoid problems, and trial and error with the cell type of interest is often the best plan. Among the most variable features are loading into nuclear or other compartments. Considerable progress in dye chemistry is being made all the time, and this is producing a range of new probes that may prove easier to work with. Apart from the scientific literature, probably the best means of staying abreast of developments is to acquire a subscription to the Molecular Probes (Eugene, Oregon) catalog and technical bulletins.

Recent developments to watch include dyes such as fura-2-labeled dextran conjugates, which can be injected and do not localize in the nucleus; multiple dye-labeled dextran conjugates; nuclear-targeted indicators, such as the single-wavelength probe calcium green; and indicators for ions that require excitation only in the visible portion of the spectrum, making them potentially easier to work with.

## Calibration of Ion-Sensitive Dyes in Living Cells and in Solution

Calibration for free ion concentration may be performed by two distinct methods, either imaging free acid solutions of the probe with various known ion concentrations or imaging the cells of interest loaded with the ester derivative of the probe under permeabilized conditions.

Although the fluorescent dyes developed for ion-sensitive measurements can also be utilized as single-wavelength excitation or emission probes, their power is only fully utilized using dual-excitation or dual-emission measurements with probes such as fura-2 and BCECF. These eliminate many of the artifacts associated with simple fluorescence intensity measurements. Fluo-3, a calcium-sensitive optical probe that changes spectral properties at only a single wavelength when $Ca^{2+}$ activity changes, is commonly used with single-wavelength confocal microscopy, but is prone to artifacts and is difficult to calibrate reliably.

In all the following methods, the probe is excited with either a single or dual wavelength and is measured, respectively, at a dual or single wavelength.

All calibration is performed in similar medium and at the same pH as used for subsequent experiments.

The following methods use as an example intracellular free $Ca^{2+}$ measurements using the dual-excitation fluorescent probe fura-2, which is available in both the free acid and acetoxymethyl ester forms (Molecular Probes). Many other probes for different ions are now available but the general methods are valid for any dual-excitation probe. For pH-sensitive dyes such as BCECF, permeabilization with nigericin could be used in place of ionomycin, as in the following example.

Data required for calibration may be entered into the TARDIS software by entering the maximum ($R_{max}$) and minimum ($R_{min}$) ratios obtained into the following equation, which is preprogrammed into the software.

$$\text{Ion concentration} = K_d\beta[(R - R_{min})/(R_{max} - R)],$$

where $K_d$ is the dissociation constant of the probe, $\beta$ is intensity at the upper wavelength at $R_{min}$ divided by intensity of the upper wavelength at $R_{max}$, $R$ is the measured ratio, $R_{max}$ is the ratio when the probe is saturated with $Ca^{2+}$, and $R_{min}$ is the ratio with no $Ca^{2+}$ present.

## *Free Acid Solution Method*

The simplest method for initial calibration is to prepare two solutions of medium, one containing 5 m*M* $CaCl_2$, which will saturate the probe with free $Ca^{2+}$ and result in the maximum ratio obtainable. The other should have the $CaCl_2$ substituted by 1–10 m*M* EGTA, which will bind all free $Ca^{2+}$ to result in the minimum ratio obtainable. Both must contain the free acid form of the probe at a concentration of about 50–100 $\mu M$, which is the approximate concentration found in cells loaded with the ester derivative.

Readings of these solutions are taken and the ratios and constants are obtained by the following method:

1. A background sample should be captured at each wavelength to correct for any background photons not attributable to optical probe.
2. After background correction, individual values are measured using the standard computer software. Raw values at each wavelength and ratio values are determined in high and low $Ca^{2+}$ concentrations. This procedure provides an estimate of $R_{min}$ and $R_{max}$, which can then be entered into the software.
3. The raw 380-nm value at minimum $Ca^{2+}$ concentration is divided by the raw 380-nm value at maximum $Ca^{2+}$ concentration. This is the $\beta$ value

that is required to be entered into the equation. Note that this value will differ with changes in the optical components of the system, and even with changes in objective or lamp.

### *Intracellular Method*

This is the preferred method of calibration. The cells are loaded with the acetoxymethyl ester derivative of the probe as previously described and the cell membrane is permeabilized to ions with a specific ionophore. For $Ca^{2+}$, ionomycin at a concentration of about 1–2 $\mu M$ has been found to be particularly suitable for $Ca^{2+}$. Another popular $Ca^{2+}$ ionophore, the antibiotic A23187, is only suitable as its halogenated analog 4-bromo-A23187, because native A23187 exhibits autofluorescence at the upper fura-2 wavelength. Some laboratories have also used low concentrations of digitonin or saponin to permeabilize cells for calibration; effective concentrations will vary from cell to cell.

The $R_{max}$ measurement should be made in elevated extracellular [$CaCl_2$]. A concentration of 10 m$M$ has been found sufficient to saturate the probe within the cell. Addition of the ionomycin causes a rapid, sustained rise in intracellular [$Ca^{2+}$] that reaches maximum within tens of seconds. This whole process should be captured and time allowed for equilibration to measure the $R_{min}$ value. This may now be determined on the same field by washing the cells two or three times with calcium-free medium containing 1–10 m$M$ EGTA. Transport across the cell membrane—resulting in binding of free $Ca^{2+}$—may require at least 15 min.

## Applications of Photometric and Imaging Technology on Single Cells

Excitable tissues, such as neurons and endocrine cells, have several systems that regulate the concentration of cytosolic free $Ca^{2+}$ ($[Ca^{2+}]_i$). On the one hand, $Ca^{2+}$ influx into the cytoplasm is largely controlled by voltage-sensitive $Ca^{2+}$ channels, receptor-operated $Ca^{2+}$ channels (Bean, 1989), second messenger-operated $Ca^{2+}$ channels (Meldolesi and Pozzan, 1987), mechanically operated $Ca^{2+}$ channels (Lansman *et al.*, 1987), background (or tonically active) $Ca^{2+}$ channels (Benham and Tsien, 1987), gap junction channels (Brehm *et al.*, 1989), and by release from internal organelle compartments (Carafoli, 1987). On the other hand, mechanisms of reducing $[Ca^{2+}]_i$ levels include plasma membrane extrusion systems, such as the $Ca^{2+}$-ATPase and the $Na^+/Ca^{2+}$-exchanger, organelle sequestration, and $Ca^{2+}$-binding proteins (for a review see Tsien and Tsien, 1990). All of these processes control both the resting $[Ca^{2+}]_i$ levels and the regulation of $[Ca^{2+}]_i$ transients occurring as a result of electrical activity and/or other environmental chemical signals such as neurotransmitters. We shall first show how the $Ca^{2+}$ influx passing through voltage-sensitive $Ca^{2+}$ channels can be recorded in voltage clamp before proceeding to show related examples for studying the link between electrical activity and $[Ca^{2+}]_i$ homeostasis.

## *Calcium Influx through Voltage-Gated Channels*

In excitable cells, $Ca^{2+}$ channels are the most extensively studied class of ionic channels. They represent a diverse class of molecules that traditionally have been grouped into two major categories according to their kinetics and voltage-dependent properties (for reviews, see Bean, 1989; Snutch and Reiner, 1992). High-voltage-activated (HVA) $Ca^{2+}$ channels first activate on depolarization to relatively positive potentials and display diverse kinetics, pharmacologies, and sensitivities to voltage (L-, N-, and P-types). The HVA $Ca^{2+}$ channels are multisubunit complexes, including a large, pore-forming $\alpha 1$ subunit that encodes many of the unique electrophysiological and pharmacological properties of these channels. Low-voltage-activated (LVA) $Ca^{2+}$ channels are available for opening only from negative membrane holding potentials and transiently activate with small depolarizations. The electrical properties of LVA $Ca^{2+}$ channels (also called T-type channels) have led to proposals for their roles in the mediation of pacemaking activity, repetitive bursting, and secretion. Some LVA $Ca^{2+}$ channels have been identified as targets of anticonvulsants, and dysfunction of HVA $Ca^{2+}$ channels has been implicated in some forms of epileptiform activity (for a review, see Tsien and Tsien, 1990).

Because ionic channel activities and agonist-induced responses are very susceptible to alterations during the whole-cell recording technique, we have used the nystatin-perforated patch method (Horn and Marty, 1988) in experiments designed to record simultaneously ionic currents and $[Ca^{2+}]_i$ levels. Moreover, this allowed us to measure $[Ca^{2+}]_i$ homeostasis without outward diffusion of cytoplasmic constituents and without introducing alien $Ca^{2+}$ buffers into the recorded cell.

We have used a prolactin-secreting cell line from rat anterior pituitary gland (GH3 cells) as a model for the study of such a relationship between ionic channel activity and intracellular $Ca^{2+}$ level. Under voltage-clamp conditions, depolarizations evoked outward potassium currents that dominated the whole-cell current recorded so that net outward currents were always seen (Fig. 9A). The pipette was filled with 55 m$M$ KCl, 75 m$M$ $K_2SO_4$, 8 m$M$ $MgCl_2$, and 10 m$M$ K-HEPES. In other experiments using a different pipette

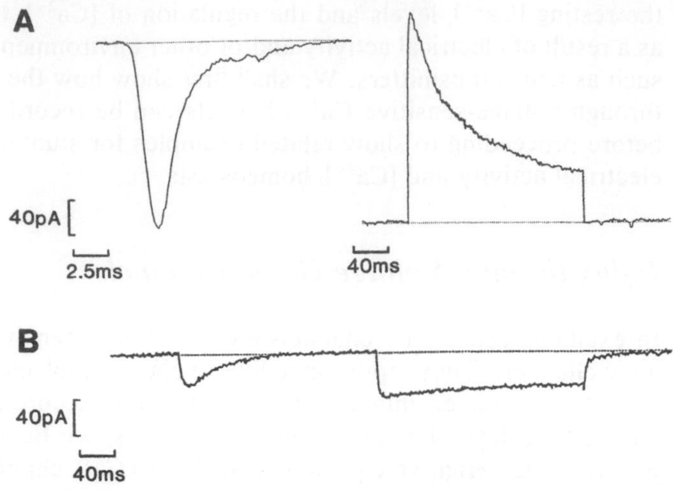


FIG. 9 Voltage-gated ionic current in GH3 Cells. (A) Left trace, voltage-dependent $Na^+$ current elicited at −10 mV from −85 mV; right trace, voltage-dependent $K^+$ current elicited at +40 mV from −85 mV. Patch pipettes were filled with a KCl/$K_2SO_4$ solution containing 150 μg/ml nystatin. (B) Separation of T- and L-type $Ca^{2+}$ currents in GH3 cells. T-Type current was activated by 200-msec pulses from −85 to −30 mV (left), whereas the L-type current was elicited following the full inactivation of the T-type current by stepping the membrane potential to +10 mV (right). External $Ca^{2+}$ was 2.5 m*M* with 1 μ*M* TTX and patch pipettes were filled with a CsCl/$Cs_2SO_4$ solution containing 150 μg/ml nystatin.

solution, for example, 55 m*M* CsCl, 75 m*M* $Cs_2SO_4$, 8 m*M* $MgCl_2$, and 10 m*M* Cs-HEPES, voltage-activated inward currents were recorded during depolarizing voltage steps. Under the presence of 1 μ*M* of tetrodotoxin in the external milieu, GH3 cells exhibit T-type and L-type $Ca^{2+}$ currents (Fig. 9B). Both currents can be reliably voltage clamped and well separated in this cell type (Matteson and Armstrong, 1986). The T-type current is activated by membrane potentials positive to about −50 mV and appears to exhibit complete inactivation over most or all of its activation range. The L-type current is activated by more positive command potentials (about −30 mV) and exhibits little or only slow inactivation (reviewed by Bean, 1989). One basic procedure we have used for separation of the two components of $Ca^{2+}$ current is shown in Fig. 9B. T-Type currents were activated by stepping to −30 mV from a holding potential of −85 mV; virtually all current activated at this potential inactivates during the voltage steps, indicating that no detectable L-type current has been activated. Thus, measurement of peak inward current activated by steps to −30 mV largely reflects the exclusive activation of the T-type current. To allow examination of the L-type current, following the full inactivation of the T-type at −30 mV, the cell membrane was stepped to +10 mV (see Fig. 9B).

## *Homeostasis of the Intracellular Free $Ca^{2+}$ Concentration*

Over the last few years, knowledge about mechanisms of $Ca^{2+}$ entry and $Ca^{2+}$ discharge from intracellular $Ca^{2+}$ stores has increased dramatically. Patch-clamp recordings and other electrophysiological methods, when combined with methods for measuring and manipulating $[Ca^{2+}]_i$, have provided new views of $Ca^{2+}$ channel activity and have led to a better understanding of the complexity of $[Ca^{2+}]_i$ regulation. Multiple pathways for $Ca^{2+}$ mobilization provide great flexibility for cellular control, but also make it more difficult to decipher mechanisms and functions. Fortunately, an array of new methods exist for measuring and manipulating $[Ca^{2+}]_i$. In this review, we first describe the combination of current-clamp recordings with measurements of $[Ca^{2+}]_i$ using photometry and then analyze the link between ionic currents monitored in voltage-clamp and $[Ca^{2+}]_i$ homeostasis.

## *Combining Current-Clamp Recording with Photometry*

The nervous system and the glandular cells of the anterior hypophysis share common properties. They both secrete peptidergic hormones, and both exhibit electrical properties such as excitability, with production of action potentials. Thus, electrophysiological techniques can be readily applied. The electrophysiological properties of these cells reveal (1) stimulus–secretion coupling and (2) modifications in membrane electrical properties brought about by the action of different regulatory factors with their receptors. These observations are used to explain the modulation of membrane properties by each factor that enhances or inhibits hormone secretion. Thus the electrical activity plays a central role in the regulation of endocrine secretion in the anterior hypophysis.

In an attempt to correlate such electrical activity with regulation of cell activation, we have combined patch-clamp recording and microfluorimetry using indo-1. Indeed, changes in intracellular $Ca^{2+}$ were monitored in single cells of the pituitary line GH3 preloaded with indo-1-AM and we have found that a single action potential leads to a marked transient increase in cytosolic free $Ca^{2+}$ (Fig. 10). $[Ca^{2+}]_i$ increased concomitantly with the action potentials to a maximum around 600 n*M* within 400 msec, and $Ca^{2+}$ was reduced to basal levels within a few seconds (Fig. 10). Interestingly, we have found that the magnitude and duration of the $[Ca^{2+}]_i$ transients depended on the number of action potentials fired but were independent of the duration of the depolarizing pulses used to triggered action potentials when cells were silent.

This typical kinetics aspect of the $[Ca^{2+}]_i$ transients, as observed in our combined experiments, was also seen in spontaneous fluctuations of $[Ca^{2+}]_i$ detected by single-cell monitoring of fura-2-loaded GH3 cells but not subjected to patch-clamp recordings. Such $[Ca^{2+}]_i$ oscillations stopped when $Ca^{2+}$ was removed from the bath, implying that these $[Ca^{2+}]_i$ spikes result from enhanced $Ca^{2+}$ influx during action potentials occurring spontaneously in these cells.

Figure 10 indicates that the fluctuations in $[Ca^{2+}]_i$ arising from spontaneous action potentials permit calcium-dependent cell activation. The dependence on $[Ca^{2+}]_i$ for prolactin secretion, determined in permeabilized GH3 cells, has been found to occur over the range 0.2–10 $\mu M$ free $[Ca^{2+}]_i$. Thus, the average rise in $[Ca^{2+}]_i$ associated with spontaneous action potentials is sufficient to trigger prolactin secretion.

## *Combining Voltage-Clamp Recording with Photometry*

We have also used ratiometric photon counting technology to detect fluorescence at two wavelengths, acquired simultaneously with electrophysiological data while the cells were voltage-clamped using the perforated patch method. From a holding potential of −80 mV, the measured basal $[Ca^{2+}]_i$ in indo-1-AM-loaded $GH_3$ cells was around 100 n*M* (Lledo *et al.*, 1992). These data are comparable to values obtained previously with the same paradigm (Benham, 1989; Mollard *et al.*, 1991) or from cells loaded with fura-2 and not subject to patch clamping (Schlegel *et al.*, 1987; Akerman *et al.*, 1991). Depolarization of cells to 0 mV caused a transient rise of $[Ca^{2+}]_i$ in cells loaded with the ester form of the $Ca^{2+}$ indicator dye indo-1-AM. Figure 11 shows the reciprocal changes in counts at 490 and 405 nm and estimated $[Ca^{2+}]_i$ derived from these values in a cell clamped to −80 mV. Thus, we were able to measure $Ca^{2+}$ entry through voltage-sensitive $Ca^{2+}$ channels and to assess simultaneously the $[Ca^{2+}]_i$ level. Several observations indicate that changes in $[Ca^{2+}]_i$ observed were due to $Ca^{2+}$ entry into the cell. First, the magnitude of the changes were closely correlated with the size of the voltage-dependent $Ca^{2+}$ current, both signals being very small during depolarizations to potentials more negative than −20 mV or more positive than +60 mV. With increasing depolarization, inward current became larger as more and more channels

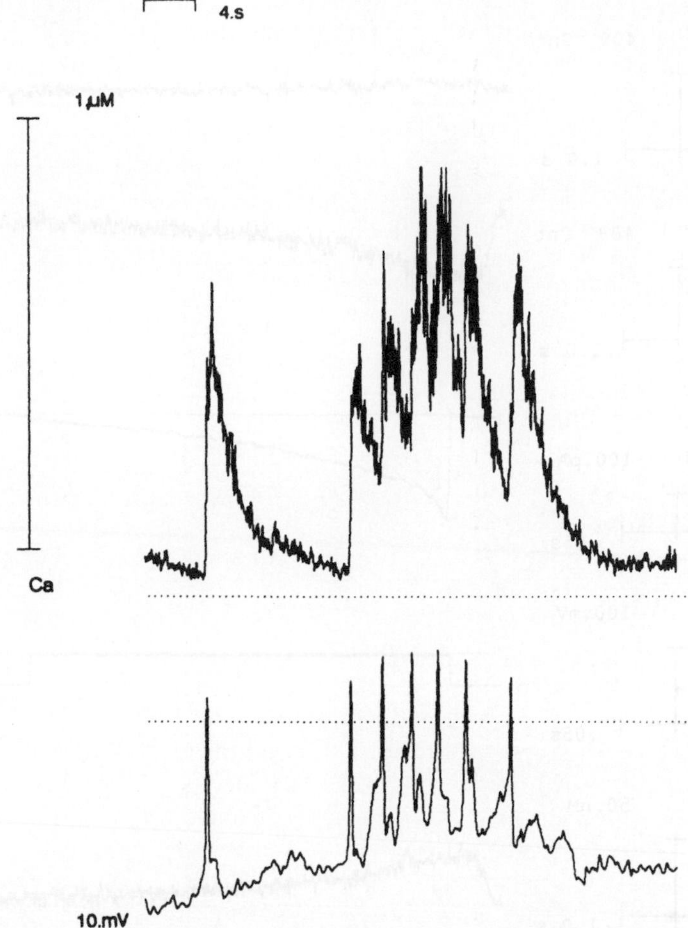


FIG. 10 Spontaneous action potential-induced rises in $[Ca^{2+}]_i$ in current-clamped GH3 cells. Intracellular free $Ca^{2+}$ concentration was monitored with the fluorescent probe indo-1 during electrophysiological recordings obtained with the patch-clamp technique in its perforated configuration using nystatin in the pipette. Action potentials occurred spontaneously (lower trace) while the ratio of the two fluorescences was measured at emission wavelengths 405 and 490 nm with a time resolution of 1 ms (upper trace).

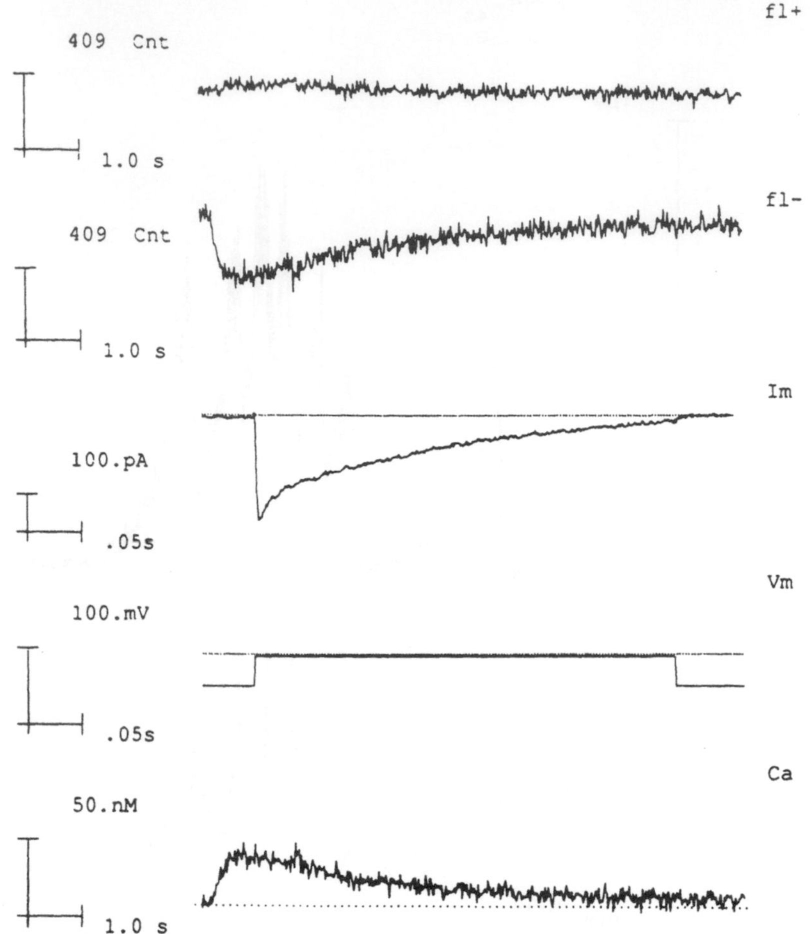


FIG. 11 $Ca^{2+}$ current triggered by voltage depolarization-induced rises in $[Ca^{2+}]_i$ in voltage-clamped GH3 cells. The two upper traces show the reciprocal changes in counts at 490 and 405 nm; the middle trace is the voltage-gated $Ca^{2+}$ current elicited from −80 to 0 mV during 250 msec and the bottom traces reflect the estimated $[Ca^{2+}]_i$ derived from the ratio of the two fluorescences values.

opened, then diminished as the driving force for $Ca^{2+}$ influx fell, and ultimately turned outward (Fig. 12). Thus, each inward current led to a transient rise in $[Ca^{2+}]_i$ whose voltage dependence closely paralleled that of the inward current (Fig. 12). The absence of $Ca^{2+}$ transients under extreme depolarization excludes the possibility that they arise, as for example in skeletal muscle,

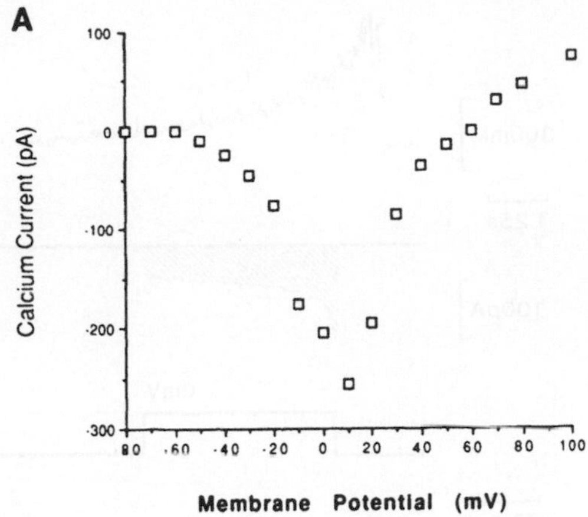


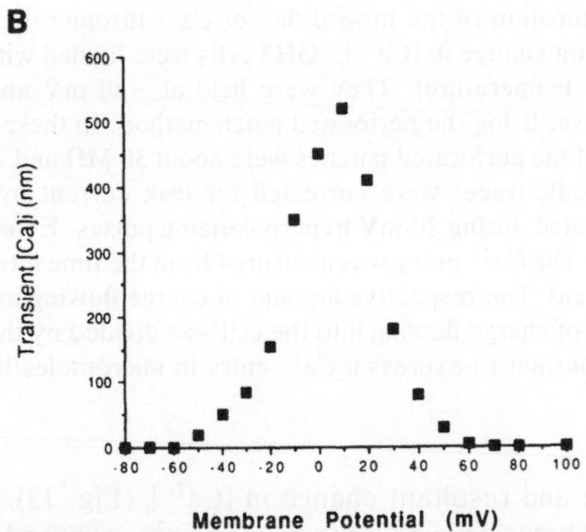


FIG. 12 Voltage dependence of $[Ca^{2+}]_i$ transients. (A) Current–voltage relationship of $Ca^{2+}$ current recorded from a holding potential of −80 mV and (B) estimated $[Ca^{2+}]_i$ versus membrane potential from the same single cell. Note the similar feature of the two relationships.

through voltage-activated release from internal store. Moreover, these $[Ca^{2+}]_i$ transients were absent in cells dialyzed with 10 m*M* EGTA.

The $[Ca^{2+}]_i$ rises induced by the voltage steps during 200 msec persist for 7–9 sec (Figs. 11 and 13). The decline of $[Ca^{2+}]_i$ can be fitted by two exponentials with both fast and slow time constants (1.5 and 4 sec, respectively at room temperature). To determine the $Ca^{2+}$ buffering capacity of the cell, we have measured simultaneously, on the same cell, the $Ca^{2+}$ entry per unit

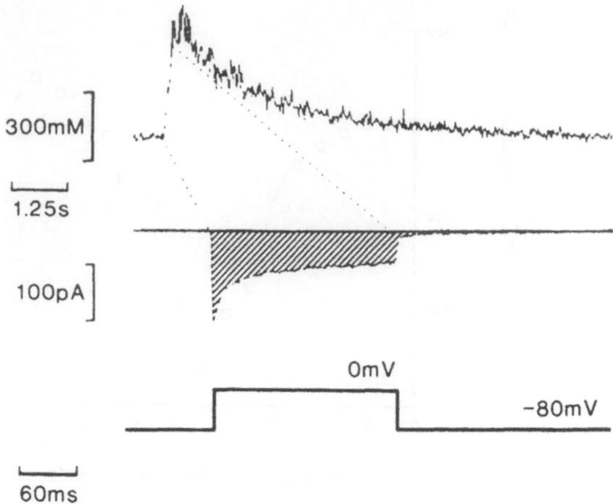


FIG. 13 Estimation of the inward flux of $Ca^{2+}$ through the plasma membrane and its concomitant change in $[Ca^{2+}]_i$. GH3 cells were loaded with indo-1-AM (5 $\mu M$, 30 min at room temperature). They were held at −80 mV and depolarized to 0 mV during 200 msec using the perforated patch method. In these experiments, the series resistances of the perforated patches were about 30 MΩ and 70% compensated. $Ca^{2+}$ currents (middle trace) were corrected for leak current by adding four times the current activated during 20-mV hyperpolarizing pulses. External $Ca^{2+}$ concentration was 2.5 m$M$. The $Ca^{2+}$ entry was measured from the time integral of the $Ca^{2+}$ current (hachured area). The respective amount of charge flowing into the cell was $10^{-11}$ C. This amount of charge flowing into the cell was divided by the cell volume and twice Faraday's constant to express a $Ca^{2+}$ entry in micromoles/liter cell volume.

cell volume and resultant change in $[Ca^{2+}]_i$ (Fig. 13). Figure 14 shows the relationship between such a normalized $Ca^{2+}$ entry and $[Ca^{2+}]_i$ changes. $Ca^{2+}$ currents were obtained from different cells held to −80 mV and subjected to voltage-clamp steps ranging from 0 to +10 mV with a duration varying from 50 to 250 msec. The $Ca^{2+}$ entry was measured from the time integral of the $Ca^{2+}$ current. This amount of charge flowing into the cell was divided by the cell volume and twice Faraday's constant to express a $Ca^{2+}$ entry in micromoles/liter cell volume, according to the relation:

$$[Ca^{2+}]_{entry} = (2FV)^{-1} \int_0^t - I_{Ca}(t)\, dt,$$

where $F$ is Faraday's constant ($0.19 \times 10^6$ $C/\mu M$), $V$ is the cell volume, and $I_{Ca}$ is the recorded $Ca^{2+}$ current. The cell volume can be estimated from the measured capacitance and the observed radius as follows. With an average

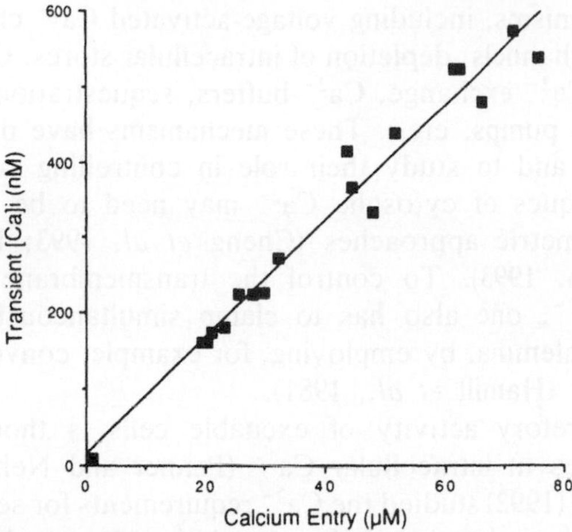


FIG. 14 Relationship between normalized $Ca^{2+}$ entry versus amplitude of $[Ca^{2+}]_i$ transients in GH3 cells. $Ca^{2+}$ currents were obtained from different cells held to −80 mV and subjected to voltage-clamp steps ranging from 0 to +10 mV with a duration varying from 50 to 250 msec. The change in $[Ca^{2+}]_i$ is clearly correlated with the $Ca^{2+}$ entry. The regression line has a slope of about 8 $nM/\mu M$ ($r$ = 0.75), which reflects the endogenous buffering capacity of these cells. Data were obtained from protocols described in Fig. 13.

initial capacitance and diameter of 10 pF and 17 μm, respectively, this gives a specific membrane capacitance of 9.8 fF/$\mu m^2$, assuming GH3 cells are spherical. Thus, cell volume could be evaluated from cell membrane capacitance measurements by $V = [(C/9.8)^3/36\pi]^{1/2}$, with $C$ in femtofarads and $V$ in $\mu m^3$. The regression line for $GH_3$ cells has a slope of 8 $nM/\mu M$ (Fig. 14). Interestingly, only about 1/100th of the $Ca^{2+}$ influx through the plasma membrane of anterior pituitary cells was seen as free $Ca^{2+}$ at the peak of the $[Ca^{2+}]_i$ transient, revealing that most of the $Ca^{2+}$ (≈99%) that enters through $Ca^{2+}$ channels binds to endogeneous $Ca^{2+}$ buffers.

## *Electrophysiology and $Ca^{2+}$ Imaging*

Imaging of optical probes is effectively a multiarray photometric measurement, and although it may have less sensitivity overall, it does provide spatial information that straight photometry cannot. Imaging cameras are essentially arrays of photosensitive elements.

An increase in cytosolic $Ca^{2+}$ activity can be effected by a number of mechanisms, including voltage-activated $Ca^{2+}$ channels, receptor-operated $Ca^{2+}$ channels, depletion of intracellular stores, $Ca^{2+}$ removal systems (i.e., $Na^+/Ca^{2+}$ exchange, $Ca^{2+}$ buffers, sequestration into organelles via ATP-driven pumps, etc.). These mechanisms have different cytosolic localizations, and to study their role in controlling cellular processes, imaging techniques of cytosolic $Ca^{2+}$ may need to be considered in addition to photometric approaches (Cheng *et al*, 1993; Rizzuto *et al.*, 1993; see Mason, 1993). To control the transmembrane electrochemical gradient of $Ca^{2+}$, one also has to clamp simultaneously the voltage across the plasmalemma, by employing, for example, conventional patch-clamp techniques (Hamill *et al.*, 1981).

Secretory activity of excitable cells is thought to be correlated to changes in intracellular $Ca^{2+}$ (Penner and Neher, 1988). Augustine and Neher (1992) studied the $Ca^{2+}$ requirements for secretion in bovine chromaffin cells and deduced that depolarizations produce substantial changes in $[Ca^{2+}]_i$ at secretory sites. These spatiotemporal gradients of $Ca^{2+}$ have been detected by employing digital imaging microscopy of fura-2-loaded cells during depolarization in voltage-clamped cells (Neher and Augustine, 1992).

The advantage of the whole-cell patch-clamp technique is that diffusible substances can be introduced into the cytosol, such as fluorescent indicators, ions, and substances playing a role in stimulus–secretion coupling. The inclusion of fura-2 in the pipette solution would not only allow the monitoring of cytosolic $Ca^{2+}$, but would also interfere with endogenous cytosolic $Ca^{2+}$ buffers. Thus the amplitude and the kinetics of depolarization-induced $Ca^{2+}$ transients are affected by loaded fura-2, and may provide a method to study intracellular $Ca^{2+}$ buffers (Augustine and Neher, 1992).

Intuitively, one would expect that $Ca^{2+}$ activity profiles develop under steady-state conditions, if the rate of $Ca^{2+}$ entry across the plasmalemma is equal to the rate of chelation by the introduced $Ca^{2+}$ buffer. Intracellular $Ca^{2+}$ profiles in the presence of a diffusible chelator were predicted earlier (Neher, 1986; Mathias *et al.*, 1990). Figure 15 shows an experiment wherein a bovine lactotroph cell was loaded via the patch pipette with a solution containing 10 m*M* EGTA and 40 $\mu$*M* fura-2. The time constant of the loading, monitored by fluorescence excitation at 340 nm, is around 80 sec, consistent with previous reports (see Pusch and Neher, 1988). The cell was voltage-clamped at −65 mV and the steady-state inward current was around 10 pA under such conditions. In Fig. 15B fluorescence images show that during loading the intensity of the signal increases (Fig. 15A), but a distinct inhomogeneity is observed in that the signal intensity is lowest beneath the mouth of the pipette, suggesting a low $Ca^{2+}$ activity. If intensity is plotted three-

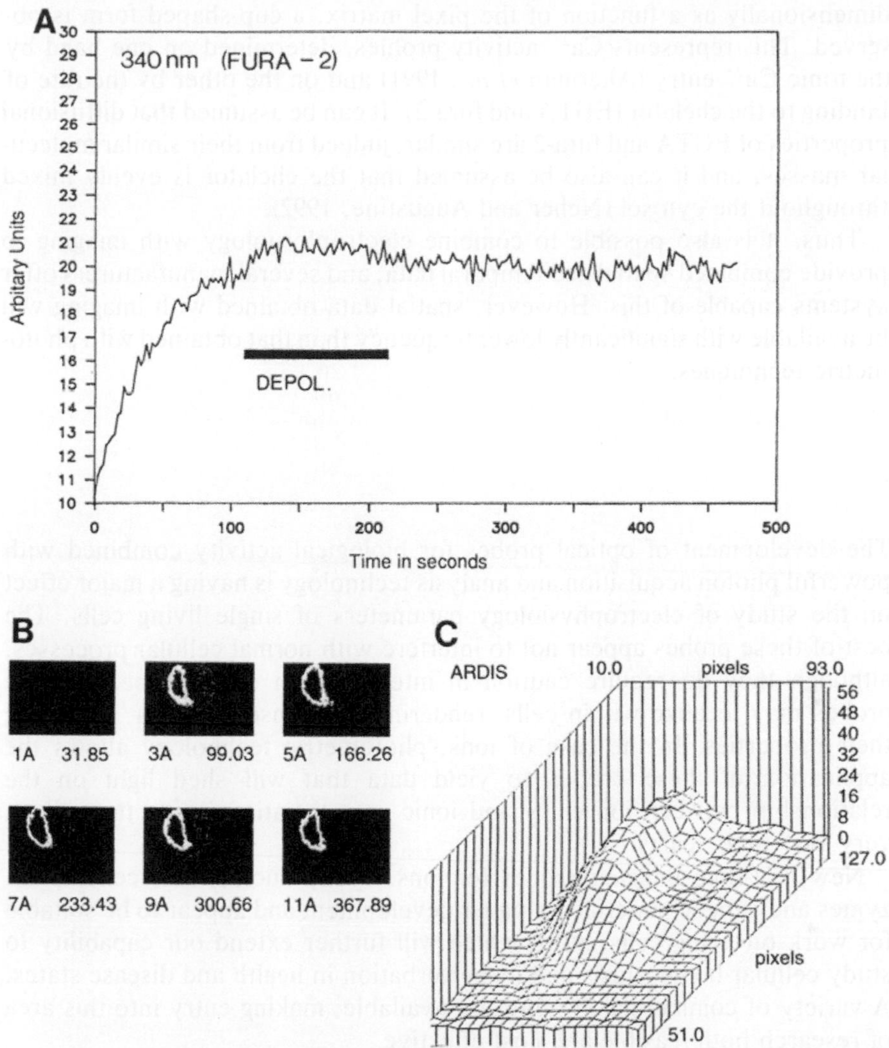


FIG. 15 (A) Fura-2 loading of a bovine lactotroph via the patch pipette. Time-dependent increase in fluorescence (arbitrary units) of fura-2 loaded into a bovine lactotroph via the patch pipette. Fura-2 was excited at 340 nm and emission was recorded at 510 nm. Bar indicates depolarization from −65 to −7 mV. (B) Fluorescence images taken at different times during loading of fura-2. Intensities are represented in grey levels, white denoting a higher intensity. The patch pipette position is indicated by the drawing. (C) Three-dimensional representation of intensity over the field of cell. Note the cup-shaped depression of intensity below the patch pipette.

dimensionally as a function of the pixel matrix, a cup-shaped form is observed. This represents $Ca^{2+}$ activity profiles, determined on one hand by the tonic $Ca^{2+}$ entry (Akerman *et al.*, 1991) and on the other by the rate of binding to the chelator (EGTA and fura-2). It can be assumed that diffusional properties of EGTA and fura-2 are similar, judged from their similar molecular masses, and it can also be assumed that the chelator is evenly mixed throughout the cytosol (Neher and Augustine, 1992).

Thus, it is also possible to combine electrophysiology with imaging to provide combined spatial and temporal data, and several manufacturers offer systems capable of this. However, spatial data obtained with imaging will be available with significantly lower frequency than that obtained with photometric techniques.

## Summary

The development of optical probes for biological activity combined with powerful photon acquisition and analysis technology is having a major effect on the study of electrophysiology parameters of single living cells. The best of these probes appear not to interfere with normal cellular processes, although they do require caution in interpretation of data, because the probes may localize within cells, rendering them insensitive or modifying their properties. In the case of ions, photometric technology allows the application of these probes to yield data that will shed light on the relationship between ion entry and ionic concentration inside the cell on very fast time scales.

New real-time imaging probes for ions, cyclic nucleotides, cellular enzymes and genetic material are under development and appear to be suitable for work on living cells, all of which will further extend our capability to study cellular homeostasis and its perturbation in health and disease states. A variety of commercial systems are available, making entry into this area of research both feasible and cost-effective.

## Acknowledgments

We thank the Biotechnology and Biological Science Research Council, Medical Research Council, Applied Imaging Ltd., Kabi Pharmacia, Wellcome Trust, Nuffield Foundation, Guggenheim Foundation, and British Heart Foundation for valuable funding, which has supported various aspects of the work discussed here.

## References*

Akerman, S. N., Zorec, R., Cheek, T., Moreton, R. B., Berridge, M. J., and Mason, W. T. (1991). *Endocrinology (Baltimore)* **129,** 475.

Augustine, G. J., and Neher, E. (1992). *J. Physiol. (London)* **450,** 247.

Bean, B. P. (1989). *Annu. Rev. Physiol.* **51,** 367.

Benham, C. D. (1989). *J. Physiol.* **415,** 143.

Benham, C. D., and Tsien, R. W. (1987). *Soc. Gen. Physiol. Symp.* **42,** 45.

Brehm, P., Lechleiter, J., Smith, S., and Dunlap, K. (1989). *Neuron* **3,** 191.

Carafoli, E. (1987). *Annu. Rev. Biochem.* **56,** 395.

Cheng, H., Lederer, W. J., Cannell, M. B. (1993). *Science* **262,** 740.

Dempster, J. (1993). "The Computer Analysis of Electrophysiological Signals." Academic Press, London.

Grynkiewicz, G., Poenie, M., and Tsien, R. Y. (1985). *J. Biol. Chem.* **260,** 3440–3450.

Hamill *et al.* (1981). *Pflugers Archiv.* **391,** 85.

Haugland, R. P. (1994). "Handbook of Fluorescent Probes and Research Chemicals." Molecular Probes, Eugene, Oregon.

Horn, R., and Marty, A. (1988). *J. Gen. Physiol.* **92,** 145.

Lansman, J. B., Hallam, T. J., and Rink, T. J. (1987). *Nature (London)* **325,** 811.

Lledo, P.-M., Somasundaram, B., Morton, A. J., Emson, P., and Mason, W. T. (1992). *Neuron* **9,** 943.

Mason, W. T. (1993). "Fluorescent and Luminescent Probes for Biological Activity. A Practical Guide to Technology for Quantitative Real-Time Analysis," Biological Techniques Series. Academic Press, London and San Diego.

Mason, W. T., Hoyland, J., Davison, I., Carew, M., Zorec, R., Shankar, G., and Horton, M. (1993). *in* "Fluorescent and Luminescent Probes for Biological Activity—A Practical Guide to Technology for Quantitative Real-Time Analysis" (W. T. Mason, ed.), p. 223. Academic Press, London and San Diego.

Mathias, Cohen, and Oliva (1990).

Matteson, D. R., and Armstrong, C. M. (1986). *J. Gen. Physiol.* **87,** 161.

Meldolesi, J., and Pozzan, T. (1987). *Exp. Cell Res.* **171,** 271.

Mollard, P., Guérineau, N., Chiavaroli, C., Schlegel, W., and Cooper, D. M. F. (1991). *Eur. J. Pharmacol.* **206,** 271.

Neher, E., and Augustine, G. J. (1992). *J. Physiol. (London)* **450,** 273.

Neher, E. (1986). *in* "Calcium Electrogenesis and Neuronal Functioning" (V. Heinemann and M. Klee, eds.), p. 80. Springer-Verlag, Heidelberg.

Penner, R., and Neher, E. (1988). *J. Exp. Biol.* **139,** 329.

*In addition to the references, the February/March 1990 issue of *Cell Calcium, Volume 11*, provides a very comprehensive treatment of additional methods and applications for work in this area.

Pusch, M., and Neher, E. (1988). *Pfluegers Arch.* **411,** 204.
Rizzuto, R., Brini, M., Murgia, M., and Pozzan, T. (1993). *Science* **262,** 744.
Sakman, B., and Neher, E. (1983). "Single-Channel Recording." Plenum, New York.
Standen, N. B., Gray, P. T. A., and Whitaker, M. J. (1987). "Microelectrode Techniques—The Plymouth Workshop Handbook." Company of Biologist, Cambridge.
Schlegel, W., Winiger, B. P., Mollard, P., Vacher, P., Wuarin, F., Zahnd, G. R., Wollheim, C. B., and Dufy, B. (1987). *Nature (London)* **329,** 719.
Snutch, P., and Reiner, P. B. (1992). *Curr. Opin. Neurobiol.* **2,** 247.
Tsien, R. W., and Tsien, R. Y. (1990). *Annu. Rev. Cell Biol.* **6,** 715.
Williams, D. A., and Fay, F. S. (1990). *Cell Calcium* **11,** 75.

# [7] Use of Chelators and Photoactivatable Caged Compounds to Manipulate Cytosolic Calcium

Alison M. Gurney and Susan E. Bates

## Introduction

Calcium is an important signaling molecule in neuronal systems, with many of the intracellular pathways that regulate neuronal activity being calcium dependent. In most mammalian neurons, the resting intracellular concentration of ionized (free) calcium ($[Ca^{2+}]_i$) is held low, in the nanomolar range, by various cellular buffering mechanisms. When neurons are activated, the $[Ca^{2+}]_i$ can rise by several orders of magnitude. The concentration may rise uniformly throughout the cell or the change may be localized to a specific region of the cell. In order to study how $[Ca^{2+}]$ is regulated in neurons, and how changes in $[Ca^{2+}]_i$ regulate neuronal activity, methods are needed that allow the $[Ca^{2+}]_i$ to be experimentally controlled. This may involve clamping the $[Ca^{2+}]_i$ to prevent it from changing in response to a stimulus, or imposing a change in $[Ca^{2+}]_i$ on the neuron that mimics the change seen in response to a normal stimulus. This is most easily achieved with $Ca^{2+}$ chelators, or buffers, incorporated into the neuron.

## Calcium Buffers

A number of small organic ligands behave as $Ca^{2+}$ buffers and can be used to control $[Ca^{2+}]$ in simple solutions. Such ligands usually also bind other di- and trivalent cations, which will interfere with $Ca^{2+}$ binding. Of particular importance are $Mg^{2+}$ ions, because these are present in cells at millimolar concentration. The main $Ca^{2+}$ buffers used in biological experiments are ethyleneglycol bis($\beta$-aminoethyl ether)-$N,N,N',N'$-tetraacetic acid (EGTA; Fig. 1a) and 1,2-bis($o$-aminophenoxy)ethane-$N,N,N',N'$-tetraacetic acid (1) (BAPTA; Fig. 1b), because they bind $Ca^{2+}$ with high selectivity over $Mg^{2+}$. Metal ions bind to these ligands primarily by coordinating with the oxygen atoms of the four carboxylate groups, although the nitrogen atoms to which they are linked also contribute to the coordination of the ion. Coordination can be improved by increasing the electron density on the nitrogen atoms,

*Methods in Neurosciences, Volume 27*

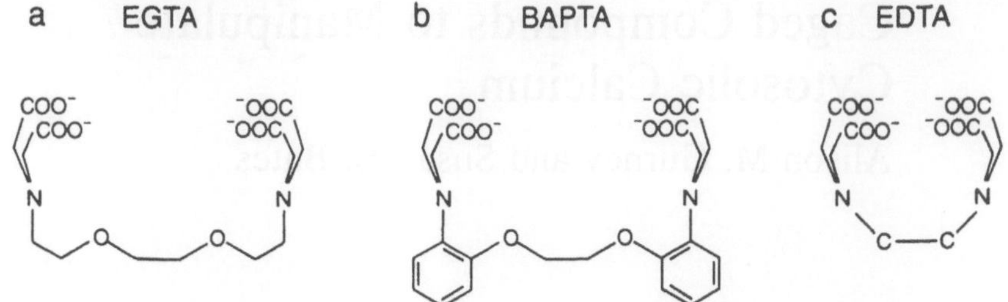


FIG. 1 Structures of $Ca^{2+}$ chelators commonly used in biological experiments.

by, for example, adding an electron-donating moiety to the ligand at an appropriate position. This has the effect of increasing ion affinity.

The binding site is in effect a "cavity" formed in the chelator molecule by the steric disposition of the carboxylate groups. Selectivity for $Ca^{2+}$ is mainly determined by having a cavity that is the right size to promote $Ca^{2+}$ binding, but which cannot constrict further to comfortably accommodate a $Mg^{2+}$ ion. Selectivity of a ligand can be improved by sterically constraining the carboxylate groups to limit the size of the cavity formed. This is apparent in comparing EGTA and BAPTA with the simpler tetracarboxylate ligand ethylenediaminetetraacetic acid (EDTA; Fig. 1c), which binds $Ca^{2+}$ and $Mg^{2+}$ equally well. Presumably the smaller ethylene linkage between the two nitrogen atoms allows the carboxylate groups in EDTA to move closer together, so that the cavity can constrict further than is possible in the larger molecules (1).

Chelator affinities for $Ca^{2+}$ are influenced by the pH of the solution in which they are dissolved, because they contain several ionizable groups that bind protons. The chelator molecules are present as several different ionic species, the proportion of which varies with the pH. For example, at alkaline pH, EGTA exists mainly in the deprotonated form ($EGTA^{4-}$). As pH decreases, the proportion of singly protonated ($HEGTA^{3-}$), doubly protonated ($H_2EGTA^{2-}$), and then $H_3EGTA^-$ species increases. The binding of protons to each of the four ionized species is defined by separate $pK_a$ values (Table I); at pH values above the highest $pK_a$, the chelator is present as a single, fully ionized species. The deprotonated, fully ionized species has the highest affinity for $Ca^{2+}$, so that the ligand becomes a more effective $Ca^{2+}$ chelator as pH is increased. The two highest $pK_a$ values measured for EGTA are 8.8 and 9.5 (Table I), so that $Ca^{2+}$ buffering by EGTA is very sensitive to pH in the physiological range. These $pK_a$ values represent proton binding to the nitrogen atoms, which is suppressed in BAPTA by the benzene rings that link the oxygen and nitrogen atoms (1). This reduces the pH sensitivity of BAPTA compared to EGTA. The two highest $pK_a$ values for BAPTA are

TABLE I Buffering Properties of EGTA and EDTA

| Property | Value[a] | |
|---|---|---|
| | EGTA buffer | EDTA buffer |
| Proton binding ($pK_a$ or log stability constant) | | |
| $K_1$ | 9.5 | 10.3 |
| $K_2$ | 8.8 | 6.2 |
| $K_3$ | 2.7 | 2.8 |
| $K_4$ | 2.0 | 2.0 |
| $Ca^{2+}$ binding | | |
| Absolute $K_d$ | $1.0 \times 10^{-11}$ *M* | $0.25 \times 10^{-11}$ *M* |
| Apparent $K'_d$[d] | $9.5 \times 10^{-8}$ *M* (pH 7.3)[b] | $5.0 \times 10^{-8}$ *M* (pH 7.0) |
| $Mg^{2+}$ binding | | |
| Absolute $K_d$ | $6.3 \times 10^{-6}$ *M* | $2.0 \times 10^{-9}$ *M* |
| Apparent $K'_d$[d] | $6.0 \times 10^{-3}$ *M* (pH 7.6)[c] | $4.0 \times 10^{-6}$ *M* (pH 7.0) |

[a] Values from R. M. C. Dawson, D. C. Elliot, W. H. Elliot, and K. M. Jones "Data for Biochemical Research" (Clarendon Press, Oxford, 1987), except where noted.
[b] Value from E. M. Adler, G. J. Augustine, S. N. Duffy, and M. P. Charlton, *J. Neurosci.* **11,** 1496 (1991).
[c] Value from R. Y. Tsien, *Biochemistry* **19,** 2396 (1980).
[d] Apparent $K'_d$ values at ionic strength 0.1–0.15 *M*, 22°C.

5.5 and 6.4, so that there is little proton binding above pH 7. Thus the affinity of BAPTA for $Ca^{2+}$ is fairly insensitive to pH in the physiological range.

EGTA and BAPTA also differ in the rates at which they interact with $Ca^{2+}$. This interaction can be written as shown in Scheme I,

$$Ca^{2+} + B \underset{k_{off}}{\overset{k_{on}}{\rightleftharpoons}} CaB,$$

SCHEME I

where $k_{on}$ and $k_{off}$ represent the forward and backward rate constants for $Ca^{2+}$ binding to the buffer (B). For simplicity, the variable charge on the buffer molecule has been ignored in the reaction scheme. When $Ca^{2+}$ is added to a solution containing buffer, the reaction approaches equilibrium at a rate that depends on these rate constants as well as on the concentrations of $Ca^{2+}$ and buffer. The buffer concentration would normally be much higher than the $[Ca^{2+}]$ and can be assumed to remain constant. In that case, the time constant ($\tau$) of the relaxation to equilibrium will be given by Eq. (1):

$$\tau = (k_{on}[B_{tot}] + k_{off})^{-1}, \tag{1}$$

where $[B_{tot}]$ represents the total buffer concentration (2). Measured values of $k_{on}$ and $k_{off}$ for EGTA (see Ref. 3) are around $2 \times 10^6 M^{-1} sec^{-1}$ and 0.8 $sec^{-1}$, respectively (pH 7.1, 22°C, 0.1 *M* ionic strength). In the presence of 10 m*M* EGTA, equilibrium would therefore be approached with a time constant of 50 μsec. BAPTA binds $Ca^{2+}$ much more rapidly than EGTA; the forward rate constant is thought to be close to the limit imposed by diffusion. The rate constants have not been precisely measured for BAPTA, although values are available for fluorescent (4, 5) and photolabile (6) derivatives, which are expected to behave in a similar way. Values for $k_{on}$ range from $5 \times 10^8$ to $8 \times 10^8 M^{-1} sec^{-1}$ and estimates of $k_{off}$ range from 60 to 130 $sec^{-1}$ under conditions similar to those used for measuring the EGTA constants. Assuming the most conservative estimates of $k_{on} = 5 \times 10^8 M^{-1} sec^{-1}$ and $k_{off} = 60 sec^{-1}$, the time constant for $Ca^{2+}$ equilibration with 10 m*M* BAPTA would be at most 0.2 μsec. Thus $Ca^{2+}$ binding would equilibrate in well under 1 μsec with BAPTA, compared to 200 μsec with EGTA. Lowering the buffer concentration would slow the reaction. Although in the presence of 1 m*M* BAPTA equilibrium would still be reached in 2 μsec, with 1 m*M* EGTA $[Ca^{2+}]$ would continue to change for several milliseconds. The slow buffering by EGTA may limit its ability to buffer $[Ca^{2+}]_i$ transients when a neuron is activated. This was found to be the case at the squid giant synapse, where even at high concentrations (80 m*M*) EGTA failed to prevent transmitter release (2). In contrast, BAPTA caused a substantial reduction of release at concentrations <1 m*M*. This is presumably because EGTA is considerably less effective than BAPTA at buffering $Ca^{2+}$ in the vicinity of $Ca^{2+}$ channels as it rapidly enters the presynaptic terminal (7).

Other factors that can affect the $Ca^{2+}$ buffering power of a ligand are the ionic strength and temperature of the solution (8). The affinity of a buffer for $Ca^{2+}$ is reduced with increasing ionic strength or decreasing temperature. Most of the published values for the $Ca^{2+}$ affinities of buffers used in physiological experiments were measured at room temperature (22°C), in conditions reflecting the normal ionic environment of mammalian cells, that is, pH 7–7.4 and 0.1–0.15 *M* ionic strength. These values are appropriate for studies performed under the same conditions. However, experiments are often carried out at more physiological temperatures, which are higher in the case of mammals and lower in the case of marine invertebrates. Also, neurobiological studies often employ marine organisms, in which the ionic strength is substantially higher than 0.15 *M*. In these cases, the effect that the environment has on buffer affinity must be quantified, and corrected values of affinity used to calculate $[Ca^{2+}]$. This can be done either by experimentally measuring the $Ca^{2+}$ affinity under the appropriate conditions (for examples of methodology see Refs. 1 and 8), or by calculating the appropriate values using equations that have been shown to predict accurately the experimental effects of temperature and ionic strength (8).

Finally, it should be noted that commercial sources of chelator vary in their purity (9), and this can influence the accuracy of calculations of $[Ca^{2+}]$. Most of the impurity seems to be water (8, 9), so it can be estimated and corrected for by drying samples of the chelator.

## Calculating the Free Calcium Concentration

When incorporated into a neuron, $Ca^{2+}$ buffers interact with cytoplasmic $Ca^{2+}$ according to the reaction given in Scheme I. The equilibrium of this reaction is characterized by the apparent dissociation constant, $K'_d$, defined as

$$K'_d = \frac{k_{off}}{k_{on}} = \frac{[Ca][B]}{[CaB]}. \tag{2}$$

Equilibrium constants are often quoted as stability constants ($K_s$), where $K_s = 1/K_d$. As long as the total Ca ($Ca_{tot}$) and buffer concentrations in the solution are known, then the free $[Ca^{2+}]$ at equilibrium may be calculated from Eq. (3):

$$[Ca^{2+}]^2 + (B_{tot} - Ca_{tot} + K'_d)[Ca^{2+}] - K'_d Ca_{tot}] = 0, \tag{3}$$

where $[Ca^{2+}]$ is given by a root of the equation according to Eq. (4):

$$[Ca^{2+}] = -(B_{tot} - Ca_{tot} + K'_d) \pm \frac{[(B_{tot} - Ca_{tot} + K'_d)^2 - 4K'_d Ca_{tot}]^{1/2}}{2}. \tag{4}$$

The $K_d$ used in this computation would be correct for the appropriate conditions of pH, ionic strength, and temperature. In most physiological experiments, corrections for pH will not be necessary when BAPTA-based buffers are used (8). Corrections will often need to be made for other types of $Ca^{2+}$ buffer. Absolute stability constants ($K_s$), measured at high pH where all of the buffer is present in the deprotonated state, have been published for EGTA and EDTA. These are listed in Table I in the form of absolute dissociation constants. From these values, and the published stability constants for proton binding to the buffer ($K_1$, $K_2$, $K_3$, and $K_4$; see Table I), the apparent $K'_s$ for $Ca^{2+}$ binding at a particular pH can be estimated according to Eq. (5):

$$K'_s = \frac{K_s}{1 + [H^+]K_1 + [H^+]^2 K_1 K_2 + [H^+]^3 K_1 K_2 K_3 + [H^+]^4 K_1 K_2 K_3 K_4}. \tag{5}$$

Table I also lists apparent $K_d$ values for binding of EGTA and EDTA to $Ca^{2+}$ and $Mg^{2+}$ at near physiological pH. More detail on these calculations and corrections for ionic strength and temperature can be found in Harrison and Bers (8).

The amount of $Ca^{2+}$ bound by an intracellular chelator will be influenced by the presence of endogenous cell buffers. The properties of these buffers are not precisely known, so their effects cannot easily be predicted. However, incorporating chelators into cells at high concentrations (millimolar) should make them the dominant buffers and largely overcome the competition from endogenous buffers. Nucleotides, which bind $Ca^{2+}$, are also found in cells, and they are often included in solutions designed to mimic the intracellular environment. Their presence will also contribute to $Ca^{2+}$ buffering. Calculations of $[Ca^{2+}]$ in the presence of nucleotides, other metal ions, and additional competing $Ca^{2+}$ buffers can be found in Bartfai (10). Computer programs are available that will calculate $[Ca^{2+}]$ in the presence of competing buffers, including nucleotides, under different conditions (see Ref. 3).

## *Choice of Buffer to Control Intracellular Calcium*

EGTA remains the most widely used $Ca^{2+}$ buffer in biological experiments, partly because it is substantially less expensive than BAPTA. However, as outlined above, BAPTA clearly has advantages over EGTA and would be the buffer of choice when pH sensitivity and speed of $Ca^{2+}$ buffering are important. Several derivatives of BAPTA (Table II) that retain its desirable properties are available. These buffers cover a wide range of $Ca^{2+}$ affinity, the main property that determines how effective a buffer will be at controlling $[Ca^{2+}]$ at different levels. Figure 2a shows how the $K_d$ for $Ca^{2+}$ binding influences the degree of saturation of a buffer at different $[Ca^{2+}]$ levels. From Eq. (2), the proportion of bound buffer is given by Eq. (6):

$$\frac{[CaB]}{[B_{tot}]} = \frac{[Ca^{2+}]}{[Ca^{2+}] + K_d}. \tag{6}$$

This relationship was used to construct the plots in Fig. 2a for $K_d$ values ranging from 10 n$M$ to 1 $\mu M$. Buffering is clearly most effective when the $[Ca^{2+}]$ is close to the $K_d$. At $[Ca^{2+}]$ well below the $K_d$, little $Ca^{2+}$ is bound by the buffer. At a $[Ca^{2+}]$ high relative to the $K_d$, the buffer becomes saturated and would be unable to respond to extra $Ca^{2+}$ added to the system. The range of $[Ca^{2+}]$ over which a chelator can act as an effective buffer is limited to around an order of magnitude on either side of its $K_d$ (Fig. 2). Thus if an

TABLE II Calcium Dissociation Constants for BAPTA Analogs[a]

| Analog | $K_d$ ($M$) |
|---|---|
| BAPTA | $1.6 \times 10^{-7}$ |
| Dibromo BAPTA | $3.6 \times 10^{-6}$ |
| Dimethyl BAPTA | $4.0 \times 10^{-8}$ |
| 5,5'-Difluoro BAPTA | $6.1 \times 10^{-7}$ |
| 4,4'-Difluoro BAPTA | $4.6 \times 10^{-6}$ |
| Dinitro BAPTA | $2.0 \times 10^{-3}$ |
| Nitro BAPTA | $9.4 \times 10^{-5}$ |

[a] Values from R. P. Haughland, "Handbook of Fluorescent Probes and Research Chemicals" (Molecular Probes Inc., Eugene, Oregon, 1992), measured at 22°C in solutions mimicking the mammalian intracellular environment, without magnesium.

experiment requires $[Ca^{2+}]$ to be tightly controlled around the resting cell level, then a $Ca^{2+}$ buffer with a $K_d$ value in the resting $[Ca^{2+}]$ range should be used.

The ability to buffer $[Ca^{2+}]$ also depends on the concentration of chelator present in the cell. The effect of chelator concentration on the ability to absorb changes in $[Ca^{2+}]$ injected into or removed from a system is illustrated in Fig. 2b. From Eq. (2),

$$[Ca^{2+}] = K_d + [CaB]/[B]. \quad (7)$$

The $[Ca^{2+}]$ can therefore be calculated if the ratio of bound to free buffer is known. As an example, we start with a 10 m$M$ solution of buffer with $K_d$ = 100 n$M$ and set the $[Ca^{2+}]$ to 100 n$M$. Under these conditions [CaB] = [B] = 5 m$M$, and in 10 ml of solution there will be 50 $\mu$mol of each species. When $Ca^{2+}$ is injected into the solution, it will be bound by some of the free buffer molecules, thereby changing the ratio [CaB]/[B]. In our hypothetical solution, adding 1 $\mu$mol of $Ca^{2+}$ will change [CaB]/[B] from 1 to 1.04, which from Eq. (7) gives a $[Ca^{2+}]$ of 104 n$M$. For comparison, adding the same amount of $Ca^{2+}$ without a buffer present would change the $[Ca^{2+}]$ by 100 $\mu M$. Figure 2 confirms that a buffer would be very effective at maintaining $[Ca^{2+}]$ constant at $[Ca^{2+}]$ levels around the $K_d$. Effective buffering exists provided [CaB]/[B] is in the range 0.1 to 10, but outside that range the capacity of the buffer is rapidly lost. Lowering the buffer concentration reduces the amount of $Ca^{2+}$ that can be added to or removed from the system before the buffering capacity is exceeded and the $[Ca^{2+}]$ becomes uncontrollable (Fig. 2b). How-

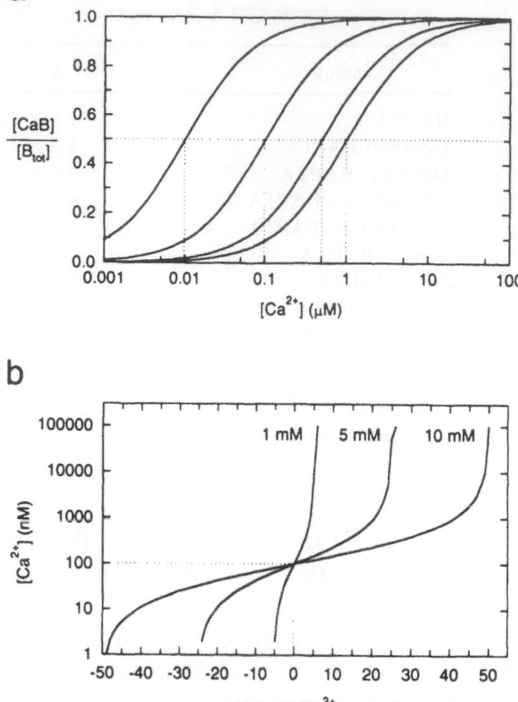


FIG. 2 Theoretical plots showing the effect of $K_d$ and concentration on the buffering properties of $Ca^{2+}$ chelators. (a) The proportion of chelator that is bound to $Ca^{2+}$ ($[CaB]/[B_{tot}]$) plotted as a function of $[Ca^{2+}]$, according to Eq. (6). Plots shown are for $K_d$ values of, from left to right, 10, 100, and 500 n$M$ and 1 $\mu M$. (b) The effect of adding $Ca^{2+}$ to, or removing it from, 10 ml of solution containing a chelator with $K_d$ = 100 n$M$ and with the $[Ca^{2+}]$ initially set at 100 n$M$. The relationship between $[Ca^{2+}]$ and micromoles $Ca^{2+}$ added or removed was calculated using Eq. (7), and is shown separately for chelator concentrations of 1, 5, and 10 n$M$.

ever, even at 1 m$M$ of our hypothetical buffer, injection of 1 $\mu$mol of $Ca^{2+}$ would only change the $[Ca^{2+}]$ by 50 n$M$.

The availability of $Ca^{2+}$ chelators with a variety of $K_d$ values means that it is possible to set $[Ca^{2+}]_i$ at levels outside the normal range encountered at rest, and still retain the ability to buffer changes in $[Ca^{2+}]_i$ concentration due to intracellular $Ca^{2+}$ release, $Ca^{2+}$ influx, or extrusion. The buffer would, however, need to be present in the cell at high concentraton. Comparison of the abilities of buffers with different $Ca^{2+}$ affinities to interfere with transmitter release at the squid giant synapse has proved helpful in localizing the site in the nerve terminal to which $Ca^{2+}$ binds and triggers release (2).

## Caged Calcium and Caged Calcium Buffers

In general, "caged" compounds are biologically inert molecules until activated by high-intensity light, when they are structurally changed to "release" active agents. This property allows the inactive compound to be preequilibrated within the cells or tissue under examination, and the active species released directly at its site of action using a pulse of light. The same principle is employed in the use of caged $Ca^{2+}$ and caged $Ca^{2+}$ chelators, except that these molecules are not inert before photolysis. The photolabile Ca probes buffer intracellular $Ca^{2+}$ before being irradiated. The light pulse changes the buffering power of the molecule, resulting either in a net release of $Ca^{2+}$ into the cell, or a reduction of $[Ca^{2+}]_i$, depending on the type of probe used. The rapidity of the light-induced reaction, which is usually complete within 1 msec, means that the $[Ca^{2+}]_i$ changes in the form of a step or "jump." This allows the study of the kinetics of $Ca^{2+}$-dependent processes at sites where speed of diffusion and $Ca^{2+}$ buffering by the cells would otherwise distort the time course of the response.

### *Why Use Caged Compounds?*

The photolabile Ca probes, which yield an active product that is either $Ca^{2+}$ or a high-affinity $Ca^{2+}$ chelator, offer several advantages over standard chelators as a means of manipulating $[Ca^{2+}]_i$. The size, intensity, and duration of the light spot used to activate the probes can be controlled; depending on the type of light source used, it may be focused to a small, intense spot, irradiating one cell or even part of a cell, or it may encompass a whole population of cells. This allows comparison of the effects of a change in $[Ca^{2+}]$ in defined areas of a cell. Because the change in $[Ca^{2+}]_i$ on photolysis depends on the intensity of the light flash, varying this intensity may produce a graded response, from which a quantitative dose–response relationship for $[Ca^{2+}]_i$ may be derived. Also, under certain conditions, by varying the duration of the flash, it is possible to produce either a brief "spike" or a reversible "pulse" of $Ca^{2+}$ (11). For example, a brief, high-intensity flash or a low-intensity flash of several milliseconds duration could induce transmitter release at the squid giant synapse (12). Both approaches provided useful information, although the response to a presynaptic $Ca^{2+}$ spike, caused by a brief, intense flash, more closely resembled the normal response to a presynaptic action potential.

## How Caged Compounds Work

In order to use caged compounds to best effect, it is necessary to understand the way in which light causes the release of $Ca^{2+}$ from caged $Ca^{2+}$ or the sequestration of $Ca^{2+}$ by caged chelators. It is also important to appreciate how experimental factors may influence the efficiency of this conversion.

The first stage in a photolytic reaction is the absorption by the caged compound of a photon of light energy ($E$):

$$E = h\nu, \tag{8}$$

where $h$ is Planck's constant and $\nu$ is the frequency of the light. The subsequent events are summarized in Scheme II:

$$B + E \rightleftharpoons B^* \rightarrow P + sp.$$

SCHEME II

As a result of absorbing the photon, the caged compound (B) gains energy and is thereby promoted to an excited state (B*). This step occurs on a time scale of <1 nsec. The molecule can then either lose the extra energy and decay back to its unexcited state, or proceed to form stable products—the desired route in flash photolysis. One of the stable products formed (P) is usually a chelator with altered $Ca^{2+}$-binding properties. The change in $Ca^{2+}$ affinity should ideally be very large. This is seen with the caged $Ca^{2+}$ molecule DM-nitrophen, which undergoes a large decrease in $Ca^{2+}$ affinity following photolysis. More usually, the photoproduct P continues to buffer intracellular $Ca^{2+}$ significantly, but at a different level to that determined by the precursor. Other side products (sp) are also formed. These may be quite innocuous or have unwanted biological effects. It is important to know what the side products of photolysis are and to determine if they have any biological activity in the system under study.

The reactions leading to the formation of photoproducts from the energized intermediate are known as the dark reactions, because they will proceed in the absence of light once the intermediate has been formed. The dark reactions are usually the rate-limiting step in photolysis and vary widely in time course—from microseconds to hundreds of milliseconds—and in efficiency.

The ratio of the precursor remaining to the product formed after photolysis determines the final $[Ca^{2+}]_i$ produced. This ratio depends on the amount of light absorbed by the caged compound and how effective the absorbed light is at inducing the dark reactions. The amount of light absorbed depends not only on the intensity of the illumination but also on the absorbance ($A$) of the caged molecule in solution, which is expressed as

$$A = \varepsilon c l, \tag{9}$$

where $\varepsilon$ and $c$ are the molar extinction coefficient and concentration of the molecule, respectively, and $l$ is the length of the light path through the solution. For efficient photolysis, $\varepsilon$ should be large at those wavelengths that trigger the photochemical reactions. However, with a thick preparation, such as the large neurons found in several molluscs (>200 $\mu$m), a high extinction coefficient may lead to poor and uneven photolysis, because the molecules at the front surface may absorb much of the light before it reaches the center of the tissue. Most mammalian neurons are sufficiently thin and transparent in the near-UV region that they absorb little of the exciting light. Thus efficient photolysis can be achieved in isolated neurons and in many thin tissue preparations. However, in pigmented cells, for example, some cell bodies in the nervous system of the marine mollusc *Aplysia*, the amount of effective light reaching the caged molecules may be reduced (13). This can interfere with the calibration of the amount of $Ca^{2+}$ released or buffered by the probe when the cell is irradiated. Nevertheless, the absorbance properties of a preparation can be measured and included in the calculation of $[Ca^{2+}]$ changes. The influence of cell absorbance and preparation thickness on the photolytic efficiency of a flash are discussed in detail in Landò and Zucker (14). They calculate the average photolysis of nitr-5 in an *Aplysia* neuron of 300–350 $\mu$m diameter, where the cytoplasm has an absorbance coefficient ($\varepsilon c$) of 25 $cm^{-1}$ at 360 nm; photolysis declines steeply as the light penetrates deeper into the cell, so that it is greatest at the front surface membrane and zero at the back of the cell. Rat cerebellar slices, containing layers of cells, have a lower absorbance coefficient of 10 $cm^{-1}$ at 320 nm (15), giving less attenuation of light flashes. Since individual cells in the slice are relatively small, photolysis would probably be fairly uniform within cells. However, before calculating photolysis-induced $[Ca^{2+}]_i$ changes, it would be important to measure the position of the cell within the slice. For similar reasons, it is advisable that the light does not travel through a deep layer of saline before reaching the preparation.

For efficient photolysis, there should be a high probability that the light-induced intermediate, once created, will form photoproducts rather than return to the unexcited state. The probability that an absorbed photon will result in the formation of photoproducts is termed the quantum yield or quantum efficiency ($Q$) of the caged compound, and is defined as

$$Q = \text{product molecules formed/photons absorbed.} \qquad (10)$$

Thus, less light is required to trigger $Ca^{2+}$ release or sequestration by caged compounds that have a high quantum yield as well as a high extinction coefficient. The values of quantum yield are low in the available photolabile $Ca^{2+}$ probes, but sufficient to allow [$Ca^{2+}$] changes in the physiological range to be produced with flashlamps or lasers.

The proportion of molecules photolyzed by a single flash may also be influenced by the lifetime of the reactive intermediate relative to the flash duration. If the lifetime of the reactive intermediate is short compared to the duration of the flash, multiple excitations may occur, resulting in a greater percent conversion from precursor to active molecule than would be predicted by the quantum yield (16).

## *Basic Structure and Activation of Photolabile Calcium Probes*

The caged $Ca^{2+}$ compounds include the nitr series (17) (Fig. 3a) and DM-nitrophen (18) (Fig. 3b). The nitr series of molecules are derived from the highly selective $Ca^{2+}$ chelator, BAPTA (Fig. 1b). DM-nitrophen is derived from EDTA (Fig. 1c), which has a higher affinity for $Ca^{2+}$ than BAPTA, but poorer ion selectivity. The metal coordinating sites of the parent chelators are retained in the photolabile derivatives. The currently available caged $Ca^{2+}$ compounds all exploit the light sensitivity of the 2-nitrobenzyl moiety (Fig. 3c) found in most other types of caged compound. Caged molecules usually contain only one 2-nitrobenzyl group, although derivatives containing more than one such group have been synthesized (19). The cation-binding parts of the molecules are linked to the photolabile group at the position labeled R in Fig. 3c.

When irradiated, the 2-nitrobenzyl group in the nitr compounds is converted to a more electron-withdrawing group (Fig. 3a), which reduces the electron density at the coordinating nitrogen atom. The result is a loss of affinity for $Ca^{2+}$ of around 40-fold. The main differences among the various members of the nitr family lie in their affinities for $Ca^{2+}$ before photolysis and the rate at which $Ca^{2+}$ is released. In contrast, photolysis of the 2-nitrobenzyl group in DM-nitrophen results in cleavage of the molecule at the cation binding site (Fig. 3b), leading to a $10^5$-fold loss of $Ca^{2+}$ affinity.

Caged $Ca^{2+}$ buffers include diazo-2 (6) (Fig. 4a), which is commercially available, related diazo derivatives (6), and caged BAPTA (20). Like the nitr

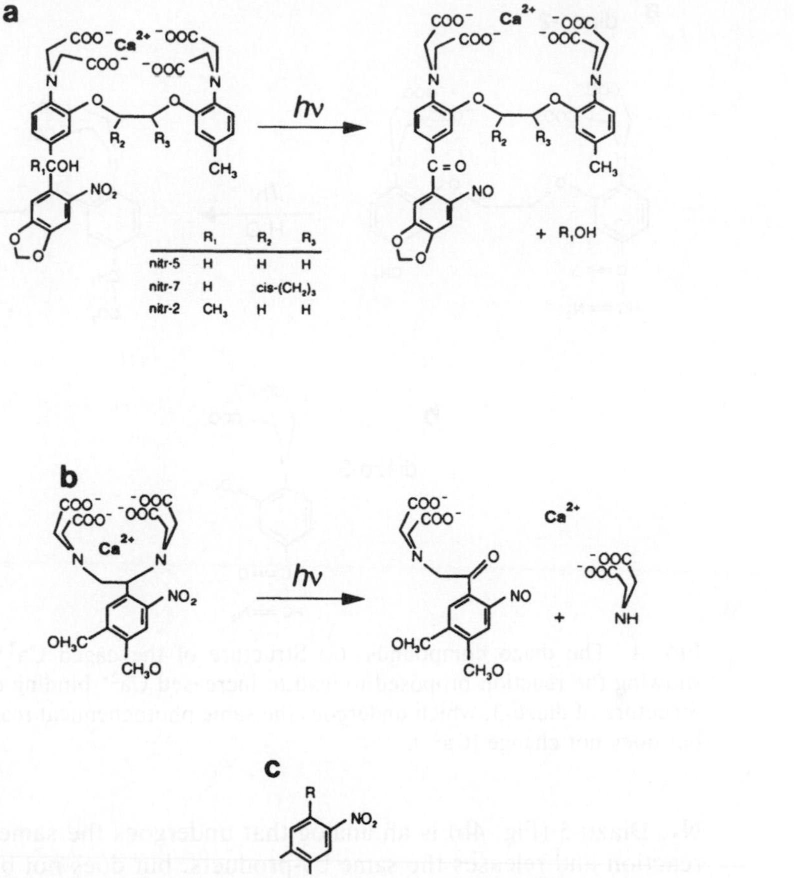

| | $R_1$ | $R_2$ | $R_3$ |
|---|---|---|---|
| nitr-5 | H | H | H |
| nitr-7 | H | cis-$(CH_2)_3$ | |
| nitr-2 | $CH_3$ | H | H |

FIG. 3 Structures of caged $Ca^{2+}$ molecules and postulated reactions leading to $Ca^{2+}$ release. (a) The nitr series of compounds; (b) DM-nitrophen; (c) the light-sensing 2-nitrobenzyl moiety that is common to all the caged $Ca^{2+}$ molecules. The letter R represents the cation-chelating part of the molecule.

group, the diazo compounds are derivatives of BAPTA, but they exploit a different photochemistry. In order to lower $Ca^{2+}$ affinity before photolysis, an electron-withdrawing diazoketone group is linked to the chelator at a site remote from the $Ca^{2+}$ binding site. This has the effect of lowering the electron density at the coordinating nitrogen. Light, in the presence of $H_2O$, converts the diazoketone to an electron-donating carboxymethyl group (Fig. 4a), which increases the electron density on the nitrogen atom and hence increases $Ca^{2+}$ affinity. Photolysis of diazo-2 results in a 28-fold increase in $Ca^{2+}$ affinity. This change occurs simultaneous with the formation of a proton and

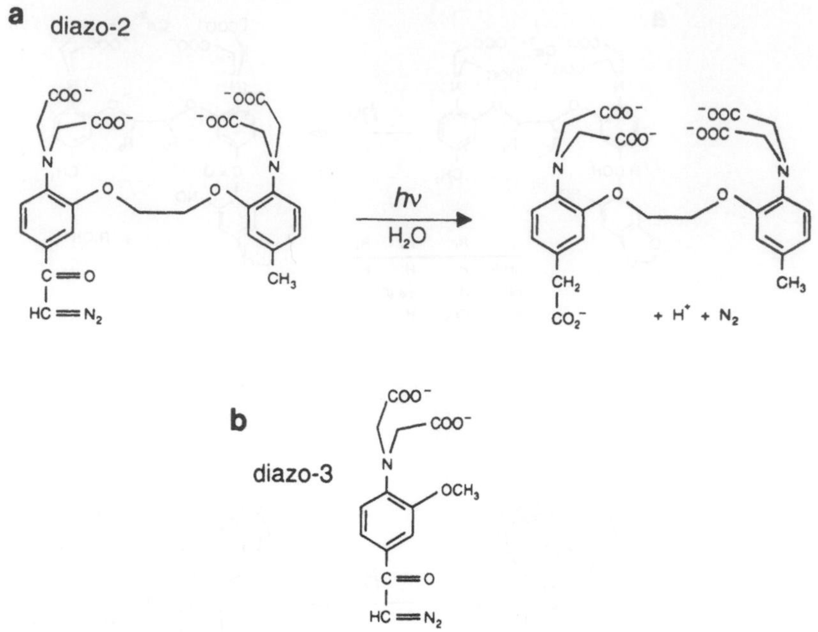

a diazo-2

$$\xrightarrow[H_2O]{h\nu}$$

$+ H^+ + N_2$

b diazo-3

FIG. 4 The diazo compounds. (a) Structure of the caged $Ca^{2+}$ buffer, diazo-2, showing the reaction proposed to lead to increased $Ca^{2+}$ binding on photolysis. (b) Structure of diazo-3, which undergoes the same photochemical reactions as diazo-2, but does not change $[Ca^{2+}]$.

$N_2$. Diazo-3 (Fig. 4b) is an analog that undergoes the same photochemical reaction and releases the same by-products, but does not bind $Ca^{2+}$ before or after photolysis.

All of the currently available photolabile $Ca^{2+}$ probes require wavelengths in the near-UV region of the spectrum for photolysis, with maximum efficiency at around 350 nm. Most caged compounds absorb little above 500 nm, and are therefore fairly stable in normal room lighting. This means that experiments need not be carried out in a dark room. However, it is good practice to prepare experimental solutions in dim light and store them in light-tight containers such as brown bottles or tubes, or wrapped in foil.

## *Choice of Caged Calcium Compound*

When studying the regulation of neuronal calcium and its role as an intracellular messenger, it may be desirable to produce a rapid increase or a rapid decrease in $[Ca^{2+}]_i$. The probes available for this differ widely in their $Ca^{2+}$

buffering and photochemical characteristics. None of them have ideal properties; they all have a number of advantages and disadvantages that determine their value in particular studies of $Ca^{2+}$-dependent neuronal pathways. In order to select the most appropriate probe for the task required, their various properties should be carefully considered when designing experiments. The properties of the different groups of photolabile $Ca^{2+}$ probe are detailed below, with particular reference to their advantages and drawbacks for different types of investigation. A summary of their properties is provided in Table III.

### *DM-Nitrophen*

DM-nitrophen shares many of the properties of EDTA, around which it was designed. In its unphotolyzed form, DM-nitrophen binds $Ca^{2+}$ with an extremely high affinity ($K_d$ = 5 n$M$). This is advantageous, because at physiological levels of $[Ca^{2+}]_i$, a high proportion of the probe would be complexed with $Ca^{2+}$, thereby providing a large reservoir of $Ca^{2+}$ that would be available for release on photolysis. Light causes the chemical disintegration of the divalent cation-binding site and yields two photoproducts (Fig. 3b), both with negligible $Ca^{2+}$ affinity ($K_d$ > 3 m$M$). The large loss of affinity also helps to generate large increases in $[Ca^{2+}]$ on photolysis. The photoprod-

TABLE III Properties of Commercially Available Caged Calcium Compounds and Caged Chelators[a]

| Property | Nitr-5 | Nitr-7 | Nitr-2 | DM-nitrophen | Diazo-2 |
|---|---|---|---|---|---|
| Before photolysis | | | | | |
| Absorbance maxima (nm)[b] | 369 | 360 | 360 | 350 | 370 |
| $\varepsilon \times 10^{-3}$ ($M^{-1}$ cm$^{-1}$)[b] | 4.8 | 4 | ~4 | 4.3 | 22.2 |
| $K_d$ $Ca^{2+}$ (n$M$) | 145 | 48 | 160 | 5 | 2200 |
| $K_d$ $Mg^{2+}$ (m$M$) | 8.5 | 5.4 | — | $2.5 \times 10^{-3}$ | 5.5 |
| After photolysis | | | | | |
| $K_d$ $Ca^{2+}$ ($\mu M$) | 6.3 | 3 | 6–10 | $3 \times 10^3$ | $73 \times 10^{-3}$ |
| $K_d$ $Mg^{2+}$ (m$M$) | 8 | 5 | — | 2 | 3.4 |
| $\tau$ $Ca^{2+}$ release/uptake (msec) | ~0.3 | ~1.8 | ~200 | <0.24 | <0.43 |
| $Q$ free cage | 0.012 | 0.011 | 0.01 | — | 0.030 |
| $Q$ Ca-bound cage | 0.035 | 0.042 | 0.05 | 0.18 | 0.057 |

[a] Values from S. R. Adams, J. P. Y. Kao, G. Grynkiewicz, A. Minta, and R. Y. Tsien, *J. Am. Chem. Soc.* **110**, 3212 (1988); S. R. Adams, J. P. Y. Kao, and R. Y. Tsien, *J. Am. Chem. Soc.* **111**, 7957 (1989); and J. H. Kaplan and G. C. R. Ellis-Davies, *Proc. Natl. Acad. Sci. U.S.A.* **85**, 6571 (1988); pH 7.1, 0.1–0.15 $M$ ionic strength, and 20–23°C.

[b] Absorbance properties of $Ca^{2+}$-free probes; absorbance of nitr compounds is little affected by addition of excess $Ca^{2+}$, but DM-nitrophen displays a 7% loss of absorbance in saturating $[Ca^{2+}]$ and $Ca^{2+}$ shifts the absorbance maximum of diazo-2 to 320 nm, with $\varepsilon = 20.2 \times 10^{-3}$ $M^{-1}$ cm$^{-1}$.

ucts illustrated in Fig. 3b are those produced by photolysis of $Ca^{2+}$-bound DM-nitrophen; irradiation of unbound chelator yields different products. This feature of DM-nitrophen complicates the design of control experiments (see below), because the omission of $Ca^{2+}$ from the preflash intracellular solution will not just prevent an increase in $[Ca^{2+}]_i$, but it will also result in a different photochemical reaction when the cell is irradiated.

The very large change in $Ca^{2+}$ binding on DM-nitrophen photolysis allows millimolar jumps in $[Ca^{2+}]_i$ to be produced. However, DM-nitrophen also has a significant affinity for $Mg^{2+}$ ($K_d$ = 2.5 $\mu M$), so that in physiological intracellular solutions an appreciable proportion of the chelator will be bound to $Mg^{2+}$. This will result in an increase in $[Mg^{2+}]_i$ being produced alongside the $[Ca^{2+}]$ jump on photolysis. This is a disadvantage when using DM-nitrophen to manipulate cytosolic $[Ca^{2+}]$, because in order for DM-nitrophen to act as a "pure" caged $Ca^{2+}$ compound, it is necessary to reduce the $[Mg^{2+}]$ in the cells under study to extremely low levels, which may alter cellular function and the response to $Ca^{2+}$. The $Mg^{2+}$-binding ability of DM-nitrophen can, however, be an advantage if changes in $[Mg^{2+}]_i$ are of interest. Reducing preflash intra- and extracellular $[Ca^{2+}]$ results in a $[Mg^{2+}]_i$ jump on photolysis, without a concomitant rise in $[Ca^{2+}]_i$; DM-nitrophen has been successfully used in this way, as a caged $Mg^{2+}$ compound, by O'Rourke *et al.* (21).

A less significant disadvantage of DM-nitrophen in its use as a caged $Ca^{2+}$ compound is the pH sensitivity of its binding to $Ca^{2+}$. Like EDTA, it binds protons in the physiological range of pH. Thus pH might be expected to influence DM-nitrophen affinity and its response to photolysis. In order to estimate how much of the chelator will be bound to $Ca^{2+}$, and thus the size of the $[Ca^{2+}]_i$ jump that will be produced on photolysis, the pH of the intracellular solution must be precisely known and the appropriate pH-adjusted $K_d$ value used in calculations of $[Ca^{2+}]_i$. The products of DM-nitrophen photolysis are also likely to interact with protons. In practice, however, varying the pH between 6.9 and 8.3 was found to have little effect on the $Ca^{2+}$ binding or rate of photolysis of DM-nitrophen in cuvettes (11).

Other factors to consider when choosing a caged $Ca^{2+}$ compound are the $Ca^{2+}$-binding ability of the photoproducts, the speed at which $Ca^{2+}$ is released from the probe, and how quickly released $Ca^{2+}$ reequilibrates with unphotolyzed chelator after photolysis. The photoproducts of DM-nitrophen have a negligible affinity for $Ca^{2+}$, therefore the postphotolysis $[Ca^{2+}]$ is determined almost entirely by its binding to any remaining unphotolyzed chelator. This makes it easier to produce large changes in $[Ca^{2+}]_i$ and to predict the size of the change. $Ca^{2+}$ is released rapidly from DM-nitrophen, with a maximum half-time of 180 $\mu$sec (22). However, $Ca^{2+}$ binding to unphotolyzed chelator is relatively slow. This can be put to advantage, because it means that the experimenter has some control over the profile of the $[Ca^{2+}]_i$ jump that can be achieved. When DM-nitrophen is well saturated with $Ca^{2+}$, a brief light pulse induces a step change in $[Ca^{2+}]$, with little unphotolyzed chelator available for rebinding. When DM-nitrophen is only partially $Ca^{2+}$ bound, continuous light of an intensity low enough to produce only partial photolysis results in a sustained elevation or "pulse" of $[Ca^{2+}]$, which can be maintained for up to 100 msec and is terminated when the light is extinguished. Such maintained $[Ca^{2+}]$ elevations were used in a study of transmitter release from crayfish motor nerve terminals (23). Under the same conditions, brief, high-energy light flashes produce a $[Ca^{2+}]$ jump that is in the form of a spike, with $[Ca^{2+}]_i$ rising to a maximum level within 1 msec, then dropping to a lower, steady level over the next several milliseconds. The generation of such $Ca^{2+}$ spikes and pulses in living cells is described in detail by Zucker (11).

In spite of the complications associated with DM-nitrophen, it may still be the caged $Ca^{2+}$ of choice if large increases in $[Ca^{2+}]_i$ are required, or the profile of the $Ca^{2+}$ jump is to be varied, especially if effects of changes in $[Mg^{2+}]_i$ are likely to be insignificant.

*The Nitr Family of Caged Calcium Compounds*

The nitr family of compounds retains the high selectivity for $Ca^{2+}$ over $Mg^{2+}$ binding that is characteristic of the parent molecule, BAPTA (17). This gives the nitr compounds a distinct advantage over DM-nitrophen for use as $Ca^{2+}$ donors. In addition, like BAPTA, their $Ca^{2+}$ affinities show little pH dependence above pH 7. Photolysis of these molecules causes around a 40-fold reduction in $Ca^{2+}$ affinity, associated with the loss of a small residue as a by-product, which in the case of nitr-5 is water (Fig. 3a). The photoproducts are the same whether the chelators are or are not initially bound to $Ca^{2+}$. Unfortunately, because the prephotolysis $Ca^{2+}$ affinity and the light-induced change in affinity are both substantially smaller than for DM-nitrophen, the currently available nitr compounds cannot generate such large $Ca^{2+}$ concentration changes. In addition, the photolyzed forms of the molecules retain significant buffering power, so that the $[Ca^{2+}]$ after photolysis is determined by both the unphotolyzed and photolyzed chelators.

Since the $Ca^{2+}$ affinities of nitr-5 ($K_d$ = 145 n$M$) and nitr-2 ($K_d$ = 160 n$M$) lie within the physiological range of $Ca^{2+}$ concentrations, these molecules will contribute to intracellular $Ca^{2+}$ buffering at concentrations at which DM-nitrophen would not. So, to prevent them from controlling the $[Ca^{2+}]_i$ before photolysis, they would have to be incorporated into cells at very low concentration, probably too low to permit significant $Ca^{2+}$ release on photolysis. Also, because they have a lower affinity for $Ca^{2+}$ than DM-nitrophen, at physiological $Ca^{2+}$ levels a smaller proportion of the probe would be complexed with $Ca^{2+}$ before photolysis, reducing the reservoir of $Ca^{2+}$ available for release. These chelators are best used to provide complete control of

$[Ca^{2+}]_i$, clamping the concentration at a known level both before and after photolysis. Nitr-7 has a higher affinity ($K_d$ = 48 n*M*) than the other nitr compounds, which confers certain advantages, although it is still substantially lower than for DM-nitrophen.

The binding of $Ca^{2+}$ to nitr-5 [time constant around 2 μsec (11)] is faster than the rate of its photolysis, so that on irradiation with a brief light flash, released $Ca^{2+}$ rapidly reequilibrates with unphotolyzed chelator. The result is always a step change in $[Ca^{2+}]_i$, rather than the spike or pulse achievable with the slower chelator, DM-nitrophen. The high $Ca^{2+}$ selectivity of the nitr compounds, along with their rapid binding characteristics, will often make them the preferred caged $Ca^{2+}$ probes, where step increases of $[Ca^{2+}]_i$ in the physiological range are required.

The different nitr derivatives vary mainly in their $Ca^{2+}$ affinities and in the rate at which they release $Ca^{2+}$ on photolysis. Nitr-7 has a higher affinity than nitr-5 or nitr-2, although it shows a decrease of similar size when irradiated. This allows greater $Ca^{2+}$ loading of nitr-7 and therefore a larger release of $Ca^{2+}$ on photolysis (see below). However, the rate of $Ca^{2+}$ release is slower than that achieved with nitr-5 (see Table III), so changes in $[Ca^{2+}]_i$ will be more gradual. The rate of $Ca^{2+}$ release from nitr-7 is still sufficiently fast that it will not necessarily be rate limiting (24), so nitr-7 may be the most appropriate caged $Ca^{2+}$ to use. Nitr-2 is similar to nitr-5 in all its properties except the speed at which $Ca^{2+}$ is released (see Table III). This is slower by nearly three orders of magnitude, making it less useful as a caged $Ca^{2+}$ compound in most investigations.

Two further molecules in the nitr family may prove useful in physiological investigations, although they are not yet available commercially (19). Nitr-8 shows a 1600-fold decrease in $Ca^{2+}$ affinity on photolysis, allowing much larger changes in $[Ca^{2+}]_i$ than is possible with nitr-5. However, this involves photolysis of two nitrobenzyl groups, requiring a higher light intensity for efficient photolysis. Nitr-9 does not change its $Ca^{2+}$ affinity on photolysis and will therefore be of use when controlling for effects of photochemical changes to other nitr molecules.

Another important advantage of the nitr probes over DM-nitrophen is that they are available in the membrane-permeant acetoxymethyl (AM) ester form, allowing them to be loaded into cells noninvasively. At present, DM-nitrophen must be introduced in ways more likely to cause damage to cells (see below).

### *Caged Calcium Chelators—The Diazo Group*

The diazo series of compounds are structurally related to the nitr family, but gain affinity for $Ca^{2+}$ on photolysis, thereby reducing $[Ca^{2+}]_i$. This change is accompanied by the release of $N_2$ gas and protons (Fig. 4). Diazo-2 (Fig. 4a and Table III), the most popular in this series, undergoes an approximately 30-fold increase in $Ca^{2+}$ affinity on photolysis, with its $K_d$ changing from 2.2 $\mu M$ to 73 n$M$. Diazo-4 undergoes a larger change in affinity, requiring photolysis of two substituent photolabile groups, but it is not currently available commercially. The structurally related molecule diazo-3 (Fig. 4b) can be used as a control molecule for experiments with any of the diazo compounds. It has photochemical properties similar to those of other members of the group, but has little affinity for $Ca^{2+}$ before or after photolysis. Thus the photochemical reaction proceeds as normal, but without producing a change in $Ca^{2+}$ concentration. Like nitr-5, the diazo compounds are available in AM ester form, allowing them to be incorporated into intact cells.

## Possible Sources of Error When Using Caged Compounds

The principal sources of error to be considered when using photolabile Ca probes are (1) the light flash used to effect photolysis, which may have direct effects on the cells under study; (2) the introduction of a caged compound that will interfere with the cell's normal mechanisms for regulating $[Ca^{2+}]_i$; (3) products of photolysis that may affect cellular functions; and (4) the time course of a response or the ability to analyze fast responses, which could be limited by the speed of photolysis. Any of these factors could cause major experimental problems; the probable influence of each should be investigated and controlled for wherever possible, or minimized by careful choice of the caged compound and other experimental conditions.

### *Direct Effects of the Light Flash*

Light of damaging wavelengths should be carefully excluded from the target cells. The 300- to 400-nm wavelengths required to photolyze caged compounds are unlikely to produce physiological responses in most types of neuronal cell. Photoreceptors are an obvious exception, therefore flash photolysis will not usually be the method of choice for investigating $Ca^{2+}$-dependent processes in these cells. Light sensitivity would not be predicted in most neuronal cell types and a variety of experiments using nitr compounds or DM-nitrophen have found no direct effect of the photolyzing light (12, 24). The possibility of such effects should not, however, be ruled out; light has been found to produce responses in giant ganglion cells of *Aplysia* (25) and flashes of the type required to photolyze intracellular caged compounds relax vascular smooth muscle (26).

## *Effects of the Unphotolyzed Caged Compound*

Photolabile $Ca^{2+}$ probes will, by their nature, affect the $Ca^{2+}$ buffering of the cells into which they are introduced. Some experiments require a caged $Ca^{2+}$ compound to be present in the cell simply as a source of photoreleasable $Ca^{2+}$, so that controlled bursts of $Ca^{2+}$ can be produced. To achieve this, the probe would have to have a high affinity and be present at a sufficiently low concentration not to contribute to buffering. This could theoretically be accomplished with DM-nitrophen and possibly nitr-7, but the affinities of the other probes are too low. More usually, the caged compound controls $[Ca^{2+}]_i$ both before and after photolysis. Even so, in some regions of the cell, buffer diffusion may be limited or the buffer may become saturated during cell activation, preventing effective control of $[Ca^{2+}]_i$. The contribution of such behavior cannot easily be predicted. It is also possible that the photolabile probes could have effects independent of their $Ca^{2+}$-buffering ability. The probes used so far appear to be inert within cells, but actions independent of $Ca^{2+}$ buffering should not be discounted.

### *Effects of Photolysis Products*

The available probes vary in the type of photolysis products that they produce and in the ease with which their possible effects can be controlled for. Photolysis of the nitr $Ca^{2+}$-releasing compounds produces a small molecule in addition to the structurally modified chelator. In the case of nitr-5 and nitr-7 this small molecule is water, which will not affect cell function. In the case of nitr-2, the by-product is methanol; potential effects of methanol could be controlled for by applying it extracellularly to cells. Because the nitr compounds are neither sensitive to, nor produce, protons on photolysis, changes in cell pH should not be a problem. However, photolysis of the diazo compounds does produce protons, along with nitrogen. These by-products can be controlled for by using diazo-3, which undergoes the same photochemical reaction but with no change in $[Ca^{2+}]$. In experiments wherein the cell contents can be controlled, acidification can be minimized by strong pH buffering of the intracellular solution.

Photolysis of DM-nitrophen could produce changes in pH, due to the pH sensitivity of the buffer and the difference in proton-binding ability between the precursor and the photoproducts. Such changes can be minimized in some experiments with strong pH buffering, but are not easily abolished. There is no analog of DM-nitrophen available that photolyzes without releasing $Ca^{2+}$, so effects of the photoproducts must be controlled for in other ways. The caged compound could be photolyzed in a test tube and then introduced into the cell. Because DM-nitrophen photolysis results in different photoproducts depending on whether it is bound to $Ca^{2+}$ or not, it is important to include an appropriate amount of Ca and test all the possible products. This approach is not appropriate for the nitr compounds, because they retain $Ca^{2+}$-buffering activity after photolysis. Thus direct actions of the chelator product would be hard to distinguish from the effects of changing $[Ca^{2+}]_i$.

A different approach to investigating effects of the photoproducts is to introduce an excess of a light-insensitive $Ca^{2+}$ buffer into the cell, so that it, rather than the caged compound, controls the $[Ca^{2+}]$. In this case, photolysis would be expected to produce the normal products without significantly changing the $[Ca^{2+}]_i$. The light-insensitive buffer could be used without added Ca to deplete intracellular Ca effectively. However, it may prove difficult to deplete cell Ca sufficiently to prevent completely the formation of caged compound–Ca complexes and hence $Ca^{2+}$ release (27). Furthermore, because the quantum yield of photolysis is $Ca^{2+}$ sensitive, the amount of photoproduct depends on the proportion of Ca-bound caged compound present in the cell before photolysis.

### *Speed of Photolysis*

With most of the caged $Ca^{2+}$ and caged chelator compounds, the rate of photolysis should not pose a problem, being complete within 1 msec. The release from nitr-7 is somewhat slower, having a time constant of 1.8 msec, but this will still be fast enough for the analysis of many $Ca^{2+}$-dependent processes. Only nitr-2 has a significantly slower rate of photolysis, with a time constant of about 200 msec. This chelator will not, therefore, be suitable for investigations of rapid $Ca^{2+}$-dependent processes, such as transmitter release produced by a $[Ca^{2+}]_i$ jump in a presynaptic nerve terminal (12).

## *Equipment*

The only specialized equipment needed for experiments involving flash photolysis of caged $Ca^{2+}$ and caged chelator compounds is a high-intensity light source, with output in the near-UV region. If brief light flashes are not required, a xenon lamp equipped with a shutter will be sufficient. However, to take advantage of the high speed of $[Ca^{2+}]_i$ changes offered by these probes, and if time resolution is important, a flashlamp or laser will be necessary.

A xenon arc flashlamp designed specifically for flash photolysis has been developed by Rapp and Güth (28) and can be obtained from Hi-Tech Scientific Ltd. (Salisbury, England). A suitable flashlamp (Strobex model 238) is also available from Chadwick Helmuth (El Monte, California); it is supplied without a lamp housing or focusing optics, but these can be obtained from most optical suppliers. These lamps produce light of a broad spectrum, from 250 to 1500 nm, and usually emit flashes of well under 1-msec duration. The flashbulb is a high-pressure tube filled with xenon gas and contains metal electrodes, the points of which are separated by a few millimeters. A 12-kV trigger pulse discharges a capacitor bank and ignites the gas, which emits high-intensity light as it is ionized. The intensity of the light is varied by adjusting the charge on the capacitors; the maximum energy stored is about 244 J for the Rapp lamp. The short-arc flash tube essentially forms a point source of light, which is collected and focused by either quartz lenses or an eliptical mirror (Fig. 5). The mirror arrangement gives an approximately threefold higher energy output than the lenses, but a larger spot (8–10 mm compared with ~3 mm), making it more appropriate for multicellular preparations. Flashlamps are often incorporated into microscopes, where flash photolysis is used in conjunction with microinjection and/or electrophysiological techniques. The light is directed onto the cells either through the optical path of the microscope, or by using a liquid light guide. Detailed descriptions of such arrangements can be found in Ref. 26.

When using flashlamps, damaging wavelengths must be removed by placing

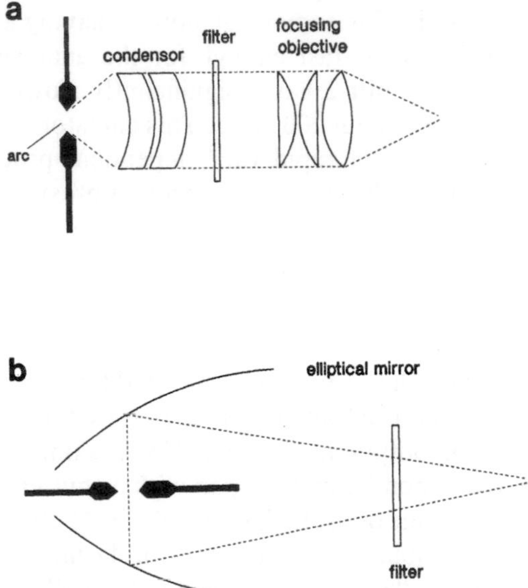


FIG. 5 Optical arrangements used with flashlamps to photolyze caged compounds. (a) Quartz lenses collect and focus the light formed by an arc between the two metal electrodes in the flashbulb. Filters are usually placed in the light path between the condensing and focusing lenses. (b) An elliptical mirror focuses the light from the arc through a filter.

filters in the light path (Fig. 5). The Schott UG 11 bandpass filter (Ealing Electro-optics PLC, Greycaine Road, Watford, England) has an appropriate peak transmission, of about 80% at 320 nm, but also has a smaller transmission peak in the near-infrared part of the spectrum, which might cause problems. The Hoya U350 bandpass filter is similar but without the transmission peak at long wavelengths. If long wavelengths are not likely to be a problem, an absorbance filter such as the Schott WG305, which cuts off sharply at around 305 nm, may be used. However, many filters of this type absorb in the order of 20% at longer wavelengths. A cheap alternative, which transmits more than 90% above 320 nm, is the Chance Propper glass coverslip, No. 0, which has been used successfully in this laboratory. In some cases, it may be desirable to use an additional heat filter, although this will also cut down the amount of transmitted light at desirable wavelengths. We have not found it necessary to employ heat filters, and they are unlikely to be required in experiments wherein light travels through bathing solution before reaching the tissue. Liquid light guides remove long wavelengths from the exciting light, but they also substantially diminish the wavelengths desirable for photolysis.

Lasers can also be used as a high-intensity light source in flash photolysis, and have some advantages over flashlamps, despite their higher cost. Laser flashes can be extremely brief and intense, and the light is monochromatic, obviating the need to filter out unwanted wavelengths. Lasers in common use are the frequency-doubled ruby laser, which emits a brief (50 nsec), intense (200 mJ) pulse at 347 nm, and a cheaper nitrogen laser, which emits a 200-$\mu$J pulse at 337 nm and can be focused through a microscope objective to give sufficient intensity for flash photolysis. Due to their price and the difficulty of incorporating them into some experimental setups, lasers are less commonly used than flashlamps, but if experiments involve irradiating only part of a cell, a laser may be necessary to produce a sufficiently small but bright light spot.

### *Localizing a Calcium Jump to Part of a Cell*

A clear advantage of light as a means of controlling the change in $[Ca^{2+}]_i$ brought about by caged compounds is that it can be localized into small areas—small enough to irradiate a single cell or even only a part of a cell. In many cases, irradiating a single cell is relatively straightforward, because the cell and the light can be simultaneously brought into focus through a microscope. Experiments have also been reported in which activation of nitr-5 was restricted to one end of a cardiac cell (29). The part of the cell irradiated can be controlled by an iris diaphragm placed in the path of the activating light (30). However, restricting photolysis to a defined part of a cell in a controlled way is not straightforward. Focusing the light to a clearly defined spot is more easily carried out with a laser than a flashlamp, because the beam from the laser is parallel and highly concentrated. Nevertheless, although the brightest light will occur in the focal plane, the planes above and below the focus will also be irradiated (30). Photolysis would be better restricted to the focal plane if the caged compound had to absorb two or more photons in quick succession. This is the idea behind the method of two-photon excitation (19, 31), in which a UV-requiring photosensitive molecule is irradiated by intense red or infrared light; simultaneous absorption of two long-wavelength photons simulates the effect of absorbing one UV photon of half the wavelength. Such methods are becoming possible, but, as discussed by Adams and Tsien (19), they require quite sophisticated technology.

## *Calibrating the Size of Light-Induced Intracellular Calcium Jumps*

### *Predicting the Jump*

In many investigations (e.g., when whole-cell patch clamping) the caged compound is the dominant buffer in the cell and its concentration is known with some degree of certainty. In these cases the intracellular concentrations of both $Ca^{2+}$ and the caged compound can be manipulated in a controlled way, and it is possible to make a reasonable estimate of how much $[Ca^{2+}]_i$ is altered by photolysis. The amount of Ca that must be added to give a predetermined $[Ca^{2+}]_i$ before photolysis can be calculated by rearranging Eq. (3). Where binding of the buffer to another ion such as $Mg^{2+}$ is likely to be significant, or the binding is highly dependent on the protonation of the buffer, the concentration of the competing ion and the pH of the solution should be included in the calculations as described earlier.

In order to estimate the change in $[Ca^{2+}]_i$, the degree of photolysis of both Ca-bound and free buffer should be known. For example, the quantum yield of Ca-bound nitr-5 is about three times greater than that of the free molecule (17). In rat sympathetic neurons, a single flash from a xenon arc flashlamp produced approximately 30% photolysis of $Ca^{2+}$-bound nitr-5 and about 10% of the unbound buffer (24). After photolysis, the photolyzed and unphotolyzed forms of nitr-5 compete for $Ca^{2+}$, and $[Ca^{2+}]_i$ is calculated by solving Eq. (10) for two competing buffers:

$$[Ca^{2+}]^3 + (K_B + K_p - Ca_{tot} + B_{tot} + P_{tot})[Ca^{2+}]^2 + [K_B P_{tot} + K_P B_{tot} - Ca_{tot}(K_B + K_P) + K_B K_P][Ca^{2+}] - K_B K_P Ca_{tot} = 0, \qquad (11)$$

where $B_{tot}$ and $P_{tot}$ are the concentrations of buffer and photolyzed buffer, respectively, and $K_B$ and $K_P$ are their corresponding dissociation constants. Calculations of this type are most easily done by computer. Again, if $Mg^{2+}$ or pH sensitivity of $Ca^{2+}$ binding is likely to be significant, a program that corrects for them should be used (3).

When pre- and postphotolysis $[Ca^{2+}]_i$ are calculated in this way, using the $K_d$ values listed in Table III, some important distinctions between different types of caged $Ca^{2+}$ compound are highlighted. Table IV shows calculated pre- and postflash $[Ca^{2+}]_i$, and the magnitude of the $[Ca^{2+}]_i$ change for a variety of photolabile Ca probes, assuming 30% of bound and 10% of free buffer are photolyzed in each case. In all of the examples it is assumed that 1 m*M* caged compound has been introduced into the cell, with or without 1 m*M* Mg added. The [Ca] that must be added to set the preflash $[Ca^{2+}]_i$ at 100 n*M* effectively corresponds to the total preflash Ca, because most of the added Ca is bound by the buffer. Where Mg is included, pre- and postflash $[Mg^{2+}]_i$ are also shown. For simplicity, $K_d$ values measured at physiological pH are used; any effect of changing pH within the physiological range will, in most cases be small.

From Table IV, it is clear that, with nitr-5 as the dominant intracellular $Ca^{2+}$ buffer, the addition of 1 m*M* Mg has little effect on the amount of Ca that must be added to set the preflash $[Ca^{2+}]_i$ at 100 n*M*, or on the size of the $[Ca^{2+}]_i$ jump produced (~40 n*M*). However, in the presence of Mg, photolysis is predicted to increase the $[Mg^{2+}]_i$ by about 1 $\mu M$. With nitr-7, which has a higher affinity for $Ca^{2+}$ than nitr-5, a greater proportion of the buffer is $Ca^{2+}$ bound, so a larger amount of Ca must be added to give an

TABLE IV Estimates of Pre- and Postphotolysis Ionic Concentrations[a]

| Parameter | Nitr-5 −Mg | Nitr-5 +Mg | Nitr-7 −Mg[+] | Nitr-7 +Mg | DM-nitrophen −Mg | DM-nitrophen +Mg | Diazo-2 −Mg | Diazo-2 +Mg |
|---|---|---|---|---|---|---|---|---|
| Before flash | | | | | | | | |
| Total [Ca] ($\mu M$) | 409 | 384 | 676 | 640 | 952 | 195 | 44 | 38 |
| $[Mg^{2+}]$ ($\mu M$) | — | 937 | — | 945 | | 204 | — | 869 |
| After flash | | | | | | | | |
| $[Ca^{2+}]$ (n*M*) | 142 | 142 | 287 | 257 | 221,850 | 340 | 27 | 26 |
| $[Mg^{2+}]$ ($\mu M$) | — | 938 | — | 944 | — | 441 | — | 863 |
| Jump | | | | | | | | |
| $[Ca^{2+}]$ (n*M*) | +42 | +42 | +187 | +157 | +221,760 | +240 | −73 | −74 |
| $[Mg^{2+}]$ ($\mu M$) | — | +1 | — | −1 | — | +237 | — | −6 |

[a] Starting conditions are 1 m*M* probe with free $[Ca^{2+}]$ set at 100 n*M*, pH 7, 0.14 *M* ionic strength, and 22°C. Calculations of $[Ca^{2+}]$ are made with and without 1 m*M* Mg present, assuming 30 and 10% photolysis of bound and free probe, respectively.

$[Ca^{2+}]_i$ of 100 n*M*. As a result, the $[Ca^{2+}]_i$ jump is about four times as great as that produced by nitr-5 under the same conditions and, when Mg is present, photolysis actually produces a decrease in $[Mg^{2+}]_i$ due to the slightly enhanced Mg-binding ability of the photolyzed chelator. Nitr-7 is thus a very useful "pure" caged $Ca^{2+}$ compound, provided that the speed of $Ca^{2+}$ release is not critical.

With DM-nitrophen, effects of Mg are more dramatic. The high affinity of DM-nitrophen means that, in the absence of added Mg, a large amount of Ca would need to be added to set the preflash $[Ca^{2+}]_i$ at 100 n*M*, and a large proportion of the buffer would be bound to Ca. A large rise in $[Ca^{2+}]_i$, of more than 200 $\mu M$, is therefore predicted after photolysis. However, when 1 m*M* Mg is present, Ca is displaced from DM-nitrophen and less Ca must be added to maintain the starting $[Ca^{2+}]_i$ at 100 n*M*; nearly 80% of the buffer is bound to Mg. The resultant $[Ca^{2+}]_i$ and $[Mg^{2+}]_i$ jumps are in the order of 200 n*M* and 200 $\mu M$, respectively. Clearly, this is not satisfactory if changes in $[Ca^{2+}]_i$ and not $[Mg^{2+}]_i$ are of interest. It should be noted that, even if Mg is omitted from both intra- and extracellular solutions, micromolar amounts of the metal are likely to be present as a contaminant.

Table IV also shows an example of the calculated drop in $[Ca^{2+}]_i$ caused by photolysis of 1 m*M* intracellular diazo-2. The preflash $[Ca^{2+}]_i$ is set at 100 n*M* by adding about 40 $\mu M$ Ca, whether Mg is present or not. A flash is then predicted to reduce $[Ca^{2+}]_i$ by about 30 n*M*, although, when Mg is present a 6 $\mu M$ decrease in $[Mg^{2+}]_i$ is also predicted. Changes in $[Mg^{2+}]_i$ of this order are small in comparison to the preflash $[Mg^{2+}]_i$ of several hundred micromolar, but their possible impact on cellular processes should always be considered.

The examples above show that neither the nitr compounds nor DM-nitrophen fulfill all the criteria for ideal caged $Ca^{2+}$ molecules. A caged compound should always be chosen with reference to the particular experimental requirements. So in experiments where changes in $[Ca^{2+}]_i$ alone are to be investigated, the influence of $Mg^{2+}$ on DM-nitrophen would tend to favor the use of one of the nitr compounds.

### *Measurement of the Jump*

It should be remembered that calculated changes in $[Ca^{2+}]_i$ can only be approximations. Whatever the method used to incorporate a probe into a cell, its intracellular concentration is not precisely known and it may not be uniformly distributed throughout the cytoplasm. Even when the caged compound is the dominant intracellular buffer, endogenous buffering mechanisms will be present, and may compete with the introduced chelator in some regions of the cell. Intracellular $Ca^{2+}$ gradients and barriers to the diffusion of buffer may affect the proportion of $Ca^{2+}$-bound chelator immediately preceding a flash. A higher proportion of the caged compound may be bound to $Ca^{2+}$ in parts of a cell where $[Ca^{2+}]_i$ is elevated (e.g., in the vicinity of the Ca channels) than in other parts of the cytoplasm. That this occurs is suggested by the finding that responses to photolysis of intracellular nitr-5 are influenced by $Ca^{2+}$ entering cells from the extracellular solution (27). Conversely, large buffer molecules may have limited access to the subplasmalemmal space.

The validity of calculations thus depends on how accurately the intracellular concentrations of probe and total Ca, as well as the extent of photolysis of the probe, are known. For this reason it is often desirable to measure the $[Ca^{2+}]_i$ directly. This is done by incorporating a $Ca^{2+}$ indicator into the cell along with the photolabile probe (32). This may be one of the fluorescent indicators, such as fluo-3, which can detect $[Ca^{2+}]$ in the submicromolar range, or a metallochromic indicator, such as antipyrylazo III, when larger changes in $[Ca^{2+}]_i$ are generated. Many factors influence the choice of $Ca^{2+}$ indicator in the absence of caged compounds (33); extra consideration must be given to the choice of indicator when photolabile Ca probes are present (34). For fluorescent indicators, important factors include the excitation wavelength and possible interactions with the caged compound, including filtering of the excitation light by the caged compound if it is present at high concentration. Zucker (35) has described effects of nitr-5 and DM-nitrophen on the fluorescent indicators, fura-2 and fluo-3, both of which have been used to monitor photolysis-induced $[Ca^{2+}]_i$ jumps. It is clear from his studies that without calibrating the fluorescence of these indicators in the presence of the photolabile Ca probe, both before and after photolysis, serious errors in the estimate of the $[Ca^{2+}]_i$ jump could result. Another approach to calibrating the photolabile probes is to use computer simulations, which combine the calculations of the $[Ca^{2+}]_i$ jump outlined above with equations that correct for light filtering by the probe and the cell, as well as the inherent $Ca^{2+}$ buffering by the cell (14).

## Incorporation of Calcium Buffers and Caged Compounds into Neurons

In order to control $[Ca^{2+}]_i$, buffers and caged compounds must be present in the cytoplasm of the neurons to be studied. However, due to their charged nature, these molecules will not readily cross membranes, so they have to be introduced into cells by other means. The least invasive method of incorporating buffers uses membrane-permeant, acetoxymethyl (AM) ester derivatives of the buffer molecules. After entering the cell, the AM ester is cleaved by intracellular enzymes to release the free buffer, which is trapped in the cell, leaving the membrane undamaged. The main problem with this method is that the levels of buffer incorporated and the degree of $Ca^{2+}$ loading are not easily controlled. This may not matter if the purpose of the buffer is to prevent or diminish intracellular $Ca^{2+}$ transients; resting $[Ca^{2+}]_i$ will probably be unchanged as a result of titration of the buffer by the cell's homeostatic mechanisms as loading proceeds. It does, however, complicate the use of caged compounds, partly because it may be difficult to achieve sufficient Ca-bound buffer in the cell to enable a $[Ca^{2+}]_i$ jump on photolysis. Furthermore, changes in $[Ca^{2+}]_i$ produced by photolysis will not be readily predictable or easily measured. Chelators that are available in the AM form include EGTA, EDTA, and most of the BAPTA derivatives, including the nitr and diazo series of caged compounds as well as fluorescent $Ca^{2+}$ buffers that are not discussed here.

Other methods of loading $Ca^{2+}$ buffers into cells include diffusion from a patch pipette in whole-cell patch-clamp recording, microinjection or reversible membrane permeabilization, for example, by hypoosmotic shock or electrophoresis. Whole-cell recording has the advantage that the concentration of buffer and the degree of $Ca^{2+}$ loading can be reasonably well controlled, because the solution in the recording pipette essentially overwhelms the cell interior. Nevertheless, even with this method, control of $[Ca^{2+}]_i$ is not complete, particularly when $Ca^{2+}$ is present in the extracellular solution (27). The internal dialysis that occurs with whole-cell recording may also result in the loss of important regulatory molecules from the cell. Microinjection, by pressure or ionophoresis, overcomes this problem, but it is less readily calibrated and is only applicable to relatively large neurons. Cell permeabilization has been widely used for studies of nonneuronal cells. Some methods are likely to result in damage to cell membranes and loss of important cytoplasmic molecules. Permeabilization with staphylococcal $\alpha$-toxin appears to overcome these problems (36) and allows molecules with molecular mass $<1000$ Da to enter cells. This should allow the concentration of buffer and the extent of $Ca^{2+}$ loading to be controlled, although while the cell is permeabilized the extracellular environment will be abnormal. A way of incorporating $Ca^{2+}$ buffers to combine the advantages of $\alpha$-toxin and whole-cell recording might be to include the toxin in a patch pipette, thereby gaining restricted access to the cell through a permeabilized patch of membrane.

## Recent Findings

Since submitting this manuscript, a new caged $Ca^{2+}$ has been described by Ellis-Davies and Kaplan (37), which appears to have near ideal properties. This compound, nitrophenyl-EGTA, discriminates well between $Ca^{2+}$ and $Mg^{2+}$, with $K_d$ values for binding these ions of 80 and 9 m$M$, respectively.

When irradiated, the chelator is cleaved to yield photoproducts with low $Ca^{2+}$ affinity and the reaction proceeds with a high quantum yield. This compound can therefore be used to generate large changes in $[Ca^{2+}]$. It is likely that nitrophenyl-EGTA will supersede DM-nitrophen and nitr-5, because it combines the desirable properties of these two compounds, apparently with few drawbacks.

## Acknowledgments

The work in this laboratory is supported by funds from the Wellcome Trust, the British Heart Foundation, and the Royal Society.

## References

1. R. Y. Tsien, *Biochemistry* **19,** 2396 (1980).
2. A. Fabiato, *in* "Cellular Calcium—A Practical Approach" (J. G. McCormack and P. H. Cobbold, eds.), p. 159. IRL Press at Oxford Univ. Press, Oxford, 1991.
3. A. P. Jackson, M. P. Timmerman, C. R. Bagshaw, and C. C. Ashley, *FEBS Lett.* **216,** 35 (1987).
4. J. P. Y. Kao and R. Y. Tsien, *Biophysic. J.* **53,** 635 (1988).
5. S. R. Adams, J. P. Y. Kao, and R. Y. Tsien, *J. Am. Chem. Soc.* **111,** 7957 (1989).
6. E. M. Adler, G. J. Augustine, S. N. Duffy, and M. P. Charlton, *J. Neurosci.* **11,** 1496 (1991).
7. M. D. Stern, *Cell Calcium* **13,** 183 (1992).
8. S. M. Harrison and D. M. Bers, *Biochim. Biophys. Acta* **925,** 133 (1987).
9. G. L. Smith and D. J. Miller, *Biochim. Biophys. Acta* **839,** 287 (1985).
10. T. Bartfai, *Adv. Cyclic Nucleotide Res.* **10,** 219 (1979).
11. R. S. Zucker, *Cell Calcium* **14,** 87 (1993).
12. K. R. Delaney and R. S. Zucker, *J. Physiol. (London)* **426,** 473 (1990).
13. J. M. Nerbonne and A. M. Gurney, *J. Neurosci.* **7,** 882 (1987).
14. L. Landò and R. S. Zucker, *J. Gen. Physiol.* **93,** 1017 (1989).
15. K. Khodakhah and D. Ogden, *Proc. Natl. Acad. Sci. U.S.A.* **90,** 4976 (1993).
16. J. F. Wootton and D. R. Trentham, *in* "Photochemical Probes in Biochemistry" (P. E. Nielsen, ed.), NATO ASI Series C, Vol. 272, p. 277. Kluwer, Dordrecht, The Netherlands, 1989.
17. S. R. Adams, J. P. Y. Kao, G. Grynkiewicz, A. Minta, and R. Y. Tsien, *J. Am. Chem. Soc.* **110,** 3212 (1988).
18. J. H. Kaplan and G. C. R. Ellis-Davies, *Proc. Natl. Acad. Sci. U.S.A.* **85,** 6571 (1988).
19. S. R. Adams and R. Y. Tsien, *Annu. Rev. Physiol.* **55,** 755 (1993).
20. M. A. Ferenczi, Y. E. Goldman, and D. R. Trentham, *J. Physiol. (London)* **418,** 155P (1989).

21. B. O'Rourke, P. H. Backx, and E. Marban, *Science* **257,** 245 (1992).
22. J. A. McCray, N. Fidler-Lim, G. C. R. Ellis-Davies, and J. H. Kaplan, *Biochemistry* **31,** 8856 (1992).
23. R. M. Mulkey and R. S. Zucker, *J. Physiol.* (*London*) **462,** 243 (1993).
24. A. M. Gurney, R. Y. Tsien, and H. A. Lester, *Proc. Natl. Acad. Sci. U.S.A.* **84,** 3496 (1987).
25. A. Arvanitaki and N. Chalazonitis, *in* "Nervous Inhibition" (E. Florey, ed.), p. 194. Pergamon, New York, 1961.
26. A. M. Gurney, *in* "Fluorescent and Luminescent Probes for Biological Activity—Biological Techniques Series" (W. T. Mason, ed.), p. 335. Academic Press, London, 1993.
27. S. E. Bates and A. M. Gurney, *J. Physiol.* (*London*) **466,** 345 (1993).
28. G. Rapp and K. Güth, *Pfluegers Arch.* **411,** 200 (1988).
29. S. C. O'Neill, J. G. Mill, and D. A. Eisner, *Am. J. Physiol.* **258,** C1165 (1990).
30. L. D. Chabala, A. M. Gurney, and H. A. Lester, *J. Physiol.* (*London*) **371,** 407 (1986).
31. W. Denk, J. H. Strickler, and W. W. Webb, *Science* **248,** 73 (1990).
32. J. P. Y. Kao, A. T. Harootunian, and R. Y. Tsien, *J. Biol. Chem.* **264,** 8179 (1989).
33. A. M. Gurney, *in* "Receptor-Effector Coupling: A Practical Approach" (E. C. Hulme, ed.), p. 117. IRL Press at Oxford Univ. Press, Oxford, 1990.
34. A. M. Gurney, *in* "Cellular Neurobiology. A Practical Approach" (J. Chad and H. Wheal, eds.), p. 153. IRL Press at Oxford Univ. Press, Oxford, 1991.
35. R. S. Zucker, *Cell Calcium* **13,** 29 (1992).
36. A. P. Somlyo and A. V. Somlyo, *Annu. Rev. Physiol.* **52,** 857 (1990).
37. G. C. R. Ellis-Davies and J. H. Kaplan, *Proc. Natl. Acad. Sci. U.S.A.* **91,** 187 (1994).

# [8] Measurement of Cytosolic Calcium by $^{19}F$ NMR

Gerry A. Smith and Herman S. Bachelard

## I. Introduction

Techniques for quantitative measurement of free intracellular calcium ($[Ca^{2+}]_i$) are vital to our understanding of its key role in a multitude of extra- and intracellular processes in the nervous system and in heart function. The techniques need to be sensitive as well as accurate and specific, given the very low intracellular calcium concentration of 0.1–0.2 $\mu M$, relative to the extracellular calcium concentration (1–2 m$M$) and to the high concentrations of other cations present in the heart and nervous system. The methods available [ion-specific electrodes, electron microscopic X-ray analysis, fluorescence, and magnetic resonance spectroscopy (MRS)] have advantages and disadvantages; none is ideal in fulfilling all requirements.

Cation-specific electrodes are very invasive, injuring the cell wall on penetration, and are limited in application and in spatial resolution, because they are not easily performed in metabolically active systems, but are sensitive and show reasonable temporal resolution. Electron microscopic X-ray analysis gives total $Ca^{2+}$, not free versus bound, but with superb subcellular localization. It is not capable of temporal resolution nor can it be performed on living tissues, because it requires fixed (i.e., dead) tissue. Fluorescence indicators give good localization, with good temporal resolution, on single cells or surface tissue, but there are problems with specificity. Problems also arise from the presence of the indicator within the cells, causing "buffering" of the $[Ca^{2+}]_i$.

$^{19}F$ magnetic resonance spectroscopic methods share problems with fluorescence methods in the danger of buffering the $[Ca^{2+}]_i$, so careful controls are essential to monitor this (see Section IV,B). The technique at present cannot give intracellular morphological localization, nor good temporal resolution. However, it can measure $[Ca^{2+}]_i$ with high specificity due to differences in the chemical shifts induced by various divalent cations and is particularly applicable to light-opaque tissue. Examples are given below in distinguishing between $[Ca^{2+}]_i$ and $[Zn^{2+}]_i$ (see Sections VI,A and VI,B). $^{19}F$ NMR indicators can be used to follow changes in $[Ca^{2+}]_i$ with metabolic or functional perturbations.

*Methods in Neurosciences, Volume 27*

Table I lists the currently usable $^{19}F$ NMR intracellular calcium indicators and their structures, calcium dissociation constants, and the calcium-induced chemical shifts.

## A. *Principles of the Use of $^{19}F$ NMR Indicators*

As for the fluorescent indicators, from which the chemically analogous $^{19}F$ indicators are derived (Tsien, 1980), they are prepared as the tetraacetoxymethyl (AM) esters, which are sufficiently lipophilic to be transported across cell membranes (Tsien, 1981). They are then hydrolyzed by intracellular esterases to the non-membrane-permeant, hydrophilic free tetraanions, which chelate divalent cations with high affinity according to their structures. Because they are intracellular and can only chelate free rather than bound $Ca^{2+}$, they are true indicators of $[Ca^{2+}]_i$. The relative concentrations of free and bound indicator are reported by their contributions to the observed spectra, either by the signal frequency (chemical shift) in the case of fast exchange or by the relative signal intensities in the case of slow exchange (see Section III,A) (Smith *et al.*, 1983).

Loading has been successfully achieved in perfused heart (Metcalfe *et al.*, 1985), in superfused brain tissues (hippocampus and cortex, Bachelard *et al.*, 1988), and in cultured cell preparations (Smith *et al.*, 1982), and is now feasible in rat brain *in vivo* (Ackerman *et al.*, 1995). The success in these *in vitro* (or *ex vivo*) preparations is providing unique information on the changes occurring in $[Ca^{2+}]_i$ in different functional states in actively metabolizing tissues.

This field of work differs from most others in that optimally it requires the collaboration of specialists from at least three fields, namely biology, NMR, and organic chemistry, with a fairly large mechanical workshop input. With this in mind this chapter has been written for the nonspecialist in each subject so that, when the collaboration possible falls short of the optimum, some reasonable attempt can be made to cover the missing requirements. In this chapter we provide precise recipes for the synthesis of the useful $^{19}F$ MRS calcium indicators, a definition of their properties (advantages and limitations), the hardware required for their use, and some examples of their applications in biomedical research.

The presentation will also facilitate each collaborator's understanding of the input of the others and their technical requirements. Most of the techniques described herein have been used on systems not directly related to neuroscience applications, but with modification are more generally applicable. Therefore the chapter is structured in a way to facilitate the finding of particular methods that can be adapted to the reader's task-in-hand.

TABLE I $^{19}$F NMR Intracellular Calcium Indicators

| Indicator | Structure | $Ca^{2+}$ affinity ($nM$ at 30°C) | $Ca^{2+}$-induced shift (ppm) |
|---|---|---|---|
| 3-FBAPTA | | 646 | 5.0 |
| 4-FBAPTA | | 2455 | 1.5 |
| 5-FBAPTA | | 537 | 5.5 |
| MFBAPTA | | 269 | 5.7 |
| DiMe-4-FBAPTA | | 155 | 2.07 |
| DiMe-5-FBAPTA | | 34 | 3.96 |
| DiMe-5-TFMBAPTA | | 724 | −0.8 |

# II. Synthesis of the Indicators

## A. *General Principles of Synthesis*

All of the $^{19}F$ NMR indicators that have been prepared and used to date are based on the EGTA analog 1,2-bis(*o*-aminophenoxy)ethane-tetraacetic acid (BAPTA), the highly selective chelator for calcium introduced by Tsien (1980). It follows, therefore, that they are all synthesized by essentially the same route as used for the synthesis of BAPTA. Compared to other spectroscopic techniques NMR is an inherently insensitive method and hence the measurement of calcium by $^{19}F$ NMR has requirements for much larger quantities of the indicators than do fluorescence methods, due to the larger amounts of tissue required, although the final intracellular concentrations are similar. In general, the commercial prices of indicators are set by market for the fluorescent indicators, and the purchase of the AM esters of the $^{19}F$ indicators becomes a hindrance to their use. To overcome this hindrance we describe the synthetic methods in a form that we hope is easily followed by those who are not specialist chemists (Tsien, 1980, Smith *et al.*, 1983). The overall synthetic route, applicable to all the variants, is shown in Scheme I.

This scheme looks lengthy to the uninitiated, but all the reagents are inexpensive and the only step requiring specialized equipment (for hydrogen addition) is the reduction of the aromatic nitro groups ($NO_2$) to the amines ($NH_2$). For those unable to acquire or make an hydrogenator we also describe the use of stannous chloride as an alternative to the use of hydrogen for this step, but recommend the hydrogenation method as the least messy.

Of prime importance in performing the preparations is the use of thin layer chromatography (TLC) to follow the course of the reactions, to ensure completion before attempting isolation of the products, and to follow the chromatography used for some of the purification steps. Samples of the reaction (a few microliters) are taken and diluted with toluene (~100 $\mu$l) in a small glass tube; the toluene solution is mixed well with either the same solution used to wash the extract in the isolation procedure or with water (~1 ml). A few microliters of the top (toluene) layer are then used for TLC. The TLC is performed on silica gel plates impregnated with an ultraviolet-absorbent material (commonly called F254) purchased as (or cut to) approximately 7 cm by 2.5 cm (miniplates). Unless one is skilled in glass cutting, plastic or aluminum-backed plates are most suitable, cut with sharp scissors. Small spots of the sample solutions to be tested are applied 5 mm from the end of the plate. To avoid misinterpretation of results if the plates do not run evenly, three spots should be applied 5 mm from the end of the plate, one with each of the starting reactants, one for the reaction sample, and a mixed spot in between. It is then obvious if the product is different from the

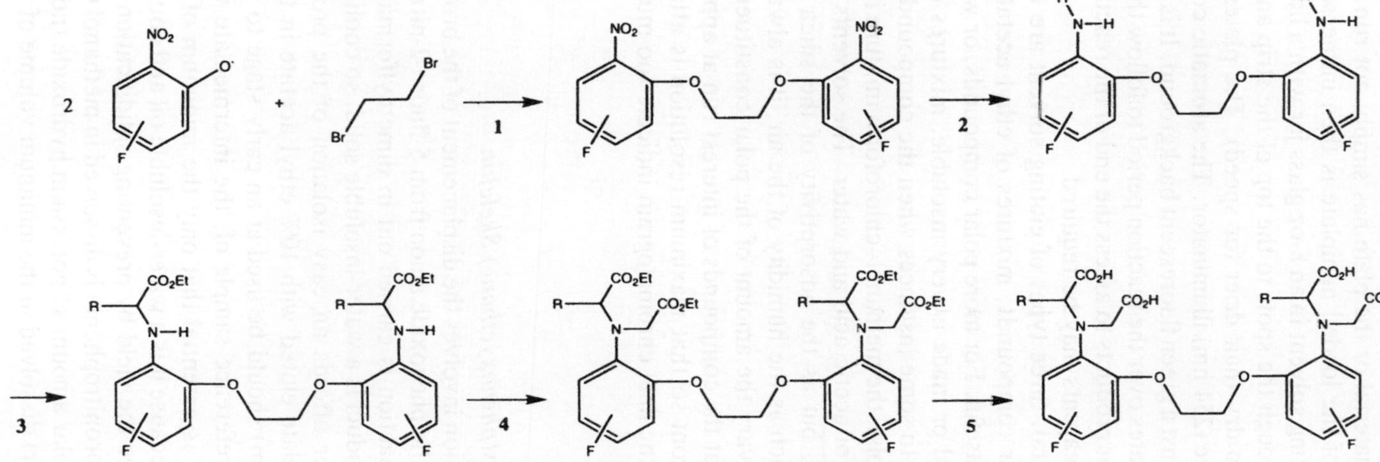

SCHEME 1 Synthesis of 5-FBAPTA and DiMe-5-FBAPTA.

starting material or the plate has simply not run the same across its width. The end of the loaded miniplate is then immersed in approximately 2 mm of the eluting solvent in an 8-oz glass jar with a lid and the solvent is allowed to rise through the spots to the top of the strip and the plate is removed and allowed to dry (hair drier for speed). The plates are then viewed under a short-wave (254 nm) illuminator. The aromatic compounds are seen as dark spots against a green fluorescent background. It is advisable to run a selection of TLC plates over the reaction period to follow the appearance of intermediates and/or products to assess the end of the reaction or whether the addition of more reagents may be required.

In general, three types of eluting solvent are used for the TLC. For the least polar compounds, mixtures of ethyl acetate in toluene are generally the most useful. For more polar compounds, or where one of the substances being used or made is very insoluble, mixtures of methanol in chloroform are used. In some instances when the compounds are ionizable (e.g., acids and phenols), the methanol–chloroform mixture is supplemented with either ammonia or acetic acid and water. The solvents required at each stage are described, but as the adsorptivity of the silica is variable, depending on factors such as the humidity of the air, it is always advisable to make trial runs and vary the amount of the polar constituent of the eluting solvent to ensure that the compounds of interest run at approximately half-way to the solvent front so that maximum resolution is attained. Very dark smeared spots on the final chromatogram indicate too much material was applied.

## *B. 5-FBAPTA*

### *1. The Bis(phenoxyethane) Skeleton*

This reaction involves the displacement of the bromine from 1,2-dibromoethane with the phenoxide anion from 5-fluoro-2-nitrophenol (step 1 in Scheme I). The reaction is carried out in dimethylformamide (DMF). Only the required product is a water-insoluble solid, so cooling the mixture and dilution with water affords an easy isolation of the product. TLC (toluene/water extract, plate eluted with 10% ethyl acetate in toluene or 2% methanol in chloroform) should be used at an early stage to show the presence of, and obtain a reference sample of, the intermediate with only one bromine replaced. It is essential that only the minimum of intermediate is present at the end because it is a water-insoluble oil and could contaminate the product and reduce the yield by preventing solidification.

The fluoronitrophenol is dissolved in methanol (~20 ml/g) and treated with an equimolar amount of potassium hydroxide (note: KOH is usually 85% + 15% water) dissolved in the minimum volume of water (3 ml/g). The bulk of the potassium salt is easily collected by precipitation on addition of 1 volume of diethyl ether and filtration; the remainder of the salt is obtained by rotary evaporation of the mother liquors and drying the residue *in vacuo* overnight.

The fluoronitrophenoxide is added to dimethylformamide (10 ml/g) in a round-bottom flask (5 times the DMF volume under a water-cooled reflux condenser with 1,2-dibromoethane (0.5 mol per mole of phenoxide). The whole is heated and stirred in an oil bath at 120°C (polyethylene glycol 400 is a very good substitute for oil at these temperatures and is easily washed away with water).

Of the dibromoethane, 10–20% is lost by side reaction; considerable starting phenol, but very little of the intermediate, will be left after 1 hr. The reaction can be worked up at this stage; however, the addition of anhydrous sodium or potassium carbonate (10 mol% of starting potassium salt) and a similar quantity of dibromoethane at 15-min intervals will ensure more efficient conversion of the phenol. This last stage is only important when dealing with other more expensive fluoronitrophenols. The reaction mixture is then cooled and diluted with two volumes of water with good mixing. The resulting solid is readily collected by vacuum filtration on a sintered glass funnel and washed well with water. The solid should then be stirred with ice-cold 50% aqueous ethanol and filtered and washed with a little more. Any traces of starting phenol may be removed by washing with dilute sodium carbonate solution (0.5 *M*), the product given a final wash with water, and dried by leaving to evacuate on the sinter. Yields will be 60–90% of chromatographically homogeneous product. If not pure, the very insoluble dinitro compound may be purified by dissolution in boiling chloroform and crystallized by the addition of a few volumes of ethanol.

*2. Reduction of Nitro Groups to Amines*

a. By Hydrogenation

Reduction of nitro groups to amines (step 2 in Scheme I) can be accomplished by hydrogenation. A simple hydrogen reservoir system is shown in Fig. 1. The dinitro compound and 5% of 10% w/w of palladium on carbon catalyst is added to ethanol (20 volumes) in a large round-bottom flask; the air is evacuated and replaced with hydrogen by judicious use of the taps. The reaction is stirred as vigorously as possible (magnetic stirrer). On a large scale (over 10 g), the reaction mixture can become very hot, which allows the dissolution of the dinitro compound and speeds the reaction; however, a cold water bath should be handy to cool the reaction flask to moderate the reaction. After absorption of the required hydrogen is achieved and uptake has ceased (6 × 22.4 liters/mol), the hydrogen is evacuated, replaced with nitrogen, and a sample of the suspension is diluted with dichloromethane and subjected to TLC (10% ethyl acetate in toluene) analysis to ensure

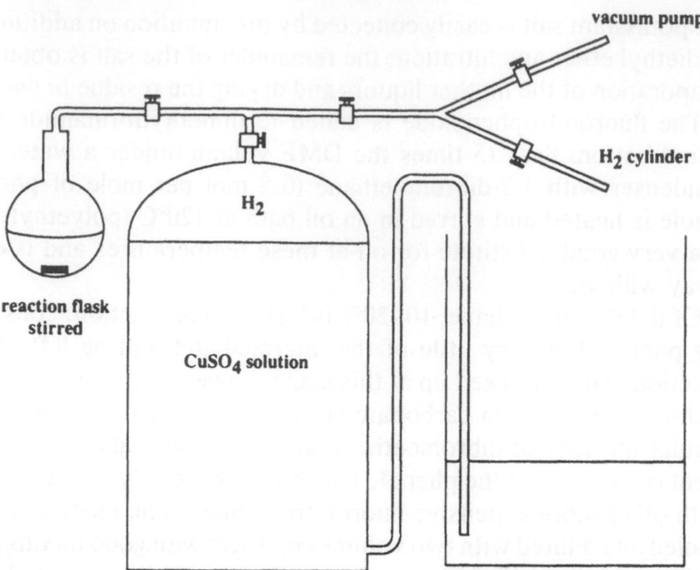


FIG. 1 Simple hydrogenator for the reduction of nitro groups. The apparatus is constructed from thick-walled glass. The flask should be charged with catalyst and reactant and all the air removed with the water pump before charging the reservoir with hydrogen (at a few psi only). Care must be taken in using the minimum vacuum to remove the air from the reservoir before hydrogen filling; an adjustable leak of air into the water pump vacuum line is required.

removal of all the dinitro compound. When the reaction is complete the product is dissolved by the addition of dichloromethane and the catalyst is removed by filtration through a pad of well-packed Hyflo Supercel on a glass Büchner filter. The solvent is evaporated to low volume and the product is collected by filtration and dried. A small second crop can be obtained by further concentration of the mother liquors, but some coloration will appear as the residues get oxidized. If further purification is required the diamine is easily dissolved in a small volume of dichloromethane and is crystallized on the addition of methanol.

b. By Stannous Chloride Reduction

For those unable to set up the much easier hydrogenation procedure, the following procedure may be used. The drawbacks are the use of very large quantities of stannous chloride dissolved in concentrated hydrochloric acid, and the very large amounts of sodium hydroxide solution required to solubilize the tin salt complex of the diamine product and subsequent extraction with dichloromethane. The dinitro compound is dissolved in 10% concentrated hydrochloric acid in tetrahydrofuran (20 ml/g) and stirred with a large excess (10 molar equivalents) of stannous chloride. Samples of the reaction are taken and added to 10% sodium hydroxide solution and extracted with dichloromethane. The organic phase is analyzed by chromatography as above until no starting material remains and a single spot is found. Sufficient sodium hydroxide to neutralize the hydrochloric acid plus at least 2 mol more per mole of stannous chloride are dissolved in water (10% solution) and cooled and mixed with the reaction mixture. The free diamine is then extracted with dichloromethane, which is separated and evaporated to dryness and the product is crystallized from methanol and dried *in vacuo*.

### *3. Addition of Acetate Groups*

Excess ethyl bromoacetate and base are added to complete the addition of all four acetate groups in one step (combined third and fourth steps in Scheme I). The reaction (sampled for TLC with toluene and molar ammonium phosphate, pH 2) can be followed by TLC in 20% ethyl acetate in toluene to follow the formation and disappearance of all the intermediates formed in this preparation. Originally the base used for the reaction was a proton sponge [1,8-bis(dimethylaminonaphthalene)], but it has since been found that the reaction proceeds just as well with the much more economical diisopropylethylamine, a much simpler nonnucleophilic organic base. The well-dried diamine is treated with 5 molar equivalents of both diisopropylethylamine and ethyl bromoacetate in a small volume of acetonitrile (equal to the bromoacetate volume). The reaction flask, under a water-cooled reflux condenser, is flushed with nitrogen from a balloon connected by a tap to the top of the condenser. The balloon is left open to the flask to avoid pressure buildup. The reaction is magnetically stirred and heated in a bath at 110°C until complete, usually overnight. The cooled reaction is diluted with toluene and washed well with molar ammonium phosphate solution, pH 2, to remove the salts and excess base, and finally with sodium bicarbonate solution, dried by stirring with a little anhydrous sodium sulfate and evaporated to dryness on a rotary evaporator (care must be taken because ethyl bromoacetate will remain and is extremely lachrymatory). This product will normally crystallize on addition of a small volume of ethanol (10 ml/g) and cooling at 0°C. If not pure enough to crystallize it may be purified by chromatography as for DiMe-5-FBAPTA (described in Section II,B).

### *4. Hydrolysis of the Esters*

The pure tetraester is dissolved in the minimum of hot ethanol and treated with five equivalents of 2 M sodium hydroxide solution in a slow stream while keeping hot. The hydrolysis (step 5 in Scheme I) may be followed by thin layer chromatography of an ethanol-diluted sample eluting with chloroform : methanol : water : 0.880 ammonia (45 : 35 : 8 : 2). When only a single low-mobility spot is attained, the cooled and stirred reaction is diluted with water (10 volumes) and acidified to pH 2.5 at 5°C with dilute HCl. The product should crystallize directly; however, the solution may need scratching to induce the gummy product to crystallize. The pH will rise as crystallization occurs and must be corrected because the partial salt will cocrystallize with the free acid. The product is removed from the pH 2.5 saline solution by suction filtration and is packed down well. The pad is washed by sucking through ice-cold 10 m*M* hydrochloric acid to ensure total removal of all the sodium chloride. Any sodium chloride present on drying the product will lead to formation of the sodium salt, which interferes with the subsequent acetoxymethyl ester formation.

## C. DiMe-5-FBAPTA

In this step (step 3 in Scheme 1) (Clarke *et al.*, 1993) the bromoacetate used in the preparation of 5-FBAPTA is substituted with $\alpha$-bromopropionate. The presence of the extra methyl group prevents subsequent reaction so only one propionyl group is added to each of the amines; the subsequent addition of the acetyl groups may then be performed in the same reaction pot. The diamine produced by reduction of the dinitro compound is treated with 2.5 molar equivalents of both ethyl $\alpha$-bromopropionate and diisopropylethylamine in acetonitrile, under nitrogen at 110°C as above. The reaction is followed until the intermediate is completely removed (a few hours or overnight) and then the 2.5 molar equivalents of ethyl bromoacetate and base are added with the required acetonitrile and the reaction is allowed to proceed for approximately 2 days until a single component product is formed. The reaction mixture work-up is exactly the same as for the tetraethyl ester of 5-F-BAPTA above.

A column of silica gel 60 (110 g per gram of product) in 5% ethyl acetate in toluene is prepared by suspending the dry silica in the solvent and applying vacuum for a few minutes to remove the gases from the silica. The silica slurry is then poured into a suitably sized glass column and allowed to settle under flow. The flat top of the column is then protected by allowing a few centimeters of acid-washed sand to settle through the solvent head. The solvent is then allowed to flow until the meniscus reaches the sand. The reaction product is then dissolved in toluene (5 ml/g) and added to the column, allowed to run into the column, and washed in with additions of 5% ethyl acetate in toluene until all is absorbed. The column head is then filled with the same solvent, with a reservoir if available, and the column is allowed to run under gravity. The eluting solvent polarity is increased by the addition of more ethyl acetate (2 to 3 column volumes of 10% and then 20%). If a reservoir is used and the column head is approximately equal to the volume of silica, then addition of each new gradient step to the reservoir, although not allowing the head to empty, will generate a smooth polarity gradient because ethyl acetate is more dense than toluene. The eluate is collected in fractions and tested for product content by spotting and drying on small pieces of TLC plate and viewing under the UV lamp. When UV-absorbing material starts to appear, the fractions are subjected to TLC analysis in 20% ethyl acetate to ascertain which are pure, and these are combined and evaporated to dryness in a rotary evaporator and finally under high vacuum.

The mixed isomers can be obtained in a solid form and free from oily contaminants by dissolution in a little diethyl ether and crystallization at 0°C by the slow addition of petroleum spirit (40–60), with scratching to initiate crystal formation, and the product is collected by filtration.

*Hydrolysis of the Esters*

This is performed by the same procedure as for 5-FBAPTA but with great care to ensure complete crystallization at pH 2.5, by scratching as before. It is advisable to remove the ethanol solvent by rotary evaporation before acidification because this will facilitate the crystallization of the mixed isomers. The solid cake of product must be washed free of salts to ensure a good yield of AM ester.

## D. Other Indicators

A variety of $^{19}$F calcium indicators are now available with varied affinities and chemical shifts; a list is given in Table I (from Smith *et al.*, 1983; Levy *et al.*, 1987; Clarke *et al.*, 1993). The use of indicators with different affinities is starting to shed more light on the buffering properties of the indicators in the heart (Kirschenlohr, 1993) and also opens the possibility of gaining information on the endogenous buffering in the cytosol. Although not yet used in a biological situation, DiMe-5-TFMBAPTA is also included in Table I to demonstrate the current position in the search for indicators with greater NMR sensitivity.

## E. The Acetoxymethyl Esters

### 1. Preparation of Bromomethyl Acetate

This reagent is sold commercially but is expensive. Making the reagent is a one-step process from readily available inexpensive materials, but does require the ability to perform a vacuum distillation. Care must be taken to ensure that all the reagents are fresh and not allowed to absorb moisture from the air. Paraformaldehyde is added portion-wise to 3 molar equivalents of acetyl bromide containing 10% by weight of aluminum bromide in a stirred round-bottomed flask cooled in a water bath. When the reaction becomes homogeneous, the flask is connected to a vacuum distillation setup with a short Vigreaux column (or air condenser set between the top of the flask and the still head) and heated in a bath at 80–90°C. The vacuum is adjusted with a screw clip closing on a piece of vacuum tubing, containing a strip of thin wire, connected to a T piece in the vacuum line. A water pump is sufficient to provide the vacuum; the adjusting air bleed protects the product from water vapor from the pump. After a small initial run is taken, the bulk of the product is collected with the vacuum set to give a constant boiling point around 60°C. Proton NMR always shows this material to be 90% pure bromomethyl acetate, with around 10% of a side product with protons at slightly higher field than those of the methylene group, probably dibromomethane; this is not removed by redistillation. The bromomethyl acetate is not stable for long periods at room temperature but can be kept for many months at −20°C in a closed Quickfit flask (greased joint).

### 2. *Acetoxymethyl Esters*

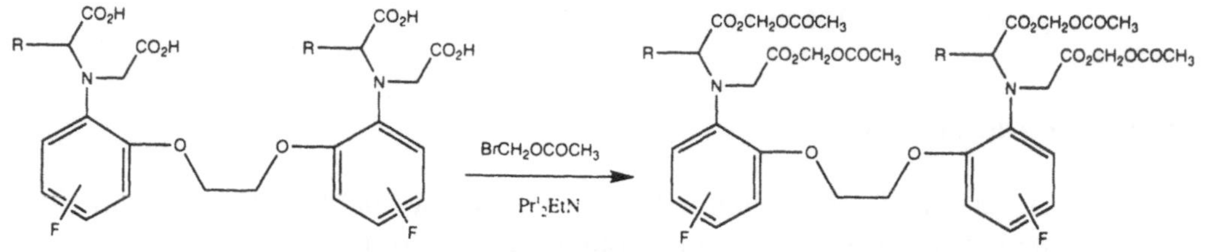

$CO_2H$, $CO_2H$, R, N, O, F → ($BrCH_2OCOCH_3$, $Pr^i_2EtN$) → $CO_2CH_2OCOCH_3$, $CO_2CH_2OCOCH_3$, R, N, O, F

SCHEME II Preparation of the acetoxymethyl esters.

For this preparation it is imperative that the free acids are absolutely dry and free from any salts (see above preparation). The free acids are suspended in stirred dichloromethane and treated with six molar equivalents of diisopropylethylamine to give a clear solution. The stirred solution is stoppered, stirred, and cooled in an ice bath, because the reaction can become quite hot on a larger scale and can compromise the yield and purity of the product. The bromomethyl acetate (5 molar equivalents) is added and the reaction is allowed to warm to room temperature. After a few hours a sample diluted with dichloromethane is spotted onto a TLC plate and developed immediately in 20% ethyl acetate in toluene. Unreacted free acid stays at the baseline; the product runs as a main spot with a much smaller contaminant just in front. When the reaction proceeds no further, some baseline will always remain. The solvent is removed on a rotary evaporator (previously cleaned and dried) with care because some sputtering will occur as the amine hydrobromide crystallizes when nearly dry. The excess bromomethyl acetate is removed under high vacuum. The gummy solid is stirred with a little ethyl acetate and the salts are removed by filtration and the ethyl acetate is rotary evaporated to give the crude product. The pure acetoxymethyl ester is isolated by rapid chromatography as detailed in the preparation of the diethyl ester of DiMe-5-FBAPTA, but increasing the ethyl acetate to 30% with only single column volumes of each gradient step (5, 10, 20, 30%). Care should be taken to avoid as much of the faster running impurity as possible. The solvent is removed again in a dry rotary evaporator and finally under high vacuum. Only the AM ester of 5-FBAPTA has been crystallized, from diisopropyl ether (mp 70–72°C). The esters are stored as aliquots of a 50 m*M* solution in pure dry DMSO at −20°C.

## III. Magnet and Probe Requirements

### A. *Field Strength and Resolution Requirements*

In general, for the type of experiments described here, when cell or tissue suspensions or whole organs (e.g., the ferret heart, a few grams) are used, it is essential to provide the sample with freshly oxygenated buffer and maintain/record the temperature and pH and sometimes supply other facilities, such as electrical stimulation. For this reason a minimum 10-cm bore is needed for the magnet.

The indicators work by reflecting the equilibrium binding of calcium in the chemical shift of the $^{19}F$ resonance signal in the case of fast exchange, and in the relative areas of the two $^{19}F$ resonance signals in the case of slow exchange. In these biological systems, the linewidths observed are considerably broadened by many factors, and hence it would be impossible to make measurements under conditions of intermediate exchange. The deesterified indicators are small molecules that diffuse freely in aqueous solution, thus the association rates are governed by the diffusion constants. The indicators are also designed to be sensitive to changes in free calcium concentration in the normal intracellular range, and the affinity is derived from the diffusion-limited association rate and the dissociation rate. Therefore, the resulting dissociation rates of usable indicators for calcium are in the range of hundreds per second, i.e., ppm, at normal NMR field strengths. Because fast or slow exchange conditions are defined by the comparison of the dissociation rate and the separation of the free and bound signals in hertz, the requirement for either pure fast or pure slow exchange sets an upper and a lower limit to the field strength, respectively, that can be used. For 5-FBAPTA and higher affinity indicators, which are in slow exchange, a minimum of around 200 MHz is required—the higher the field strength the more sensitive the measurement. Because intracellular calcium is in the 0.1 $\mu M$ range, no work has as yet been performed with fast-exchange $^{19}F$ indicators for calcium, although magnesium has been measured with fluorocitrate; pH, with FQUENE (Kirschenlohr *et al.*, 1988); and almost certainly in the future, sodium, with fluorocrown derivatives (Smith *et al.*, 1993); the intracellular concentrations are in the millimolar range and the indicators are in fast exchange. The fast-exchange condition poses another problem in that a reference signal is also required to determine the exact chemical shift and hence the free ionic concentration.

## B. Service from Below

### 1. The Transceiver Coil

Not many commercial fluorine probes of suitable physical dimensions are available, and these often sacrifice sensitivity for resolution. Home-made dedicated probes are preferable and not too difficult to build. Here we describe our version for a 10-cm bore magnet, up to 376 MHz (9.4 T).

A diagrammatic representation of the coil made from copper sheet is shown in Fig. 2. The only magnetically active elements of this coil are the two

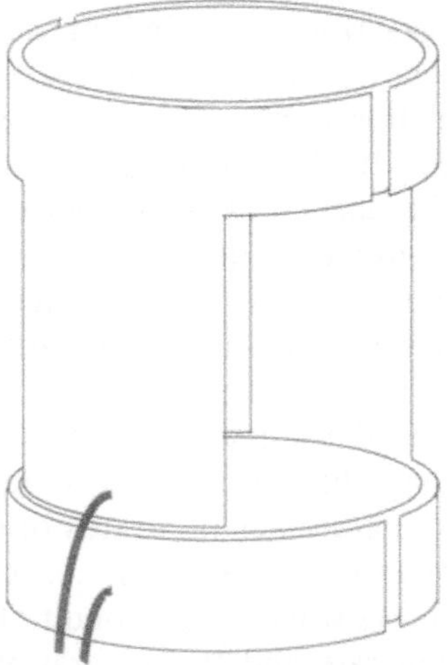

FIG. 2 Diagrammatic representation of the transceiver coil.

vertical poles; all currents in the top and bottom rings are cancelled by an equal and opposite current generated through the high-capacitance coupling to the guard rings. The 1-mm gaps on each side, and top and bottom, generate virtually no magnetic field when the whole coil is tuned to the required frequency. The shortness of the two active elements allows easy tuning of this very sensitive coil up to around 400 MHz, when very low-residual-capacitance tuners are used.

The former for the coil is a Pyrex glass tube, diameter 3.5 cm, length 5.5 cm. Two strips of 0.07-mm gauge copper (99.99%), 1 cm by $3.5\pi$ cm (circumference), are cut and rolled smooth with a solid steel rod on a hard surface. The strips are cleaned and polished and rolled again on a soft card until they bend to form a ring that fits the glass former. The lengths are adjusted to close exactly around the former and are positioned, with sticky tape, on the former with a 2.00-cm gap between. The butted ends are then soldered and the excess solder is removed by careful filing. Their tight fit on cooling temporarily holds the rings when the tape is removed. The whole is then given a uniform coat of clear acetate lacquer (spray can from any car shop). The layer of lacquer forms the dielectric for the high-capacitance link between these guard rings and the remainder of the coil. Two H-shaped pieces of copper are then cut 4.0 cm wide and $1.75\pi$ cm (half-circumference) high with a crosspiece of 2.0 cm and sides 1 cm wide. One piece has a 1-mm strip removed at the end of the crosspiece to form the gap for connecting the drivers. The three pieces are cleaned and rolled to fit the former as before and the lengths of the side pieces are adjusted to leave a 1-mm gap on either side of the guard rings. The unit is assembled and is held together with a little contact adhesive or is tied with cotton. The whole coil is placed in its final position on the probe head and copper wire is cut and shaped to make the shortest connections to the tuning circuit. The tuning circuit should already be mounted directly next to the bottom of the coil and fixed to the plastic unit on which the glass former will be finally attached with Araldite (epoxy). The coil is then removed and the connecting wires are soldered in place as shown. The coil is then given a few coatings of acetate lacquer to fix the copper strips in place and give good protection from spillage of perfusate. The ends of the copper wire connectors are cleaned and the coil is mounted on the probe head with Araldite. After soldering the connections to the tuning capacitors the whole is given a light coat of lacquer to protect the capacitors and connectors from spillage. The coating should be very light so as not to increase the minimum capacitance of the tuning circuit.

The field generated in this coil is reasonably uniform across the whole coil and is constrained to the volume between the guard rings, i.e., between the open windows. The uniformity of the field generated in the coil can be checked with a probe formed from two turns of wire (inner cable, 5 mm in diameter) attached to the end of a length of coaxial cable connected to the two-channel high-frequency oscilloscope and sweep generator used to tune the probe. This field probe should be used frequently during the life of the coil because any breakdown of the dielectric coupling to the guard rings will result in a very inhomogeneous field.

Note that the glass former extends well below and above the active coil to allow the use of a sample chamber that has the top and bottom of its liquid content outside the guard rings, to minimize the effects of changes in volume on the field uniformity. This is particularly important for a moving system such as a beating heart. The use of good-quality copper for the coil and connections, and gold tuning capacitors, which are situated close to the coil, causes minimum disturbance to the field uniformity in 10-cm bore 200- to 500-MHz magnets such that linewidths of less than 50 Hz can be achieved for the water proton signal, well within the requirements of the indicator experiments where total fluorine chemical shifts of 4–6 ppm are used. For those not familiar with magnetic resonance spectroscopy, a linewidth of 50 Hz is equivalent to 0.1 ppm in a 500-MHz magnet. For example, in the superfusion system used in studies on cerebral $[Ca^{2+}]_i$ in a 500-MHz magnet (Section VI,C), a linewidth of the $^1H$ resonance of the water of the superfusing medium of <5 Hz increases to between 10 and 12 Hz in the presence of the tissues. The linewidth of the $^{19}F$ resonance (tuned to 470.5 MHz) is normally greater between 25 and 40 Hz, arising from the fluorine proton coupling.

*2. Physical Dimensions and Structure of the Probe for the 10-cm Magnet*

A diagram of the top end of the probe for servicing from below is shown in Fig. 3, with the coil mounted. Those parts shaded are made of aluminum or copper. The main structural part of the top end is a flat, hollow cylinder through which the main temperature control bath is circulated, being fed by wide-bore (>1 cm) PVC tubing 3 m in length to remove all the service equipment, peristaltic pumps, and water baths, etc., away from the strong field region of the magnet. The top end is connected to the base by four hollow tubes, of appropriate length to center the resonance coil in the magnet. These tubes, carrying the signal cable and other services, pass through the heating tank to allow them to reach the coil, as do three smaller tubes inserted in the tank to allow connection of the adjusting rods to the tuning capacitors. The radius of the tank is such that when the tight-fitting cover is placed over the whole assembly it fits snugly inside the room-temperature shim coils of the magnet. With the spinner housing still in the magnet, the space above the probe is limited. To minimize the chances of kinking the servicing connections to the sample as they fold over the side of the coil, the removable probe head cover incorporates an additional cylindrical space on its top that fits into the spinner housing. The top section of the cover is also made

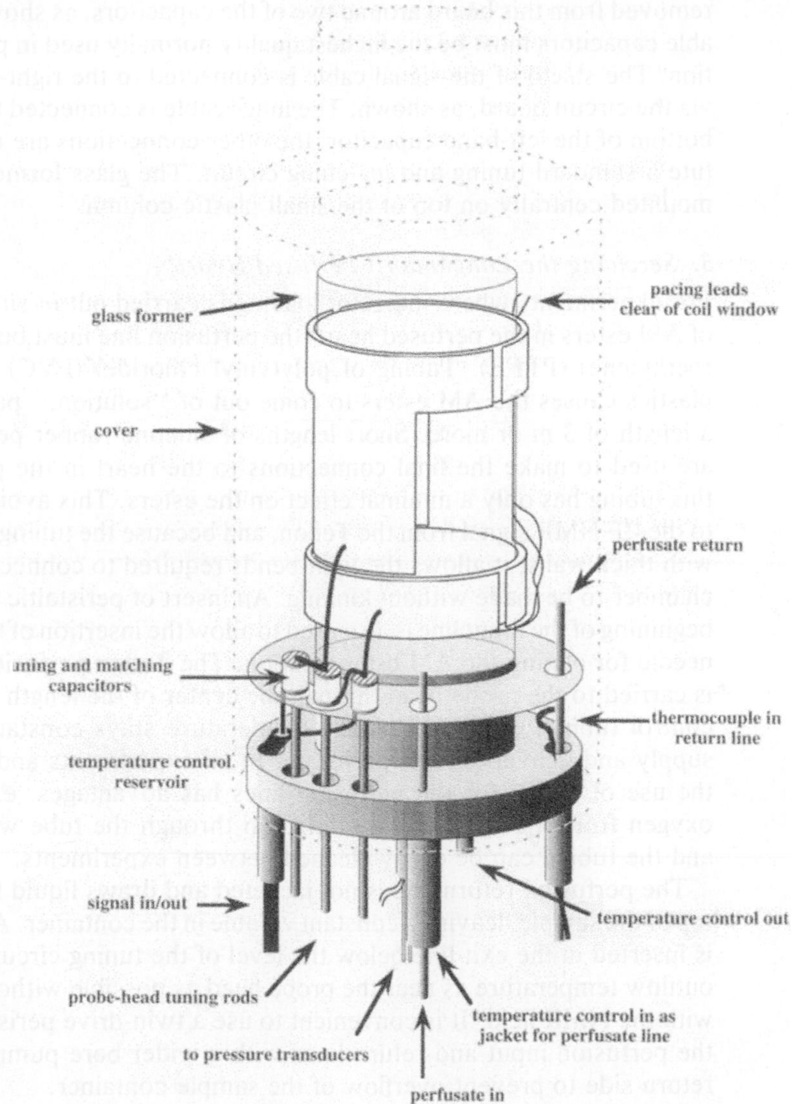


FIG. 3 Diagrammatic representation of the probe for the 10-cm magnet.

removable for ease of assembly when the heart is in place. The tubing requires careful positioning.

Mounted on a smaller plastic column on top of the tank is a ring of circuit board that carries the tuning and matching circuit. The signal conductor is removed from this board around two of the capacitors, as shown. The adjustable capacitors must be the highest quality normally used in probe construction. The shield of the signal cable is connected to the right-hand capacitor via the circuit board, as shown. The inner cable is connected to the insulated bottom of the left-hand capacitor; the other connections are made to constitute a standard tuning and matching circuit. The glass former of the coil is mounted centrally on top of the small plastic column.

### *3. Servicing the Langendorff Perfused Heart*

For experiments where indicator loading is carried out *in situ*, e.g., the use of AM esters in the perfused heart, the perfusion line must be poly(tetrafluoroethylene) (PTFE). Tubing of poly(vinyl chloride) (PVC) and other soft plastics causes the AM esters to come out of "solution," particularly over a length of 3 m or more. Short lengths of silicone rubber peristaltic tubing are used to make the final connections to the heart in the probe, because this tubing has only a minimal effect on the esters. This avoids interference to the $^{19}$F NMR signal from the Teflon, and because the tubing is very flexible with thick walls, it allows the tight bends required to connect to the sample chamber to be made without kinking. An insert of peristaltic tubing near the beginning of the input line is also used to allow the insertion of the hypodermic needle for adding the AM ester solution. The Teflon perfusion input tubing is carried to the probe head through the center of the length of temperature control tubing, ensuring that the temperature stays constant between the supply and delivery of the perfusate. In all experiments and for flexibility, the use of PTFE for the perfusion lines has advantages, e.g., the loss of oxygen from the perfusate by diffusion through the tube walls is minimal and the tubing can be easily cleaned between experiments.

The perfusion return line is not jacketed and draws liquid from above the top of the sample, leaving a constant volume in the container. A thermocouple is inserted in the exit line below the level of the tuning circuit to record the outflow temperature as near the probe head as possible without interference with the NMR field. It is convenient to use a twin-drive peristaltic pump for the perfusion input and return lines with a wider bore pump tubing on the return side to prevent overflow of the sample container.

For the heart experiments, where pacing is required, the leads from the stimulator are taken via a high-frequency rejection loop into the probe head but are kept clear of the windows of the transceiver coil by taping to a plastic post incorporated for that purpose. The leads cannot enter the coil so connection to the heart is made via two salt bridges constructed with narrow-gauge PVC tubing filled with 5 *M* sodium chloride in agar. The salt bridge ends are sutured to the left ventricle of the heart or are inserted to different depths into the right ventricle to achieve electrical contact (Fig. 4).

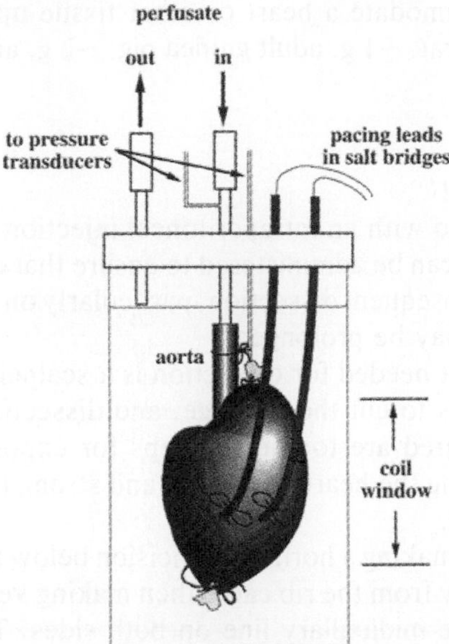


FIG. 4 Pacing the Langendorff perfused heart.

The perfusate is drawn by peristaltic pump from a large reservoir that is heated to the operating temperature by standing in a circulating water bath and is oxygenated with the appropriate gas mixture, normally 5% $CO_2$ in oxygen for the heart. It is essential that the gas equilibration is performed at the operating temperature to avoid subsequent bubble formation in the perfusion line. The flow from the peristaltic pump is passed through a coarse pad filter and then a 5-$\mu$m filter, both of which are duplicated with appropriate taps to allow easy replacement during a prolonged experiment. The filtered medium then passes through a bubble trap and via a short piece of silicone tubing into the PTFE line. The input line immediately enters the temperature control line to act as a water jacket maintaining the perfusate temperature over the 3 m to the probe head. The temperature control line is driven from the water bath circulator. The two connections to the pressure transducers, recording the perfusion pressure at the aorta entry and the pressure developed by the latex balloon inserted in the left ventricle, are made of hard-walled wide-bore catheter tubing and are completely filled with liquid to ensure minimal pressure loss over the 3-m length to the transducers. The pressures and changes at the transducers are recorded on a Gould bridge and differential recorder and tape drive for future study.

The perfusion chamber above has an internal diameter of approximately

30–33 mm and will accommodate a heart or other tissue up to 8 g. Heart sizes are as follows: adult rat, ~1 g, adult guinea pig, ~2 g, and adult ferret, ~6 g.

### *4. Preparation of the Heart*

The animal is anaesthetized with an intraperitoneal injection of pentobarbitone (250 mg/kg). Heparin can be administered to ensure that coronary blood clots do not form during subsequent dissection, particularly on larger animals, for which the procedure may be prolonged.

The minimum equipment needed for dissection is a scalpel for incision, a pair of heavy duty scissors to cut the rib cage, and dissection scissors for fine dissection. Also required are toothed forceps for exposing the heart, small flat forceps for holding the heart and aorta, and strong thread for tying the aorta onto the cannula.

The heart is exposed by making a horizontal incision below the diaphragm, cutting the diaphragm away from the rib cage, then making vertical incisions through the rib cage at the midaxillary line on both sides. The rib cage is held back so the heart can be excised. Care should be taken to leave adequate length of the great vessels to ensure cannulation. The excised heart is then placed in ice-cold perfusion solution (containing heparin) and gently agitated to remove blood from the ventricular cavities. The cooling will stop the heart contracting and minimize ischemic damage while extraneous tissue is removed to expose the aorta for cannulation. The pericardial sac is incised near the apex and peeled away to expose the pulmonary artery and aorta. When the aorta has been identified the excess tissue is removed. The aorta is slipped onto the cannula (glass or hard plastic with a slightly flanged end) with a low flow of perfusate (to displace air) and tied with the thread. Care must be taken to ensure that the cannula is not insected beyond the coronary artery branch points. The perfusion is started by raising the pump speed over a couple of minutes. The perfusion pressure in the ventricle closes the valve and the perfusate can exit via the coronary arteries only when the flow is increased to 4 ml/g.

The atria should be incised to ease drainage of the coronary effluent and can be removed almost completely, taking care not to damage the arteries near the atrio–ventricular junction. For experiments where pacing is required, it is convenient at this stage, when the heart has resumed beating, to crush the atrio–ventricular node, with a pair of blunt tweezers, to slow the heart rate to its minimum to allow well-synchronized capture by the pacing voltage pulse (10–20 msec, 10–20 volts, or three times the threshold for capture). The natural heart frequency is temperature dependent and we have used 30°C as our operating temperature for ferret and pig hearts to achieve a slow enough natural frequency to maintain capture of the heart beat with pacing in the range of 0.2 to 2 Hz.

The agar pacing electrodes are sutured in place as shown. The intraventricular balloon is inserted by tying a needle and thread to the end of the balloon and passing the needle through the cavities and out via the apex of the heart. The thread may then be used to tie the balloon in place. The balloon should be inflated to register a diastolic pressure of 5–10 mm Hg. This level will minimize the risk of endocardial ischemia while maintaining an adequate systolic pressure.

### 5. *Servicing Cell Suspension*

Two systems have been used for maintaining the viability of cells in suspension. The first successful system involved the use of bundles of dialysis fibers (ID 190 $\mu$m; OD 200 $\mu$m; cutoff, 1500 Da) to deliver the oxygen and nutrients and to remove the products of metabolism from the sample in the probe. Cell suspensions of up to 5–10% hematocrit can be maintained this way for some hours (Gillies *et al.*, 1983, 1984).

Preparation of the dialysis fiber bundle is a simple task (Fig. 5). The first step is to obtain a dry, use-by-date-expired kidney dialysis cartridge from the local hospital or suppliers. One end of the unit is removed by cutting the casing with a fine saw, taking care not to damage the fibers, and the fibers are cut with a sharp pair of scissors. The casing is then removed by cutting the other end with a saw and sliding it away without bending the dry fibers. The result is a bundle of fibers attached at one end from which the required number can be cut as required. A bunch of fibers is removed, around 50 for a 25-ml sample of cell, and without kinking fed through a pair of 2-cm pieces of heat-shrink tubing (from any electronics shop). The ends of the bundle are spread out and curable silicone rubber (clear bath sealant) is

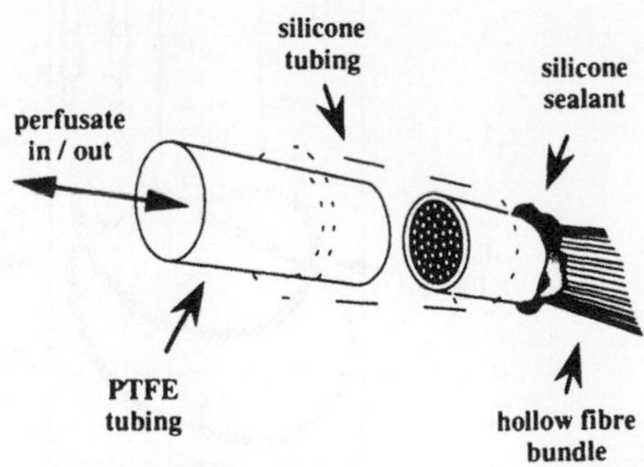


FIG. 5 The dialysis fiber bundle.

applied evenly near the fiber ends. The ends of the bundles are smoothed together and the heat-shrink tubing slid along to cover the silicone rubber. Heat is carefully applied to shrink the tubing, avoiding the fibers, most easily done by holding near a 5-mm gas flame from the end of a Pasteur pipette. After both ends have been treated the dialysis bundle is left to cure for 24 hr. After curing the bundles are cut through the middle of each of the seals with a razor blade or scalpel, to leave a clean surface with the ends of the hollow fibers open.

The dialysis bundle is connected to the in and out perfusion lines with suitable diameter soft tubing and tested for leaks by pumping. When the fibers are wet they can be bent easily to fit into the sample chamber (Fig. 6). Because the dialysis tubing is a continuous system the outflow does not need pumping. However, because the input pressure can drive water across the fibers and it is difficult to adjust the height of the sample above the end of the return line to counteract this flow, it is advisable to retain the pumped outflow tube as used in the heart system to prevent overflow of the sample during long experiments. The cells are kept in suspension by a suckblow reservoir of up to 5-ml capacity for a 25-ml suspension. The air-mediated drive to lift and drop the cells is provided most easily with a reciprocating

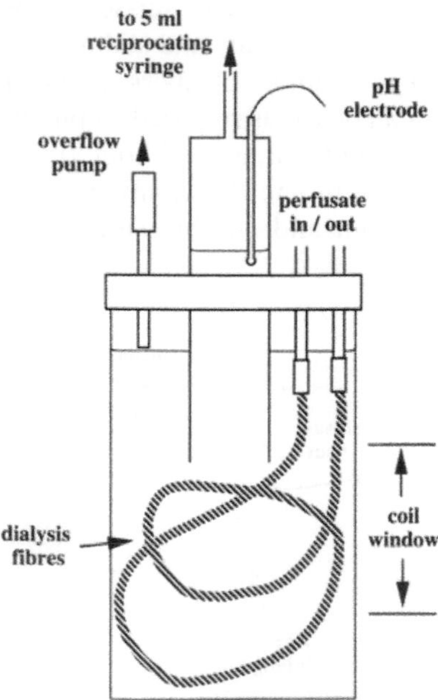


FIG. 6 The sample chamber.

syringe. The volume used has to be adjusted to minimize the field perturbation caused by the moving liquid, which is minimal if the moving surfaces are outside of the guard-ring coil. Any perturbation is easily seen on the free induction decay of the water signal used for shimming the magnetic field. An added bonus when using this stirring system is that a small pH electrode can be incorporated well outside of the coil being bathed with sample once every stir cycle. The efficiency of the dialysis fibers is thus monitored. The fibers will maintain the pH very efficiently if a $CO_2$/bicarbonate buffer is used.

The alternative method of maintaining the cells in suspension is to embed them in extruded agar to retain them in the chamber while superfusing (Chresand *et al.*, 1986). The system detailed below for the superfused brain slices is suitable in this case.

#### 6. *Adherent Cells*

These may be used in NMR experiments in either of two ways. The simplest method is to grow the cells on Cytodex beads maintained in suspension. Schanne *et al.* (1990) have used this methodology to measure the cytosolic free calcium and lead by $^{19}$F NMR in osteoblastic bone cells loaded with 5-FBAPTA. The cells-on-beads suspension was maintained in the magnet using a superfusion system similar to that applied to brain slices (Sections III,C and VI,C). The alternative is to grow the cells on fibers in small cylindrical units marketed for the collection of antibodies and other metabolites from cell culture (Gillies *et al.*, 1993). These units are packed longitudinally with hollow fibers that are a few hundred micrometers in diameter and are extremely porous, allowing molecules of several million daltons to pass freely. The dialysate medium is passed through the fibers and the cells grow around and into the interstices of the fibers. Effective hematocrits in the 1% region are attainable, but this system has not yet been used for $^{19}$F NMR studies.

### C. *Service from Above*

The superfusion system that has been applied to studies on brain slices (Section VI,C) is an example of servicing from above. The apparatus of Fig. 7 (Bachelard *et al.*, 1985) consists of a 25-mm NMR tube fitted with a Teflon [poly(tetrafluoroethylene)] insert plugged into the base of a long glass cylinder that contains the inlet and outlet tubing carrying the superfusing medium. The superfusion medium is warmed and gassed in a reservoir in a water bath a few feet from the magnet. The inlet tube is surrounded by a jacket through which warm water is circulated to maintain the temperature

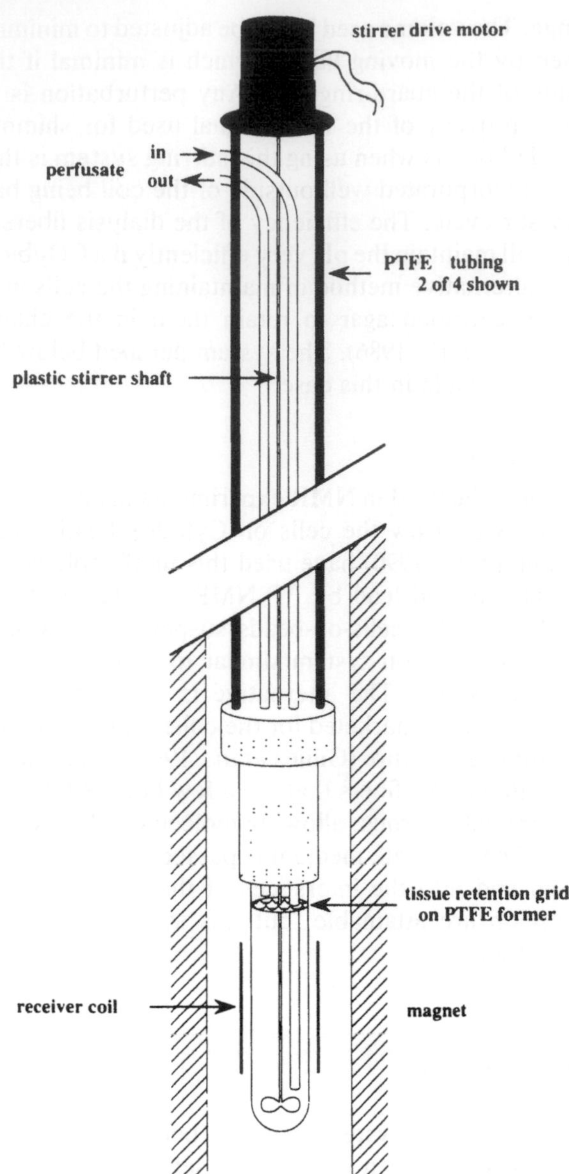


FIG. 7 Superfusion system for brain slices.

of 37°C in the NMR tube. The cylinder also carries the shaft of the glass paddle stirrer, operated by an electric motor at the top of the cylinder protruding above the magnet. The rate of stirring and the rate of superfusion (usually 25–35 ml/min) are adjusted in each experiment to maintain constant gentle movement of the tissues. The stirrer was found to be necessary to maintain adequate oxygenation of, and uptake of glucose to, the 4- to 5-g tissues. The tumbling movement of the slices has surprisingly little effect on homogeneity, as can be seen in the $^{31}$P spectra (Section VI,C), and metabolism can be maintained for 12 hr or more. This apparatus is appropriate for a wide-bore magnet, such as the Bruker WM 400 with a commercial 25-mm probe. For use in a smaller bore magnet (e.g., the Bruker AMX 500), a similar piece of equipment has a sample tube of 15-mm diameter (instead of the 25-mm tube of Fig. 7). The smaller volume has the advantage that much less brain tissue (0.5 g) can be used, which may be necessary for studies on parts of the brain such as the hippocampus. In this case we have found that the stirrer of Fig. 7 can be discarded—the tissues are maintained moving and tumbling nicely by the stream of perfusion fluid entering at rates of between 30 and 40 ml/min. Our experience suggests that 15 mm is the minimum size compatible with maintenance of adequate perfusion, and therefore metabolism, of the tissues.

Though the fluorine content of the Teflon insert is sufficiently far from the coil not to affect $^{19}$F measurements, we now use the non-fluorine-containing Delrin (an acetal polymer). The apparatus of Fig. 7 (or the smaller 15-mm version) is lowered from above into the bore of the magnet (the spinner is removed) easily and routinely as the collar of the Teflon or Delrin insert locates into the spinner seat of the shim-set within the bore.

## D. *Larger Aperture, Horizontal Magnet with Side Service*

In theory it should be possible to achieve a better signal-to-noise ratio for $^{19}$F NMR measurements of $Ca^{2+}$ by using a larger sample. At present the largest magnets available with high enough field strength (4.7 T) have a 40-cm bore. This is large enough to fit human hearts or even whole animals up to 30 kg of body weight for *in vivo* studies. Most larger magnet spectrometers have imaging capabilities but at present the signal-to-noise ratio obtainable with the $^{19}$F indicators is too low to make use of this.

### *1. The Transceiver Coil*

There are two types of transceiver used for studies in large magnets. The best review is by Styles (1991). J. I. Vandenberg and A. A. Grace in Cambridge have used birdcage resonators (Hayes *et al.*, 1985) and surface coils for measurements of calcium in Langendorff perfused pig hearts and more recently in pig hearts *in vivo*.

For simple spectroscopy measurements surface coils are easier to make and use, consisting of a simple high-quality copper loop coil with tuning capacitors. The coil and capacitors are coated with silicone gum for protection and are placed on the tissue in the magnetic field so that movement is minimized. For the beating heart it is best placed between the ventricle and the supporting stretched sheet of latex rubber. There are limitations involving the depth of signal penetration of the tissue, and the relevant review (Styles, 1991) should be seen on this point. In theory, birdcage resonators and other similar coils have higher sensitivity and should be used if available.

A representation of a birdcage coil is shown in Fig. 8. The resonator is constructed from high-quality 0.07-mm-gauge copper strips 1 cm wide by 20 cm long, and two rings of copper are assembled on a plastic tube former of suitable diameter (~15 cm). The strips are soldered to one ring and joined to the other through the high-quality gold tuning capacitors. The whole birdcage is enclosed in a tube of thicker gauge copper sheet of the same length and ~5 cm larger diameter connected to the drive shield. The resonator is driven through a tuning capacitor as shown.

### 2. *Servicing the Heart*

The perfusion rig for large hearts is essentially the same as described above for ferrets. The main changes are in size of tubing, and it is convenient to use gravity to supply the perfusion pressure (1–1.5 m head). Flow rates between 2 and 4 ml/g are sufficient for a pig heart.

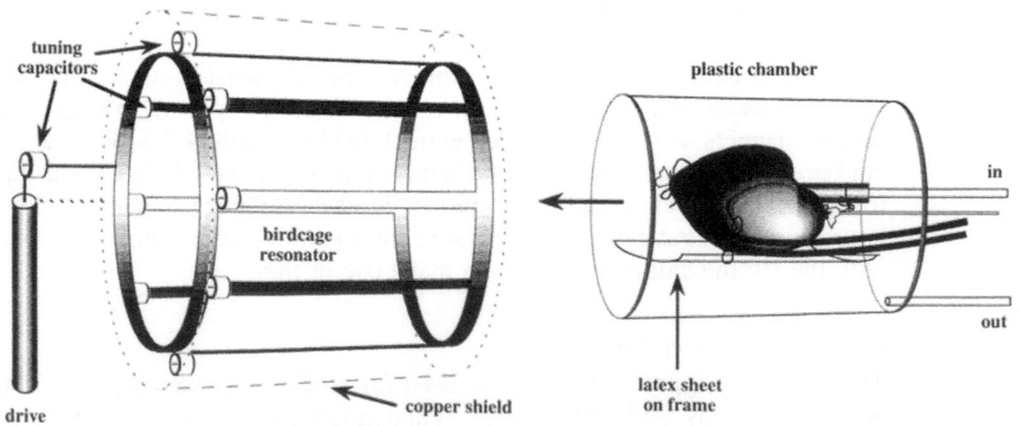


FIG. 8 Diagrammatic representation of a birdcage coil.

The major change required when working with larger animals is that it is best to intubate the animal to ensure adequate oxygenation throughout the longer operation (10–15 min). The excision of larger animal hearts is similar to that above except heavier duty shears are required to open the ribcage and retractors are also required. To avoid major blood spillage it is preferable to make a midsternum incision rather than midaxillary, because the latter will transect the intercostal arteries.

To avoid problems of reduced field penetration into the tissue, the heart is not floating in perfusate but is centered in the chamber sitting on a sheet of stretched latex, with the area of most interest, i.e., the left ventricle, underneath to minimize movement in the magnetic field. The effusate is withdrawn from the bottom of the chamber by suction, taking care to minimize the suction of cold air over the surface of the heart because this can impair function. The other requirements are as for the smaller heart.

### 3. *Work on Intact Animals*

The major consideration for these experiments is an anesthetic one. The anesthetic must not contain fluorine and must be minimally perturbing to the system under study. This is beyond the scope of this chapter.

## IV. Loading the Indicators

The acetoxymethyl esters of the indicators are designed to be hydrophobic and permeable to cell membranes and hence are virtually completely insoluble in aqueous medium. For successful loading of the indicators into the cytoplasm, the acetoxymethyl esters have to be added at about 10–25 $\mu M$ concentration to the medium the cells are suspended in or the tissue is being perfused with. This is achieved by dissolving the esters in dimethyl sulfoxide (DMSO) at 10 to 50 m$M$ and adding this solution to the medium (1000–5000 volumes) as a slow stream from a hypodermic syringe with a good flow of medium over the end of the needle. If the ester solution is added this way to a buffer that is smoothly stirred without vortex, an opalescent "solution" (colloidal suspension) results that is stable for some hours. If the solution is bubbled with a gas (oxygen) or even stirred too vigorously, the esters are immediately precipitated and are no longer available for uptake into the cells or tissue. These problems can be partially overcome by the addition of 10% by weight of a surfactant, Pluronic F-127 (Molecular Probes, Inc., Eugene, Oregon), to the DMSO solution of the esters. The addition of surfactants to cellular systems introduces further possible perturbation to the cells, such as altered membrane permeability.

## A. Cell Suspensions

The loading of the indicators into a suspension of free floating cells or cells on beads is the simplest to do. The cells will absorb the hydrophobic acetoxymethyl esters into their plasma membranes if the DMSO solution is added directly to the cells as opposed to predilution. This prevents the precipitation of the esters when used at concentrations at the limit of "solubility."

In most cell types the indicators will load under normal conditions of pH and temperature. The pH of maximum activity of the cytoplasmic esterases is around 8.4 and the rate drops off with acidification until virtually no activity is present at less than pH 7. The indicators have a tendency to stimulate the production of lactate and will in some instances cause a lowering of the intracellular pH and subsequently inhibit the esterase activity. This is best overcome by raising the extracellular pH to a level the cells are known to withstand. In one instance, namely, the loading of muscle cell line, it was found that removal of the source of lactate, i.e., glucose, was required to achieve reasonable indicator loading.

After incubating the cells with the esters for a suitable period, in most instances around 1 hr, the cells are recovered by centrifugation and washed well to remove the side products, particularly the formaldehyde. After a short further incubation, normally the time it takes to put the cell in the magnet, shim the field, and start the spectral acquisition, the cells are ready for use.

## B. The Langendorff Perfused Heart

To deliver oxygen at a rate sufficient to maintain function, the perfused ferret heart, as set up above, requires a minimal perfusate flow rate of 4 ml/min/gram wet weight. For an average size ferret heart this gives a total flow rate of between 20 and 30 ml/min. Because the esters in "solution" cannot be bubbled with gas or suffer other mechanical disturbances, the DMSO solution has to be added to the perfusion line just before its entry into the temperature control jacket. This is best achieved by inserting a hypodermic syringe needle into a short piece of silicone rubber peristaltic tubing inserted in the line. The syringe is driven automatically by a variable-speed insulin pump at around 10 $\mu$l/min to give greater than 2000-fold dilution into the perfusate, with 25 $\mu M$ maximum final concentration. The perfusate is delivered via tubing of approximately 2 to 4 mm internal diameter, and flow rates of the order of 1 m/min, with very little turbulence, sweep the DMSO solution smoothly from the syringe needle tip. Any deposition of AM ester caused by too rapid or turbulent addition can be easily seen near the tip of the needle. The presence of in-line connections after the infusion point should be avoided because the ester will tend to collect around the edges of the connectors. We have always allowed the return perfusate to go to waste during the loading procedure and for a while after the infusion ceases. Between experiments the whole perfusion apparatus should be rinsed through with water followed by ethanol to remove any deposits of ester and render the apparatus sterile.

## C. Neuronal Tissue

Preliminary experiments using $^3$H-labeled indicators can be valuable in determining the optimal conditions for loading. For the superfused brain slices described in Section VI,C, it is essential to minimize any period of anoxia, ischemia, or change in pH, because irreversible metabolic damage may result, thus rendering any studies relating $Ca^{2+}$ to function invalid. Therefore the indicator must be loaded under optimal conditions for ensuring full metabolic viability (adequate supplies of oxygen and glucose, and maintenance of pH between 7.0 and 7.4). In practice the tissues are incubated (at least 20 ml/g fresh tissue) in the normal media required for metabolic studies (containing NaCl, 124 m*M*; KCl, 5 m*M*; $KH_2PO_4$, 1.2 m*M*; $MgSO_4$, 1.2 m*M*; $CaCl_2$, 2.4 m*M*; $NaHCO_3$, 26 m*M*; and glucose, 10 m*M*) with constant gentle swirling and gently bubbled with the appropriate oxygen mixture (typically $O_2/CO_2$, 95/5%) at 37°C. The tetra ester (50 or 100 m*M* in DMSO) is delivered to a final concentration of 100 $\mu M$ from a Hamilton syringe in periodic aliquots of 10 $\mu$l. The gas is removed for each addition and immediately after the 10 $\mu$l has been injected into the preparation, gassing is reintroduced with swirling. Studies using the $^3$H-labeled indicator showed that a further 20-min incubation is required for deesterification to yield the free tetra anion, which is almost entirely in the cell cytoplasm, with little detectable in organelles. The final concentration within the tissue was calculated to be between 100 and 140 $\mu M$ (Bachelard *et al.*, 1988; Badar-Goffer *et al.*, 1990). The loading is therefore necessarily time-consuming; any effects on tissue viability are routinely monitored by $^{31}$P spectra. In our hands the $^{19}$F spectra can be followed for 4–5 hr, at which time the signal-to-noise ratio has deteriorated due to slow wash-out of the indicator.

## D. Whole Animals

Little $^{19}$F NMR indicator work has been done on intact animals, although the pH indicator FQuene, related to 5-FBAPTA, has been used in rat liver (Beech and Iles, 1991). Recently, as this review went to press, Ackerman *et al.* (1995) reported success in loading 5F-BAPTA into rat brain by slow direct intraventricular infusion of the tetraester. The loading of indicators into whole animals has some special considerations, the most serious of which is the presence of blood esterases that will hydrolyze the acetoxymethyl esters before they reach the cytoplasm and prevent their efficient uptake into the organ of interest. Future success in experiments *in vivo* may depend on the development of specific delivery vehicles analogous to those being developed by the pharmaceutical industry.

To achieve loading of indicators it is necessary therefore to infuse the solution of ester into an artery just prior to its entry into the organ of interest. The above considerations relating to solubility and the final concentration in the perfusing medium apply *in vivo*, although somewhat higher concentrations may be possible and required because the esters will bind to the serum albumin and blood cell membranes. Some knowledge of the blood flow through the delivery artery is therefore required to choose a reasonable infusion rate for the DMSO solution of the esters. For larger animals, in which the artery can be reached only with a catheter, e.g., the coronary artery for heart loading, the use of an angioplasty catheter is recommended; however, because the plastic used for this equipment is incompatible with DMSO, the ester solution should be delivered with a 20-gauge Teflon syringe needle of appropriate length (Aldrich) fed through the catheter. The destination of the esters and their hydrolysis products can be followed by using tritium-labeled 5-FBAPTA AM ester and using chloroform methanol extraction and water partitioning to analyze the tissues postmortem.

## V. Resonance and Signal Acquisition

### A. *Tuning the Probe Head and Magnet Shims in Situ*

Because all these experiments are working at the detection limit of the spectrometer, next to the biological requirements, tuning and shimming are probably the most important steps of the procedure. It is therefore most convenient and reliable to tune the loaded probe to the correct $^{19}F$ resonance frequency, *in situ*, using a sweep generator coupled to the probe and a two-channel high-frequency oscilloscope via a reflection bridge. The resonance coils are large and hence are affected by the servicing tubing, the casing, and other objects in their vicinity, thus it is always preferable to carry out the tuning with the sample in place in the magnet. This process takes only a few minutes. The frequency to tune the probe to is 94.1% of the proton-operating frequency of the spectrometer.

The uniformity of the magnetic field, and hence the sharpness of the resonance lines are maximized by shimming the magnet using the free induction decay of the proton signal from the water in the sample. The proton concentration in the sample is extremely high (110 *M*) and will thus give sufficient signal to observe, even though the coil is detuned to 94.1% of the correct proton-operating frequency. Because other resonance frequencies may be present there is no guarantee that the field distribution generated by a proton frequency applied to the coil, when tuned to the fluorine frequency, is that generated by the fluorine frequency. It is therefore advisable to carry out an experimental shimming of the magnetic field on a sample of 200 m*M* sodium fluoride, or preferably trifluoroacetate, using both the proton and the fluorine resonances to check if they respond to shimming in the same manner. Better still, the field uniformity on the fluorine signal can be checked with an NMR imaging system if available; this is particularly useful when constructing purpose-built and/or large coils.

The magnet shimming process needs to be performed manually as rapidly as possible because it occurs at the beginning of the experiment. This does take some time even when the operator is practiced, particularly with a beating heart in the coil, and is probably the longest step of setting up any experiment. Hence a good starting point is required, usually the best previous experiment. In general, with the indicators in use at this time, the linewidths of the signals arising from the intracellular fluorine resonances are of the order of 150 to 200 Hz. This means that once the signal for the water protons has been reduced by shimming to 60 to 70 Hz (less than about a third of the fluorine linewidth), one is well into the region of diminishing returns, because any improvements will be achieved only slowly and little difference will be seen between spectra of 150- (the minimum) and 200-Hz linewidth (see comments on linewidth above).

## B. Acquisition Parameters

### 1. The 90° Pulse for the Coil and Relaxation Time for the Loaded Indicator

The 90° pulse for the loaded coil is measured on a 200 m*M* sample of trifluoroacetate of the same volume and in the same container to be used in the experiment; it is advisable not to use much higher concentrations because the high ionic strength can affect the electrical properties of the coil. The need to allow a time between pulses (many minutes) that is much longer than the relaxation time ($T_1$) of the sample precludes the use of a weaker signal. Simply, the applied pulse width is varied from zero until the signal observed passes through a maximum and finally becomes inverted. A plot of signal strength against pulse width will find the exact point at which the signal is zero, the 180° pulse. The maximum signal occurs at half this value, the 90° pulse. This procedure gives a value well within the experimental requirements.

As a result of the low signals obtained in biological experiments and the slow leakage of indicator from cells, it is not possible to obtain a value for the relaxation time of the intracellular indicator under the exact experimental conditions required. The simplest procedure to estimate the relaxation time involves the gross overloading of the indicator, more than enough to completely inhibit heart function, and application of a two-pulse sequence separated by a variable time and a fixed data acquisition time. From the angle of the pulse width and the variation of the signal with time, the $T_1$ can be calculated. For the perfused Ferret heart, the $^{19}F$ $T_1$ is around 700 msec.

For the indicators that are in slow exchange, the dissociation rates, the relaxation times, and the pulse and acquisition rates are important considerations. Because the calcium measurements are dependent on two signals that may have differing relaxation times, hence the acquisition of data at a rate that does not allow the complete relaxation of the magnetization between pulses could lead to preferential saturation of one of the signals and distort their ratio and the calcium value reported. For the fluorinated indicators the dissociation rate is at least 10 times faster than the relaxation rate measured in the heart. A direct consequence of this is each molecule undergoes switches from the bound to free state many times during the relaxation period and thus any signal saturation will be equilibrated between the two states and distortion of the ratios will not arise. This has been confirmed by us and others by varying the pulse rate for data acquisition when partial saturation is present.

### 2. *Static Systems*

These are the simplest systems for data collection. After finding the 90° pulse width, tuning and matching the coil, and shimming the magnetic field, the data collection can begin. Sensitivity is the main concern because we need to keep the concentration of indicator at levels that will interfere the least with any changes that are being measured. We need to obtain the maximum signal-to-noise ratio in the shortest possible time. This is normally achieved by a pulse rate of around 5 to 10 Hz, very much shorter than the relaxation time, and a pulse angle of 15 to 30°, adjusted to minimize the signal saturation. The time taken to reach a signal-to-noise ratio of 10 : 1 varies with the cell concentration and indicator loading, from around 2 to 3 min for cultured murine mast cells at 5% hematocrit with 5 m*M* intracellular indicator, to 15 min for pig mesenteric lymphocytes with much less than 1 m*M* internal concentration and higher hematocrit (10 to 15%). In the brain slice superfusion system, with its limitations on the amounts of indicators that can be loaded without metabolic damage, at least 5 min is required to achieve a signal-to-noise ratio of 10:1 in the first spectra acquired after loading; subsequent spectra may take longer because the indicator slowly washes out of the tissue.

### 3. *Cyclic Processes Linked the Data Acquisition*

All NMR spectrometers have facilities for output of the low DC voltage used to control the rf pulses. For the experiments on the perfused heart it is simplest to collect data that are averaged over the cardiac cycle, with no coupling to the pacing, because there is an order of magnitude difference in the optimal data collection frequency (5 to 20 Hz) and that of the cardiac cycle (0.5 to 2 Hz). To gain the more valuable information on cytosolic calcium levels at various stages of the cardiac cycle the heart stimulus is coupled to every *N*th rf pulse (Fig. 9). For studying the calcium during the whole cycle we have used values of *N* from 8 to 32 with cyclical collection of the NMR data in *N* separate bins corresponding to the *N* stages of the cardiac cycle (Harding *et al.*, 1993). Illustrated in Fig. 9 are the pulses used to produce the 16 point data in Section VI,B (see later, Fig. 15). The shortening of the acquisition time to approximately 50 msec and the consequent reduction in the rf pulse (~22°) required to produce the data necessitate a considerably longer total time for data collection. If one is interested only in the calcium level at a fixed time in the cycle, e.g., diastole, it is convenient to use the same regime but to deliver only one longer rf pulse to the heart during each cycle, allowing much longer for relaxation of the magnetization, and to collect the signal (free induction decay) over the short time period of interest (40° and 55 msec, respectively for the data in Section VI,B (see later, Fig. 14).

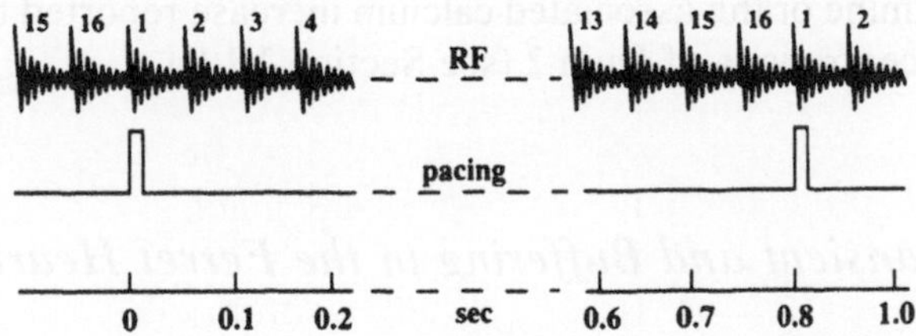


FIG. 9 Pacing the cardiac cycle. The top trace represents the train of free induction decays from the heart produced at a spectrometer pulse rate of approximately 18 Hz (30-$\mu$sec pulses at 55-msec intervals). The free induction decays are collected in 16 separate bins, and the 17th pulse is relayed to trigger the heart stimulator and recycle to the first collection bin. The computer takes a few milliseconds to operate the controlling macro, so times are approximate.

## VI. Selected Results

### A. *2H3 Cells, Ligand-Stimulated Histamine Release*

In NMR terms, these results were the simplest and quickest to produce because the cells were extremely tolerant to high intracellular concentrations of 5-FBAPTA (~5 m*M*), loaded as in Section IV,A but using a low cell density for the loading at an AM ester concentration of around 25 $\mu M$. The low cell density ensured a relatively large total quantity of AM ester to achieve high intracellular indicator concentration. 2H3 Cells are a murine mast cell line that can be grown in reasonably large numbers. After the many cycles of culture required to produce sufficient cells to fill the 25-ml NMR chamber at around 5 to 10% hematocrit, the cells show increased differentiation, reflected in higher histamine granule content. The cells have surface Fc receptors for IgE, and like other murine mast cells they can be primed with IgE directed against a specific antigen. Cross-linking of the IgE with antigen-labeled protein, or of the Fc receptors directly with the lectin concanavalin A, leads to degranulation and release of histamine. The histamine release is readily blocked by zinc in the medium. The time course of these effects on the $^{19}F$ NMR spectra and the accompanying reported changes in cytosolic calcium $[Ca^{2+}]_i$ and zinc are shown in Figs. 10, 11, and 12, respectively. It is noted that the presence of zinc salts in the medium prevents both the rise in intracellular calcium and the release of histamine.

Because the appearance of the zinc-bound form of 5-FBAPTA does not arise until much later than the time course of histamine release, it would appear that the action of zinc is extracellular, unless the intracellular site of action of calcium in histamine release has a much higher affinity for zinc relative to calcium than does 5-FBAPTA. It is noted that the presence of high intracellular 5-FBAPTA did not have a deleterious effect on the release of histamine or the associated calcium increase reported by lower intracellular concentrations of Quin 2 (see Section VI,B).

### B. *Calcium Transient and Buffering in the Ferret Heart*

In our earlier $^{19}F$ NMR studies of the perfused ferret heart (Kirschenlohr *et al.*, 1988) and rat heart (Metcalfe *et al.*, 1985) with 5-FBAPTA, we had found that the heart contractions and relaxation were particularly sensitive to the loading of the indicator (Fig. 13). The amount of cytosolic indicator required to give an adequate signal-to-noise ratio was around 100 $\mu M$ and this resulted in a loss of most of the strength of contraction. The loading

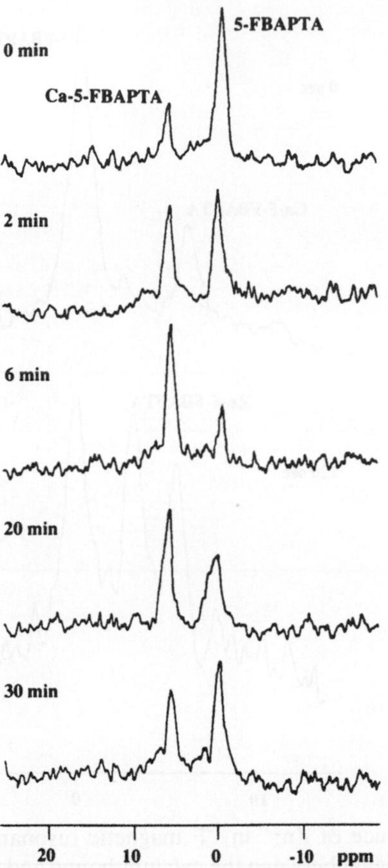


FIG. 10 The $^{19}F$ magnetic resonance spectra showing the time course of the increase in $[Ca^{2+}]_i$ in 2H3 cells stimulated with antigen at time zero.

time and conditions had to be very carefully controlled to ensure the maximum degree of reversibility of the deleterious effect (Harding *et al.*, 1993).

The average cytosolic calcium levels reported by the 5-FBAPTA were much higher than those found by other means—ion-selective microelectrodes and fluorescent indicators. The possibility that these effects of the indicator were related to its associated calcium buffering power prompted the search for a more spectroscopically sensitive $^{19}F$ NMR indicator. During these studies an indicator with higher calcium affinity was prepared, DiMe-5-FBAPTA (Section II,C). The results with this new indicator in the ferret heart did indeed confirm that the calcium buffering was the major contributor to the effects (Kirschenlohr *et al.*, 1993). The indicators had similar effects

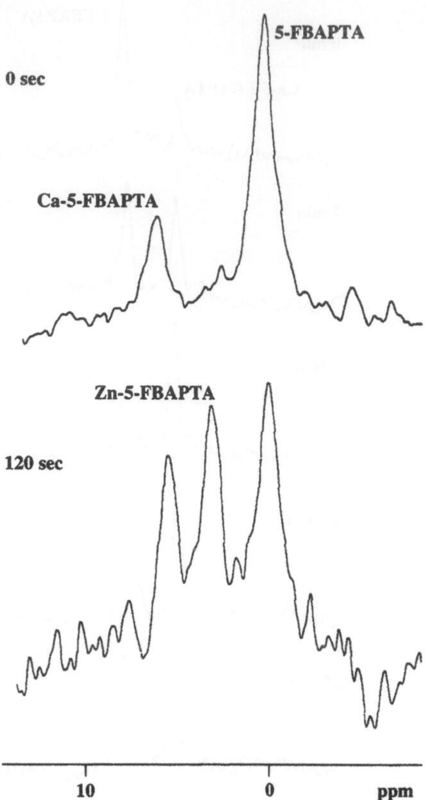


FIG. 11 Appearance of $Zn^{2+}$ in $^{19}F$ magnetic resonance spectra of 2H3 cells. The zinc-5-FBAPTA signal between the calcium-bound and free signals appears only after prolonged incubation in the presence of zinc, not seen in Fig. 10.

on the contraction, but the reported calcium levels, average and diastolic (Fig. 14a), and through the cardiac cycle (Fig. 15), were much lower with the higher affinity indicator. The higher affinity indicator also allowed a much more detailed study of the effects of a calcium channel blocker, diltiazem, on the cytosolic calcium levels, because these are out of the titration range of 5-FBAPTA (Fig. 14). These studies showed that even in the presence of total blockade of the calcium channels, the application of pacing to the heart caused a rise in intracellular free calcium (Fig. 14c and d), which we have interpreted as possibly arising from the continuation of electrically stimulated sodium influx and consequent rise in calcium via the sodium/calcium exchanger.

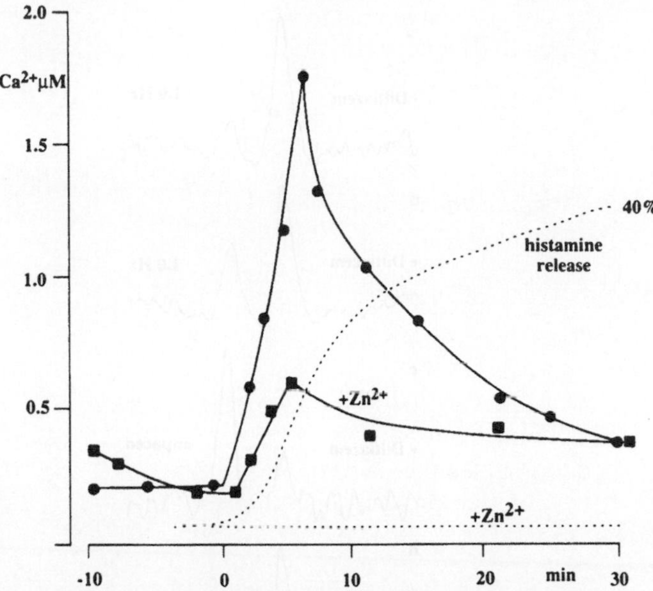


FIG. 12 Changes in $[Ca^{2+}]_i$ with the release of histamine in 2H3 cells. The changes in $[Ca^{2+}]_i$ ($\mu M$, solid line) and the histamine release (%, dotted line) are shown during control conditions and after addition of antigen stimulus at time zero. When the experiment was repeated in the presence of zinc ($+Zn^{2+}$), the large rise in $Ca^{2+}$ is largely prevented, as is the histamine release. No $Zn^{2+}$-5-FBAPTA signal is seen over this time course.

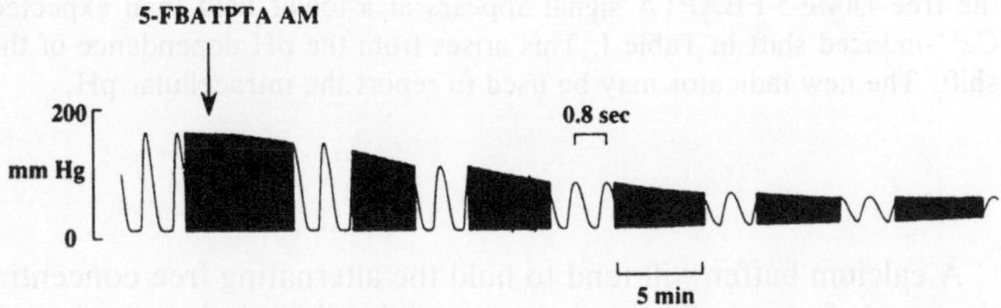


FIG. 13 Effects of 5-FBAPTA on heart contractions. The pressure trace is from the balloon catheter inserted in the left ventricle during loading of 5-FBAPTA into the heart. Note the decrease in developed pressure and the prolongation of relaxation time.

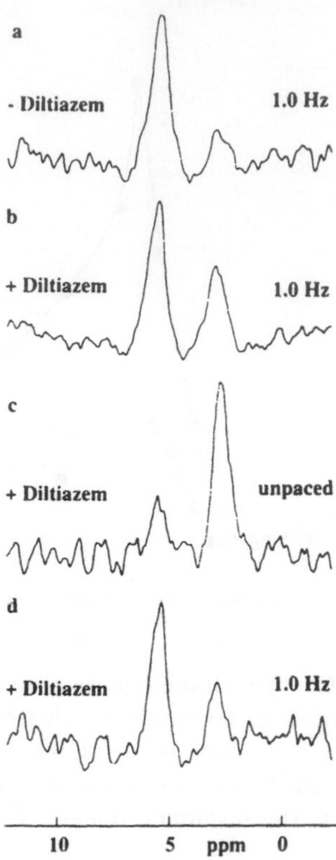


FIG. 14 Use of DiMe-5-FBAPTA and the effects of diltiazem. The spectra show the effects of diltiazem and pacing on the calcium levels in the perfused ferret heart loaded with DiMe-5-FBAPTA. The chemical shifts are given relative to 5-FBAPTA at 0 ppm. The bound forms of both indicators appear at the same shift. Note that the free DiMe-5-FBAPTA signal appears at a lower field than expected from the $Ca^{2+}$-induced shift in Table I. This arises from the pH dependence of the chemical shift. The new indicator may be used to report the intracellular pH.

A calcium buffer will tend to hold the alternating free concentration near the point of most buffering, i.e., around the dissociation constant. The lower overall cytosolic calcium seen with the higher affinity indicator is also associated with a much improved long-term recovery of the heart function on loss of the intracellular indicator by leakage over time. The range of indicators now available (Section II) with varying affinities opens the possibility of measurements of the buffering capacities of the heart cytosol and its relation to the strength of contraction.

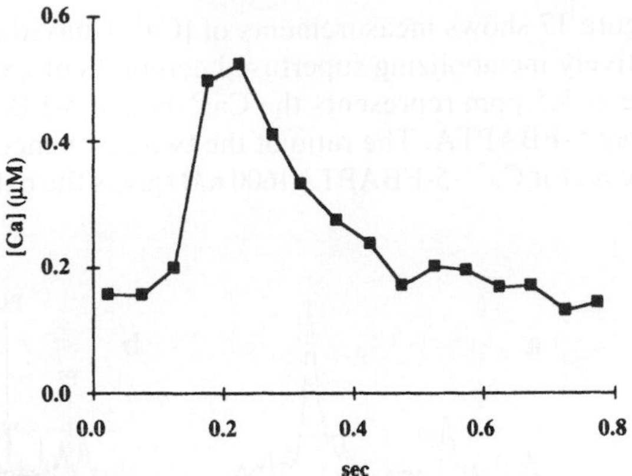


FIG. 15 Use of DiMe-5-FBAPTA in studies on the cardiac cycle. The intracellular calcium levels were found on applying the regime in Fig. 9.

## C. *Studies on Calcium in Superfused Brain Slices*

As noted in Section V,B,2, the indicator slowly washes out of the tissues (Fig. 16). The signal-to-noise ratio becomes limiting after about 4 hr, but the calculated value for $[Ca^{2+}]_i$ remains relatively constant, thus confirming that the value is not dependent on the amount of indicator present (Badar-Goffer *et al.*, 1990). Similar results were obtained for the perfused heart (Harding *et al.*, 1993).

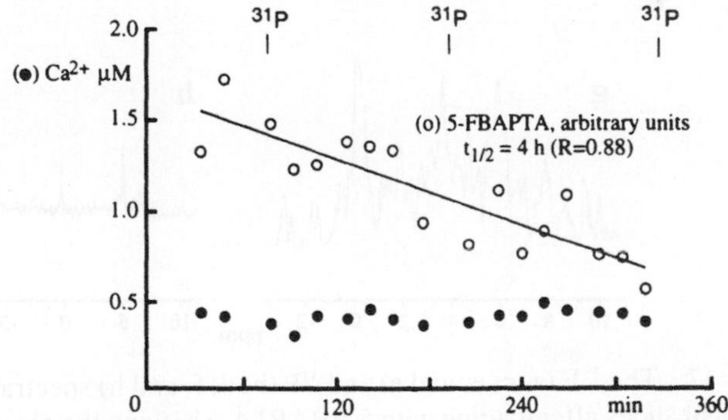


FIG. 16 Time course of washout of total 5-FBAPTA ($Ca^{2+}$-bound plus free) and calculated $[Ca^{2+}]_i$ in superfused cerebral slices. $^{31}$P gives the times when interleaved $^{31}$P spectra were used to check the energy state, which remained normal throughout.

Figure 17 shows measurements of $[Ca^{2+}]_i$ based on the use of 5-FBAPTA in actively metabolizing superfused cerebral cortex slices at 37°C. The resonance at 5.5 ppm represents the $Ca^{2+}$-bound 5-FBAPTA and that at 0 ppm, the free 5-FBAPTA. The ratio of the two resonances (bound/free) multiplied by the $K_d$ for $Ca^{2+}$-5-FBAPTA (600 n$M$) gives the calculated $[Ca^{2+}]_i$ (normally

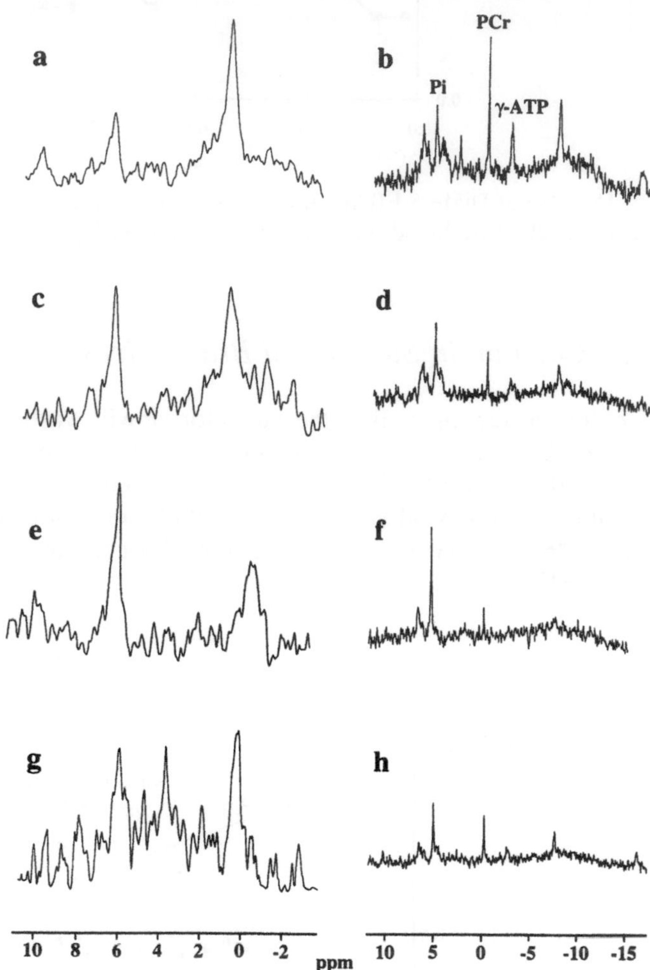


FIG. 17 The $^{19}F$ (a, c, e, and g) and $^{31}P$ (b, d, f, and h) spectra of superfused cerebral cortical slices after loading with 5-FBAPTA, showing the changes from control (a, b) caused by depolarization (c, d), decreased glucose and oxygen (e, f), and exposure to 1 m$M$ glutamate (g, h).

in the range 100–200 n*M*). Increases in $[Ca^{2+}]_i$ are caused by depolarization (Fig. 17c), after decreases of both glucose and $O_2$ in the medium (Fig. 17e) or after exposure to an excitotoxic concentration of glutamate (Fig. 17g). However, the resonance attributable to $Zn^{2+}$-5-FBAPTA (at 3.5 ppm) is detectable only after exposure to glutamate. The effects of these insults on the energy state are demonstrated (Badar-Goffer *et al.*, 1993, 1994) by the

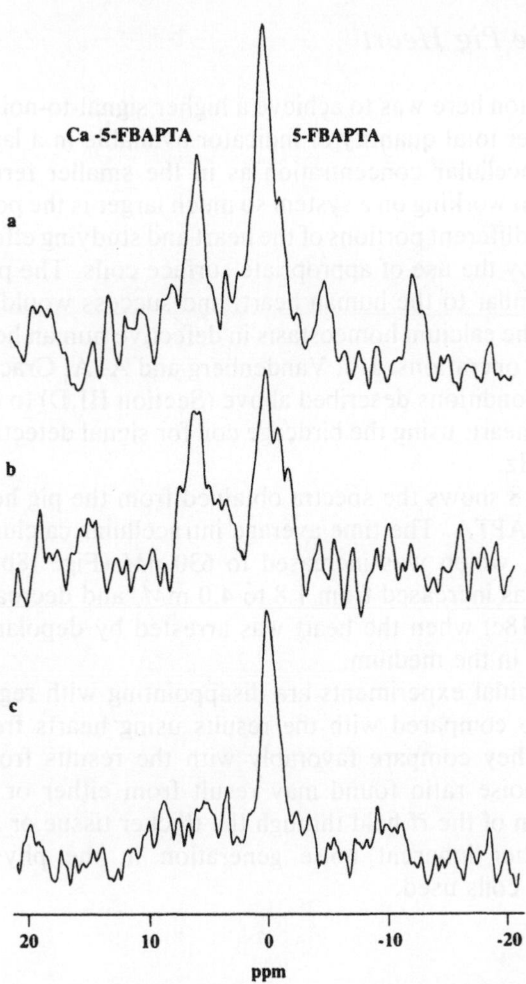


Fig. 18 Effects of (b) increased external calcium and (c) depolarization (30 m*M* KCl) on $[Ca^{2+}]_i$ in perfused pig heart. (a) Control spectrum with 1.8 m*M* external calcium.

decreases in phosphocreatine (PCr) and ATP in the interleaved $^{31}P$ spectra (Fig. 17d, f, and h). In such studies, if too much 5-FBAPTA is loaded (a temptation to achieve an improved signal-to-noise ratio), this is clearly shown by an inability of the PCr to recover on return to control conditions after depolarization, although the presence of the indicator had no perceptible effect on control PCr (Bachelard *et al.*, 1988).

### *D. Calcium in the Pig Heart*

The intention here was to achieve a higher signal-to-noise ratio based on the much larger total quantity of indicator available in a larger heart, using the same intracellular concentration as in the smaller ferret heart. An added bonus from working on a system so much larger is the possibility of acquiring data from different portions of the heart and studying effects such as localized ischemia by the use of appropriate surface coils. The pig heart is also more closely similar to the human heart, and success would lead to a method of studying the calcium homeostasis in defective human hearts removed during transplant operations. J. I. Vandenberg and A. A. Grace in Cambridge have used the conditions described above (Section III,D) to load 5-FBAPTA into a porcine heart, using the birdcage coil for signal detection at a field strength of 188 MHz.

Figure 18 shows the spectra obtained from the pig heart, at 34°C, loaded with 5-FBAPTA. The time average intracellular calcium was 310 ± 50 n*M* (Fig. 18a), which was increased to 630 n*M* (Fig. 18b) when the external calcium was increased from 1.8 to 4.0 m*M*, and decreased to less than 100 n*M* (Fig. 18c) when the heart was arrested by depolarization with 30 m*M* potassium in the medium.

These initial experiments are disappointing with regard to the signal-to-noise ratio compared with the results using hearts from smaller animals, although they compare favorably with the results from brain slices. The signal-to-noise ratio found may result from either or both the decreased penetration of the rf field through the thicker tissue or more likely from the much higher inherent noise generation in the physically much larger tranceiver coils used.

## References

Ackerman, J. J. H., Hotchkiss, R., Hsu, C. Y., Li, Y., Neil, J., Shornack, P., and Song, S-K. (1995). *SMR Workshop*, (*Woods Hole, Mass.*) p. 14.

Bachelard, H. S., Cox, D. W. G., Feeney, J., and Morris, P. G. (1985). *Biochem. Soc. Trans.* **13,** 835.

Bachelard, H. S., Badar-Goffer, R. S., Brooks, K. J., Dolin, S. J., and Morris, P. G. (1988). *J. Neurochem.* **51,** 1311.

Badar-Goffer, R. S., Ben-Yoseph, O., Dolin, S. J., Morris, P. G., Smith, G. A., and Bachelard, H. S. (1990). *J. Neurochem.* **55,** 878.

Badar-Goffer, R. S., Thatcher, N. M., Morris, P. G., and Bachelard, H. S. (1993). *J. Neurochem.* **61,** 2207.

Badar-Goffer, R. S., Morris, P. G., Thatcher, N. M., and Bachelard, H. S. (1994). *J. Neurochem.* **62,** 2488–2491.

Beech, J. S., and Iles, R. A. (1991). *Magn. Reson. Med.* **19,** 386.

Ben-Yoseph, O., Bachelard, H. S., Badar-Goffer, R. S., Dolin, S. J., and Morris, P. G. (1990). *J. Neurochem.* **55,** 1446.

Clarke, S. D., Metcalfe, J. C., and Smith, G. A. (1993). *J. Chem. Soc., Perkin Trans.* *2,* 1187.

Chresand, T. J., Dale, B. E., and Gillies, R. J. (1986). *Am. Chem. Soc.* **192** (Sept.), P105 (abstract).

Gillies, R. J., Smith, G. A., and Metcalfe, J. C. (1983). *J. Cell Biol.* **97** (abstract).

Gillies, R. J., Powell, D. A., and Drury. (1984). *Biophys. J.* **45,** 87a.

Gillies, R. J., Galons, J. P., McGovern, K. A., Scherer, P. G., Lien, Y. H., Job, C., Ratcliff, R., Chapa, F., Cerdan, S., and Dale, B. E. (1993). *NMR Biomed.* **6,** 95.

Harding, D. P., Smith, G. A., Metcalfe, J. C., Morris, P. G., and Kirschenlohr, H. L. (1993). *Magn. Reson. Med.* **29,** 605.

Hayes, C. E., Edelstein, W. A., Schenck, J. F., Mueller, O. M., Eash, M. (1985). *J. of Magn. Reson.* **63,** 622–628.

Kirschenlohr, H. L., Metcalfe, J. C., Morris, P. G., Rodrigo, G. C., and Smith, G. A. (1988). *Proc. Natl. Acad. Sci. U.S.A.* **84,** 9017.

Kirschenlohr, H. L., Grace, A. A., Clarke, S. D., Shachar-Hill, Y., Metcalfe, J. C., Morris, P. G., and Smith, G. A. (1993). *Biochem. J.* **293,** 407.

Levy, L. A., Murphy, E., and London, R. E. (1987). *Am. J. Physiol.* **252,** C441.

Metcalfe, J. C., Hesketh, T. R., and Smith, G. A. (1985). *Cell Calcium* **6,** 183.

Schanne, F. A. X., Dowd, T. L., Gupta, R. J., and Rosen, J. F. (1990). *Environ. Health Perspect.* **84,** 99.

Smith, G. A., Hesketh, T. R., and Metcalfe, J. C. (1982). "Ions Cell Proliferation and Cancer," p. 65. Academic Press, New York, ISBN 0-12-123050-3.

Smith, G. A., Hesketh, T. R., Metcalfe, J. C., Feeney, J., and Morris, P. G. (1983). *Proc. Natl. Acad. Sci. U.S.A.* **80,** 7178.

Smith, G. A., Kirschenlohr, H. L., Metcalfe, J. C., and Clarke, S. D. (1993). *J. Chem. Soc., Perkin Trans.* **2,** 1205–1209.

Styles, P. *in* "Physics of NMR Spectroscopy in Biology and Medicine" (C. Corso, ed.), pp. 412–439. Bologna Societa Italiana di Fisica, 1991.

Tsien, R. Y. (1980). *Biochemistry* **19,** 2396.

Tsien, R. Y. (1981). *Nature* (*London*) **290,** 527.

# [9] Fluorescence Measurement of Cytosolic pH in Cultured Rodent Astrocytes

S. Jeffrey Dixon and John X. Wilson

## Introduction and Scope

Glial cells, like other cell types, regulate their intracellular (cytosolic) pH ($pH_i$) by the active transport of protons and proton equivalents across their plasma membrane. In addition, glia are thought to participate in the physiological regulation of extracellular pH in the brain. Pathological changes in glial $pH_i$ may play a role in a number of disorders, including cytotoxic edema (1) and AIDS dementia complex (2).

In this chapter, we present the methods used in our laboratories for the (1) isolation and primary culture of rodent astroglia, (2) measurement of $pH_i$ in populations of glial cells using a pH-sensitive fluorescent probe, (3) calibration of the fluorescence signal, and (4) manipulation of $pH_i$, as means of investigating the proton transport mechanisms present in the plasma membrane of these cells. These methods can easily be adapted for use with other types of cultured or freshly isolated cells.

## Isolation and Primary Culture of Rodent Astroglia

### *Principle*

Highly enriched preparations of type-1 astrocytes can be obtained by primary culture of cells from the neopallium (cerebral cortex) of neonatal rats or mice. The method depends on the differential rates of maturation of glial and neuronal cells. In the neonate, neuronal cells are relatively well-differentiated and tend not to survive mechanical dissociation and culture. In contrast, less differentiated glial cell precursors are able to survive dissociation and proliferate in culture. The use of serum-containing medium favors the growth of type-1 astrocytes over oligodendrocytes (3).

### *Reagents and Solutions*

Horse serum: catalog number 16050-064, Gibco Laboratories.

Minimum essential medium (MEM) is prepared in 10-liter batches by combining the following reagents in the order listed: 5 liters deionized water; 26


*Methods in Neurosciences, Volume 27*

ml HCl (1 *M*); 0.48 g L-cystine; 0.62 g L-histidine; 1.05 g L-isoleucine; 1.05 g L-leucine; 0.96 g L-threonine; 0.94 g L-valine; 2.53 g L-arginine; 1.46 g L-lysine; 3.65 g L-glutamine; 0.3 g L-methionine; 0.66 g L-phenylalanine; 0.2 g L-tryptophan; 0.72 g L-tyrosine; 8.8 ml NaOH (1 *M*); 500 ml of an aqueous solution containing 40 mg D-pantothenic acid, 40 mg choline chloride, 80 mg inositol, 40 mg niacinamide, 40 mg pyridoxal HCl, and 40 mg thiamine HCl; 500 ml of an aqueous solution containing 4 g KCl, 68 g NaCl, 1.4 g $NaH_2PO_4 \cdot H_2O$, and 13.5 g D-glucose; 2 g $MgSO_4 \cdot 7\ H_2O$ in 500 ml water; 40 mg folic acid in 250 ml water; 4 mg riboflavin in 250 ml water; 200 mg phenol red in 1 liter water; water to a total volume of 10 liters; 22 g $NaHCO_3$; $1.5 \times 10^6$ units penicillin G; and 0.32 g streptomycin sulfate. The solution is equilibrated with 5% $CO_2$/95% air, adjusted to pH 7.2, sterilized by vacuum filtration (pore size 0.2 $\mu$m), and stored in glass bottles at 4°C in the dark.

Dibutyryl cAMP (*N*-6,*O*-2′-dibutyryladenosine-3′,5′-cyclic monophosphate): 10 m*M* in MEM, sterilized by filtration.

## Procedure

The procedure is based on that of Booher and Sensenbrenner (4) and the modifications of Hertz and co-workers (5). The neopallium is obtained from 1-day-old Sprague–Dawley or Wistar rats or Swiss CD-1 mice. Following decapitation, the areas superficial to the lateral ventricles of the cerebral hemispheres are dissected out under sterile conditions using a binocular dissecting microscope. Ten neopalliums are placed in a 60-mm petri dish containing MEM (6 ml) and the neopallial meninges are removed. Next, the tissue is minced with scissors and washed three times in MEM. The tissue pieces are further disrupted by repeated passage through a serological pipette and by vortex mixing. This preparation is then passed twice through sterile nylon sieves of pore size 10 $\mu$m to remove remnants of blood vessels and meninges. MEM is equilibrated in 5% $CO_2$, supplemented with serum (20%, v/v), and used to dilute the resulting cell suspension (30 or 12 ml for each rat or mouse neopallium, respectively). Aliquots (3 ml) of this suspension are then introduced into 60-mm culture dishes and incubated at 37°C in 5% $CO_2$/95% air with 90% relative humidity. For immunofluorescence studies, cells are cultured on glass coverslips coated with polylysine.

Examination by phase-contrast microscopy of 1-day-old cultures reveals floating clumps of cells tethered by processes to the bottom of the dish. Cultures should not be handled during the first 2 days, to avoid detaching these explants. Starting on day 4, the medium is replaced twice weekly with MEM supplemented with serum (10%).

### *Discussion*

Under the conditions described above, cultures become confluent 2 weeks after plating. At confluence, cell density is approximately $3 \times 10^6$ cells per 60-mm dish. The identity of cells can be confirmed by immunofluorescence staining for the astroglial marker, glial fibrillary acidic protein (GFAP). Using the methods described by Wilson and co-workers (6), we find that ≥85% of cells are intensely immunoreactive for GFAP.

Treatment of confluent cultures with dibutyryl cAMP (0.25 m*M*) for approximately 1 week changes the morphology of the cultured astrocytes from polygonal to stellate (process bearing). Dibutyryl cAMP-induced stellation has been used as a model of reactive astrogliosis, a condition that occurs following brain injury and is associated with elevated levels of cerebral cAMP (7, 8).

## Measurement of Cytosolic pH Using the pH-Sensitive Fluorescent Probe BCECF

### *Principle*

The measurement of $pH_i$ using pH-sensitive dyes offers several advantages over alternative methods, such as determining the distribution of weak acids or bases or microelectrode techniques. Fluorescent probes allow continuous monitoring of $pH_i$ and can be used in small or motile cells that cannot be impaled with ion-selective microelectrodes (e.g., platelets, sperm, or protozoa). Polygonal astrocytes are flat and thus difficult to impale with microelectrodes. Thomas and co-workers (9) pioneered the use of pH-sensitive fluorescent probes to measure cytosolic pH. In these early studies, cells were loaded with esterified membrane-permeant fluorescein derivatives (fluorescein diacetate or 5- or 6-carboxyfluorescein diacetate), which on entering the cell are rapidly hydrolyzed by endogenous cytosolic esterases and trapped intracellularly (as fluorescein or 5- or 6-carboxyfluorescein). The fluorescence spectra of these dyes are markedly dependent on pH, with fluorescence intensity increasing as pH is increased.

The fluorescein derivative 2′,7′-bis(carboxyethyl)-5- or 6-carboxyfluorescein (BCECF) was introduced by Tsien and co-workers in 1982 (10). BCECF offers a number of advantages over earlier fluorescein-based indicators, accounting for its current popularity as a probe for measurement of $pH_i$. First, BCECF has four to five negative charges at physiological $pH_i$ (Fig. 1), compared to only two or three for carboxyfluorescein. These additional ionic

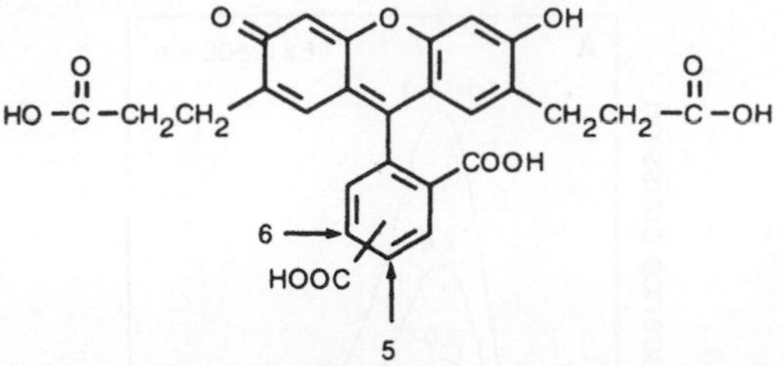

FIG. 1 Structure of BCECF free acid form. All carboxyl groups on BCECF are ionized at physiological $pH_i$ and the compound is therefore highly impermeant. The carboxyl group on the lowest ring is attached at either the 5 or 6 position. Commercially available preparations consist of a mixture of the two resulting isomers. The phenolic group is partially ionized at physiological $pH_i$, with a $pK_a$ of approximately 7.0. Deprotonation of this group is accompanied by a dramatic increase in the quantum efficiency of the chromophore (see Fig. 2). Reproduced with permission of Molecular Probes, Inc.

charges account for decreased leakage of BCECF across the plasma membrane as well as reduced accumulation within alkaline subcellular compartments, such as mitochondria (9, 11). Second, the $pK_a$ of BCECF, of approximately 7.0, is closer to physiological $pH_i$ than that of carboxyfluorescein ($pK_a \approx 6.4$). Alkalinization results in deprotonation of the phenolic group on BCECF, which dramatically increases the amplitude of its emission spectrum without a shift in wavelength (Fig. 2A). In contrast, alkalinization causes both an increase in the amplitude of the excitation spectrum and a shift toward longer wavelengths. This gives rise to an "isosbestic" point in the excitation spectrum, the point at which the spectra cross and the excitation efficiency of BCECF is independent of pH (Fig. 2B). This feature permits the use of BCECF for microscope-based measurements of $pH_i$ in single cells. The ratio of BCECF fluorescence at a pH-sensitive excitation wavelength (490–500 nm) to that at the isosbestic point ($\approx$440 nm) provides a measure of $pH_i$ independent of dye concentration and pathlength, thereby avoiding errors arising from changes in cell shape, dye leakage, or photobleaching.

The cell-permeant ester form of BCECF is known as BCECF-AM (BCECF-acetoxymethyl ester). BCECF-AM was originally thought to possess two acetate ester and three acetoxymethyl ester groups (Fig. 3A). Molecular Probes has recently reported, however, that their preparation of BCECF-AM is primarily a mixture of two lower molecular weight forms. In these forms, one or two of the carboxyethyl groups are cyclized with the adjacent phenolic hydroxyl groups (Fig. 3B and C; Ref. 11). Like the open-chain triester form, these mono and dilactone forms of BCECF-AM are highly

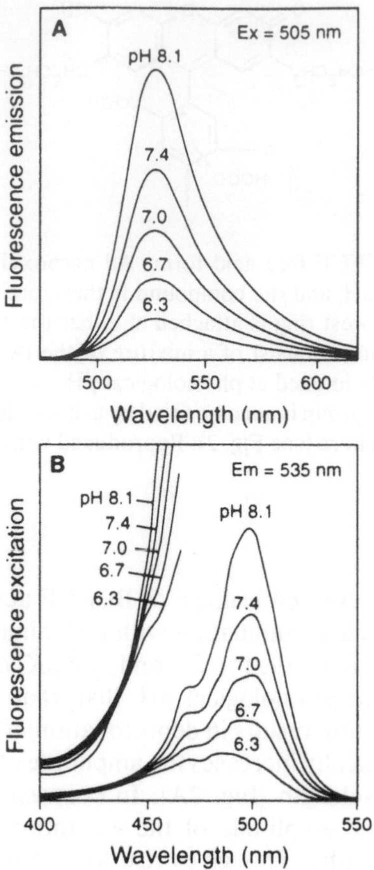


FIG. 2 The pH-dependent spectra of BCECF. (A) Emission spectra; alkalinization dramatically increases the fluorescence intensity without a shift in the emission spectrum. (B) Excitation spectra; alkalinization increases fluorescence intensity and shifts the excitation spectrum to longer wavelengths. This shift gives rise to an isosbestic point (439 nm), at which all spectra intersect and the fluorescence intensity is therefore independent of pH (shown at higher gain on the left of panel B). Reproduced with permission of Molecular Probes, Inc.

permeant and readily hydrolyzed by cytosolic esterases to yield the pH-sensitive free acid, BCECF.

## *Reagents and Solutions*

BCECF-AM: Commercially available (Molecular Probes, Eugene, Oregon) in 50-$\mu$g aliquots that are stored dessicated in the dark at $-20$°C. Stock solutions are prepared by dissolving aliquots in 50 $\mu$l dimethyl sulfoxide.

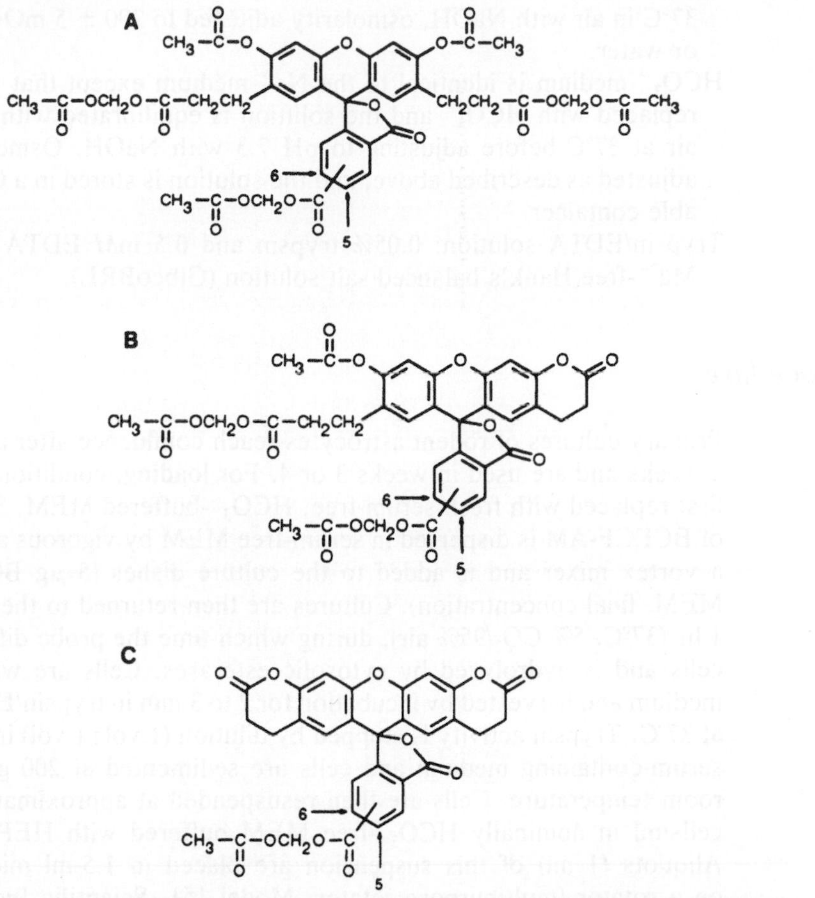

FIG. 3 Structure of BCECF-AM. All forms of BCECF-AM are membrane permeant and are readily hydrolyzed by cytosolic esterases to yield the free acid BCECF (see Fig. 1). At least some commercially available preparations of BCECF-AM are composed of a mixture of three different forms of the molecule. Molecular Probes report that their preparation contains little of the triester form (structure A, molecular mass 820.7 Da), but consists primarily of the monolactone (structure B, molecular mass 688.6 Da) and dilactone (structure C, molecular mass 556.5 Da) forms. Reproduced with permission of Molecular Probes, Inc.

Unused portions of this stock solution can be stored in the dark at −20°C for several weeks without detectable deterioration. Aqueous solutions are unstable and should be used immediately.

$Na^+$ medium contains 135 m*M* NaCl, 5 m*M* KCl, 1 m*M* $MgCl_2$, 1 m*M* $CaCl_2$, 10 m*M* glucose, and 20 m*M* Na-HEPES adjusted to pH 7.30 ± 0.02 at

37°C in air with NaOH, osmolarity adjusted to 290 ± 5 mOsM with NaCl or water.

$HCO_3^-$ medium is identical to the $Na^+$ medium except that 12 m*M* $Cl^-$ is replaced with $HCO_3^-$ and the solution is equilibrated with 5% $CO_2$/95% air at 37°C before adjusting to pH 7.3 with NaOH. Osmolarity is then adjusted as described above, and the solution is stored in a $CO_2$-impermeable container.

Trypsin/EDTA solution: 0.05% trypsin and 0.5 m*M* EDTA in $Ca^{2+}$- and $Mg^{2+}$-free Hank's balanced salt solution (GibcoBRL).

## *Procedure*

Primary cultures of rodent astrocytes reach confluence after approximately 2 weeks and are used in weeks 3 or 4. For loading, conditioned medium is first replaced with fresh serum-free, $HCO_3^-$-buffered MEM. Stock solution of BCECF-AM is dispersed in serum-free MEM by vigorous agitation using a vortex mixer and is added to the culture dishes (5 $\mu$g BCECF-AM/ml MEM, final concentration). Cultures are then returned to the incubator for 1 hr (37°C, 5% $CO_2$/95% air), during which time the probe diffuses into the cells and is hydrolyzed by cytosolic esterases. Cells are washed in $Na^+$ medium and harvested by incubation for 2 to 3 min in trypsin/EDTA solution at 37°C. Trypsin activity is stopped by dilution (1 vol : 1 vol) in conditioned, serum-containing medium and cells are sedimented at 200 g for 5 min at room temperature. Cells are then resuspended at approximately $1.5 \times 10^6$ cells/ml in nominally $HCO_3^-$-free MEM buffered with HEPES (25 m*M*). Aliquots (1 ml) of this suspension are placed in 1.5-ml microfuge tubes on a rotator (multipurpose rotator, Model 151, Scientific Industries, Inc., Bohemia, New York; set at lowest speed) to maintain cells in suspension. At room temperature, cells remain viable with little leakage of BCECF for several hours.

For fluorescence measurements, cells are first sedimented using a microfuge (10 to 15 sec at slow speed). Cells are then resuspended in appropriate medium directly in the fluorescence cuvette at a final concentration of approximately $7 \times 10^5$ cells/ml. Cells are kept in suspension by gentle magnetic stirring and maintained at 37°C using a cuvette holder equipped with a water jacket. Fluorescence is monitored at 495 mm (5-nm slit) excitation and 525 nm (10-nm slit) emission, giving a continuous recording of $pH_i$ with time. A variety of conventional fluorescence spectrophotometers are suitable for such studies; two that we have used are Model LS-5 (Perkin-Elmer Corp.) and Model F-4010 (Hitachi Ltd.). BCECF-AM, which is weakly fluorescent, becomes brightly fluorescent on hydrolysis. Thus, the fluorescence intensity of the cell preparation provides a convenient measure of the effectiveness with which cells have been loaded with BCECF. Similarly, if leakage of BCECF is a problem, it will be apparent as a decrease in the initial fluorescence intensity as loaded cells are stored for increasing periods of time in MEM, prior to sedimentation and fluorescence determination.

For studies examining $HCO_3^-$-independent transport systems, nominally $HCO_3^-$-free solutions are used (e.g., $Na^+$ medium; see above). To examine $HCO_3^-$-dependent transport systems, we utilize an $HCO_3^-/CO_2$-containing medium such as the $HCO_3^-$ medium described above. Prior to suspension of cells, $HCO_3^-/CO_2$-containing media are prewarmed to 37°C and equilibrated with 5% $CO_2$/95% air. During $pH_i$ measurement in the presence of $HCO_3^-$, a stream of 5% $CO_2$/95% air is directed toward the surface of the solution in the cuvette via a tube inserted through the injection port on the sample compartment. To avoid light scattering, care must be taken to prevent the liquid/gas interface or bubbles from impinging on the light path.

## *Discussion*

One is able to use a single wavelength to monitor average $pH_i$ in suspensions of BCECF-loaded astrocytes, because the pathlength remains constant and there is negligible dye leakage or photobleaching over the time course of a single experiment (10 to 20 min). For single-cell studies, these issues become critical and can be overcome, at least in part, through dual-wavelength ratioing techniques. In addition to BCECF, which can be used in the dual-excitation mode, a series of seminaphthofluorescein (SNAFL) and seminaphthorhodafluor (SNARF) dyes have been introduced (12). Like BCECF, these dyes are available as membrane-permeant acetoxymethyl and acetate esters for loading into cells. SNAFL and SNARF dyes can be used in either dual-excitation or dual-emission modes. Furthermore, their relatively long-wavelength spectra allow them to be used in combination with other fluorescent probes for simultaneously monitoring $pH_i$ and cytosolic free $Ca^{2+}$ or $Na^+$ concentration.

Preliminary experiments indicated that loading astrocytes with BCECF-AM in conditioned medium, or medium containing fresh serum, gave rise to compartmentation of BCECF within subcellular organelles, possibly components of the endocytic pathway. In contrast, predominantly cytosolic loading was obtained when cells were incubated with BCECF-AM in protein-free medium. Compartmentation can be assessed by direct observation using a fluorescence microscope, in which case it may be apparent as a punctate pattern of fluorescence often localized to the perinuclear region. On the other hand, cytosolic loading is typified by uniform fluorescence throughout the cytoplasmic and nuclear compartments. Alternatively, compartmentation can be assessed by comparing the amount of dye released following selective permeabilization of the plasma membrane (e.g., with low concentrations of digitonin), to the total amount of dye released following complete permeabilization (e.g., with Triton X-100).

A limitation of the technique described above is that cells are suspended by brief exposure to trypsin. This method would not be suitable if the responses under study involved trypsin-sensitive cell surface receptors or transporters. In such a situation, responses are better studied using cells cultured on coverslips that can be mounted in a modified fluorescence cuvette. A further advantage of a coverslip system is that the extracellular medium can be rapidly exchanged, while continuously monitoring the fluorescence signal from the adherent cells. Disadvantages of a coverslip system include (1) a poorer signal-to-noise ratio because fewer cells (only a monolayer, in most cases) are in the light path and (2) greater photobleaching because the same population of cells remains in the light path throughout the experiment.

Last, during studies of $pH_i$, care must be taken to avoid manipulations that might activate proton transport mechanisms that would otherwise remain quiescent. For example, the osmolarity of all extracellular media must be checked and adjusted, because osmotic shrinkage is an effective activator of the $Na^+/H^+$ exchanger in many cell types (13). During our studies on $pH_i$ regulation in astrocytes, we also noticed that $Na^+/H^+$ exchange could be activated by excessive acceleration during sedimentation of cells. Such phenomena may underlie the activation of $Na^+/H^+$ exchange observed by others in sedimented and resuspended samples of a related cell type (14).

## Calibration of the Fluorescence Signal

### *Principle*

Over the physiologically relevant range of $pH_i$ (6.4 to 7.6), the fluorescence intensity of BCECF is approximately a linear function of pH. Following an experiment, direct calibration can be performed by lysing the experimental cell sample to release BCECF into the medium. The pH of the medium is then stepped to various values and the corresponding fluorescence intensities are recorded. Using this approach, correction must be made for differences between the behavior of the dye in the cytosol and the extracellular medium. At the same pH, the fluorescence of BCECF in the cytosol is slightly less than that in the extracellular medium following lysis, presumably because of quenching of the dye when it is confined intracellularly. To compensate for this effect, a correction factor is obtained by measuring the fractional increase in fluorescence observed when cells are lysed under conditions where $pH_i$ equals extracellular pH ($pH_o$). $pH_i$ can be clamped equal to $pH_o$ using nigericin, as described in the following paragraph. The correction factor, obtained using this technique, is relatively independent of pH over the range of interest (6.4 to 7.6).

A more convenient method of calibration can be used when an experiment is performed on a series of cell samples, loaded with BCECF under identical conditions. In this situation, samples need not be calibrated individually. For such calibration, a parallel sample of loaded cells is suspended in medium with $K^+$ concentration ($[K^+]_o$) approximately equal to cytosolic $K^+$ concentration ($[K^+]_i$). Addition of the electroneutral potassium/proton exchange ionophore nigericin sets

$$[K^+]_i/[K^+]_o = [H^+]_i/[H^+]_o$$

(Ref. 9). When $[K^+]_o$ equals $[K^+]_i$, $pH_i$ will follow changes in $pH_o$. $pH_o$ is stepped to various values and the corresponding fluorescence intensities recorded. The calibration curve that is produced applies to all the samples, and no correction is required for quenching, because the BCECF remains trapped in the cytosol. However, the calibration applies only to samples loaded with the same concentration of BCECF and, therefore, the procedure must be repeated each time a new batch of cells is loaded.

## *Reagents and Solutions*

High $[K^+]$ medium contains 140 m*M* KCl, 1 m*M* $MgCl_2$, 1 m*M* $CaCl_2$, 10 m*M* glucose, and 20 m*M* Na-HEPES, adjusted to pH 7.30 ± 0.02 at 37°C in air with NaOH, osmolarity adjusted to 290 ± 5 mOsM with NaCl or water.

MES [2-(*N*-morpholino)ethanesulfonic acid]: saturated solution in water.

Tris [tris(hydroxymethyl)aminomethane]: saturated solution in water.

Nigericin: 1 m*M* in absolute ethanol (Calbiochem, San Diego, California).

Triton X-100: 10% in water (v/v).

## *Procedure*

For direct calibration following an experiment, the cell sample is first lysed with Triton X-100 (0.1% v/v) and the fluorescence intensity is recorded. The pH of the medium is then measured using a pH meter and combination electrode of sufficiently small diameter so that the probe can easily be inserted into the fluorescence cuvette (e.g., Model GK2421 C combined electrode, Radiometer, Copenhagen, Denmark). Fluorescence intensity and medium pH are again measured following the addition of a small volume of MES or Tris solution. This procedure is repeated until a series of measurements have been obtained over the range of fluorescence intensities encountered during the preceding experiment. These measurements are used to construct a standard curve of fluorescence intensity versus pH, which is approximately linear over the range of interest and can be fit by least-squares linear regression. To calculate the $pH_i$ of the unlysed experimental sample, its fluorescence intensity is first multiplied by the correction factor. The $pH_i$ is then obtained from the standard curve digitally, using a computer algorithm, or manually, by interpolation. The correction factor is measured as follows. A cell sample is suspended in high $[K^+]$ medium and nigericin (1 $\mu M$) is added. The fluorescence intensity will equilibrate (as $pH_i$ approaches $pH_o$) and the cells are then lysed with Triton X-100 (0.1%). The correction factor is the ratio of the BCECF fluorescence following lysis to that before lysis.

Indirect calibration is performed using a sample of cells loaded with BCECF together with the experimental samples. This parallel sample is suspended in high $[K^+]$ medium, nigericin (1 $\mu M$) is added, and, following equilibration of the fluorescence signal, medium pH is measured. Using this same sample, a series of determinations are then performed in which concentrated MES or Tris is added, fluorescence is allowed to equilibrate, and medium pH is measured. These measurements are used to construct a standard curve of fluorescence intensity versus $pH_i$. The cytosolic pH values for experimental samples are obtained directly from the standard curve, without need for a correction factor. It should be noted that, in the absence of detergent, it takes several minutes for $pH_i$ to equilibrate with $pH_o$ following the addition of MES or Tris to the extracellular medium.

## *Discussion*

A limitation of the indirect calibration procedure is that $[K^+]_i$ is not measured experimentally. Because $pH_i = pH_o$ only if $[K^+]_o = [K^+]_i$, error in estimating $[K^+]_i$ will alter the absolute values obtained for $pH_i$. Therefore, the indirect calibration procedure is more appropriate for assessing relative changes in $pH_i$. Calibration is also complicated if, during an experiment, agents are introduced into the cuvette that are autofluorescent or quench the fluorescence of BCECF. Last, it should be remembered that indirect calibration is appropriate only if the starting $pH_i$ of each sample is identical. Differences in starting $pH_i$ would occur if some samples were pretreated with inhibitors or substitution of extracellular ions that lead to alteration of $pH_i$. In such cases, direct calibration should be used. Reproducibility of $pH_i$ measurement in astrocyte populations using BCECF is within 0.05 units. The sensitivity of the measurement (i.e., the ability to detect changes in $pH_i$) is ≈0.02 units.

## Manipulation of Cytosolic pH

### *Principle*

A number of systems that transport protons or proton equivalents across the plasma membrane are involved in the physiological regulation of $pH_i$. Therefore, a useful strategy for investigating these systems is to alter $pH_i$ and monitor its recovery by fluorimetry. Although several transport systems may be activated simultaneously, the contribution of individual transporters can be assessed through use of ion substitution and selective blockers.

We present below a protocol for acid loading of cells, and examples of the use of this procedure for studying proton transport kinetics in astrocytes. Cells are acidified by exposure to nigericin while suspended in a special extracellular medium in which $Na^+$ and $K^+$ are replaced by a large, nontransported organic cation, *N*-methyl-D-glucamine$^+$ ($NMG^+$). Under these conditions, nigericin mediates the exchange of intracellular $K^+$ for extracellular $H^+$. Within 2 to 3 min of addition of nigericin, reproducible cytosolic acidification to pH 6.4–6.6 is observed (Figs. 4 and 5). Nigericin is removed by washing the cell suspension. Alternatively, the time course of acid loading can be monitored fluorimetrically and terminated at the desired $pH_i$ by addition of bovine albumin, which binds nigericin and removes it from the cell membranes (15).

The mechanisms responsible for recovery of $pH_i$ following acid loading can be investigated by monitoring the rate of alkalinization in the presence and absence of various extracellular ions and transport inhibitors. Using this approach, at least two transport systems can be shown to participate in the recovery of $pH_i$ following acidification of astrocytes. The first is dependent on the presence of extracellular $Na^+$ and inhibited by amiloride (Figs. 4–6), properties consistent with those of the $Na^+/H^+$ exchanger (16). The second system requires the presence of extracellular $Na^+$ and $HCO_3^-$ and is inhibited by 4,4′-diisothiocyanostilbene-2,2′-disulfonic acid (DIDS; Fig. 4), properties consistent with both $Na^+$-dependent $HCO_3^-/Cl^-$ exchange and $Na^+/HCO_3^-$ cotransport (13).

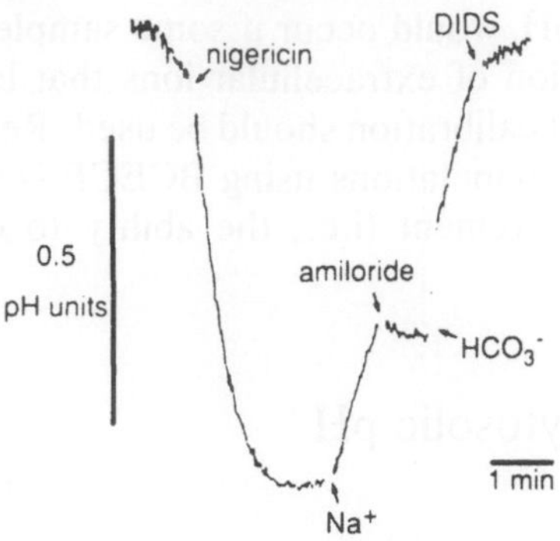


FIG. 4 Cytosolic acidification and recovery. The cytosolic pH ($pH_i$) of rat astrocytes was monitored fluorimetrically using BCECF as described in the text. Cells ($7 \times 10^5$/ml) suspended in $Na^+$-, $HCO_3^-$-, and $K^+$-free $NMG^+$ medium were acid loaded by exposure to nigericin (1 $\mu M$). This acid-loading regimen reduced cytosolic pH to 6.4–6.6. Where indicated, 50 m*M* NaCl was added, resulting in cytosolic alkalinization that was inhibited by 0.5 m*M* amiloride. Subsequent addition of 50 m*M* $NaHCO_3$ resulted in renewed alkalinization that was inhibited by 0.5 m*M* DIDS (4,4′-diisothiocyanostilbene-2,2′-disulfonic acid). For clarity, the trace was offset following addition of amiloride to compensate for quenching of BCECF fluorescence by this agent.

## *Reagents and Solutions*

$NMG^+$ medium (nominally $Na^+$ and $K^+$ free) contains 140 m*M* *N*-methyl-D-glucamine chloride, 1 m*M* $MgCl_2$, 1 m*M* $CaCl_2$, 10 m*M* glucose, and 20 m*M* HEPES, adjusted to pH 7.30 ± 0.02 at 37°C in air with *N*-methyl-D-glucamine free base, osmolarity adjusted to 290 ± 5 mOsM with *N*-methyl-D-glucamine chloride or water.

Nigericin, $Na^+$, and $HCO_3^-$ media: see above.

Bovine albumin (crystalline): 100 mg/ml in $NMG^+$ medium.

Concentrated salt solutions: 2 *M* aqueous stocks of NaCl, $NaHCO_3$, LiCl, KCl, choline chloride, and NMG chloride (to prepare NMG chloride, NMG free base is titrated to pH 7.3 with HCl).

DIDS: 10 m*M* in water, prepared fresh for each experiment and stored on ice, in the dark, to minimize degradation.

## *Procedure*

Astrocytes are loaded with BCECF as described above. Cells are suspended in $NMG^+$ medium directly in the fluorescence cuvette at a final concentration of approximately $7 \times 10^5$ cells/ml. Nigericin (0.5 to 1.0 $\mu M$) is added to the

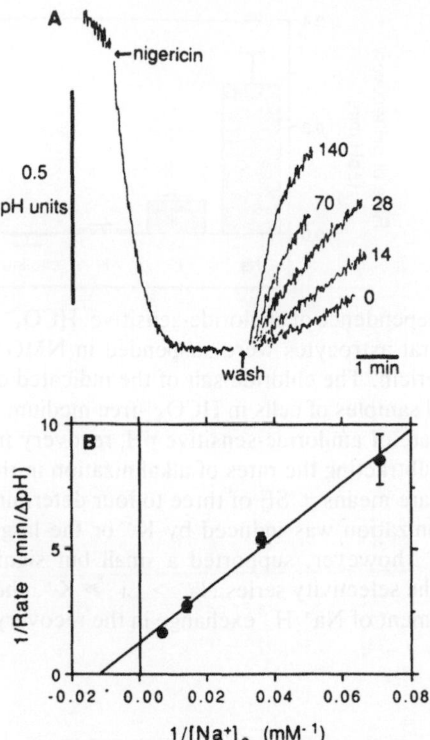


FIG. 5 Use of fluorescence techniques to measure proton transport kinetics. These data illustrate the dependence of $pH_i$ recovery on extracellular $Na^+$ concentration in nominally $HCO_3^-$-free, isoosmotic medium. Parallel samples of BCECF-loaded rat astrocytes were acidified in $NMG^+$ medium by exposure to nigericin (as described above). Cells were then sedimented and immediately resuspended in medium containing the indicated concentration of $Na^+$ (in millimoles/liter). $NMG^+$ was used for osmotic balance. (A) Individual traces from a representative experiment illustrate the time course of acid loading, and $pH_i$ recovery in parallel samples that were acid loaded under identical conditions and then resuspended in media containing $Na^+$ (0 to 140 m*M*). (B) Lineweaver–Burk plot illustrates the relationship between the initial rate of $pH_i$ recovery and the concentration of extracellular $Na^+$ ($[Na^+]_o$). Data are the means ± SE from three individual experiments. The line was fitted by least squares and had a correlation coefficient of 0.99. The negative inverse of the *X* intercept yields an apparent $K_m$ of 87 m*M* for stimulation of $pH_i$ recovery by extracellular $Na^+$.

suspension and the time course of acidification is monitored fluorimetrically (Figs. 4 and 5). Acid loading can be terminated at the required $pH_i$ by addition of albumin solution (5 mg/ml, final).

Following acid loading, addition of NaCl induces, without apparent delay,

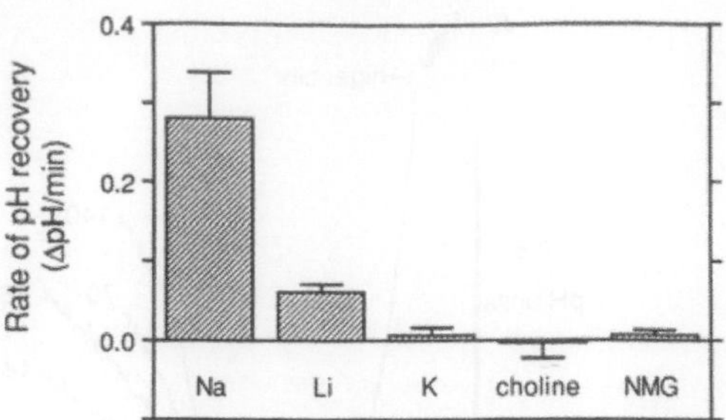


FIG. 6 Cation dependence of amiloride-sensitive, $HCO_3^-$-independent $pH_i$ recovery. BCECF-loaded rat astrocytes were suspended in $NMG^+$ medium and acidified by exposure to nigericin. The chloride salt of the indicated cations (100 m*M*, final) was added to parallel samples of cells in $HCO_3^-$-free medium with and without amiloride (0.5 m*M*). The rate of amiloride-sensitive $pH_i$ recovery induced by each cation was determined by subtracting the rates of alkalinization in the absence and presence of amiloride. Data are means ± SE of three to four determinations for each cation. No significant alkalinization was induced by $K^+$ or the large organic cations $choline^+$ and $NMG^+$. $Li^+$, however, supported a small but significant rate of cytoplasmic alkalinization. The selectivity series $Na^+ > Li^+ \gg K^+$, $choline^+$, $NMG^+$ is consistent with the involvement of $Na^+/H^+$ exchange in the recovery of astrocytes from acidification.

rapid alkalinization that is inhibited by amiloride (Fig. 4), a blocker of $Na^+/H^+$ exchange. Amiloride-sensitive alkalinization is induced by $Na^+$ and to a lesser extent by $Li^+$, but not by $K^+$, $choline^+$, or $NMG^+$ (Fig. 6). The $Na^+/H^+$ exchanger shows a similar selectivity for $Na^+$ and $Li^+$ over $K^+$, $choline^+$, and $NMG^+$ (17), supporting a role for this transporter in the recovery of astrocytes from acidification. In the presence of amiloride, addition of $NaHCO_3$ induces an additional alkalinization that is blocked by the anion transport inhibitor DIDS (Fig. 4). These findings suggest the involvement of a second transport mechanism in recovery from acidification, either $Na^+$-dependent $HCO_3^-/Cl^-$ exchange or $Na^+/HCO_3^-$ cotransport. It should be noted that DIDS may react with NMG or Tris. Thus, media containing these agents are not recommended for studies examining the dependence of $pH_i$ recovery on DIDS concentration.

The procedures described above require addition of concentrated salt solutions to the acid-loaded cells, thus inducing osmotic imbalance and cell shrinkage. To avoid this problem, parallel samples of BCECF-loaded astrocytes can be acid loaded in a microfuge tube, sedimented, and then introduced into a fluorescence cuvette containing isoosmotic extracellular medium of the desired composition. This procedure avoids changes in cell volume and ensures effective removal of nigericin from the cell suspension. An example of this procedure is illustrated in Fig. 5. One sample of cells was acid loaded in a cuvette and the time course of acidification monitored fluorimetrically. A series of parallel cell samples were then acid loaded under identical conditions in microfuge tubes. Cell samples were individually sedimented, resuspended in small volumes of $NMG^+$ medium, and added to fluorescence cuvettes containing varying concentrations of $Na^+$. In the absence of $HCO_3^-$, initial rates of alkalinization primarily reflect $Na^+/H^+$ exchange activity. The dependence of transport activity on $[Na^+]_o$ is consistent with Michaelis–Menten kinetics, and yields an apparent $K_m$ of 87 m*M* for the interaction of extracellular $Na^+$ with the exchanger (Fig. 5B).

## *Discussion*

Rates of alkalinization are obtained from the procedures described above. Flux of $H^+$ (or $H^+$ equivalents) can be calculated by multiplying the rate of change of $pH_i$ (pH units/minute) by the buffering power of the cells (moles/liter cells/unit pH). Buffering power can be estimated using several techniques, including direct titration of cell lysates or measurement of the changes in $pH_i$ arising from acute exposure to known concentrations of $NH_4Cl$. Assuming that $NH_3$ rapidly equilibrates across the cell membrane and $NH_4^+$ does not, one can calculate the acute change in $[NH_4^+]_i$. The buffering power of the cytoplasm is then given by $\Delta[NH_4^+]_i/\Delta pH_i$ (18). It should be kept in mind that total buffering power is increased when cells are suspended in media containing $CO_2/HCO_3^-$.

Besides the nigericin technique for acid loading described above, a number of other strategies are available for manipulating $pH_i$. For example, cells can be pulsed with weak acids or bases (e.g., propionic acid or $NH_3$). These rapidly enter the cell by nonionic diffusion, acidifying or alkalinizing the cytosol on ionization. Another widely used technique for acid loading is to preload cells with $NH_4^+$ and then resuspend in $NH_4^+$-free medium. Under these conditions, rapid loss of $NH_3$ lowers $pH_i$ acutely (for further details, see Chapter 12, this volume).

## Concluding Remarks

Procedures have been described for functionally studying plasma membrane transport of $H^+$ and $H^+$ equivalents by astrocytes. These techniques require use of a conventional fluorimeter and are easily applicable to many cell types. Newer pH-sensitive dyes, such as the SNARF and SNAFL indicators, have improved optical properties and readily permit quantification of $pH_i$ using ratiometric techniques. Such techniques allow microspectrofluorimetry and fluorescence imaging of pH at the single-cell and subcellular levels. With appropriate choices of indicator dyes and excitation and emission wavelengths, simultaneous monitoring of multiple ion species is feasible. Furthermore, it is now possible to monitor both membrane currents and $pH_i$ by simultaneous patch-clamp recording and microspectrofluorimetry. These developments will undoubtedly lead to a better understanding of the physiological and pathological roles of $H^+$ transporters in glia and other cell types.

## Acknowledgments

The original work from the authors' laboratories included in this chapter was supported by the Natural Sciences and Engineering Research Council and the Medical Research Council (MRC) of Canada. S. J. Dixon is supported by a Development Grant from the MRC.

## References

1. H. K. Kimelberg, E. J. Cragoe, Jr., K. D. Barron, L. R. Nelson, R. S. Bourke, A. J. Popp, D. Szarowski, J. W. Rose, S. K. Easton, O. W. Woltersdorf, and A. M. Pietruszkiewicz, *in* "The Biochemical Pathology of Astrocytes" (M. D. Norenberg, L. Hertz, and A. Schousboe, eds.), p. 337. Alan R. Liss, New York, 1988.
2. D. J. Benos, B. H. Hahn, J. K. Bubien, S. J. Ghosh, N. A. Mashburn, M. A. Chaikin, G. M. Shaw, and E. N. Benveniste, *Proc. Natl. Acad. Sci. U.S.A.* **91,** 494 (1994).
3. N. Sakellaridis, D. Mangoura, and A. Vernadakis, *Dev. Brain Res.* **27,** 31 (1986).
4. J. Booher and M. Sensenbrenner, *Neurobiology* **2,** 97 (1972).
5. L. Hertz, B. H. P. Juurlink, H. Fosmark, and A. Schousboe, *in* "Neuroscience Approached through Cell Culture" (S. E. Pfeiffer, ed.), Vol. 1, p. 174. CRC Press, Boca Raton, Florida, 1982.
6. G. A. R. Wilson, S. Beushausen, and S. Dales, *Virology* **151,** 253 (1986).
7. S. Fedoroff, W. A. J. McAuley, J. D. Houle, and R. M. Devon, *J. Neurosci. Res.* **12,** 15 (1984).
8. F. Wandosell, P. Bovolenta, and M. Nieto-Sampedro, *J. Neuropathol. Exp. Neurol.* **52,** 205 (1993).
9. J. A. Thomas, R. N. Buchsbaum, A. Zimniak, and E. Racker, *Biochemistry* **18,** 2210 (1979).
10. T. J. Rink, R. Y. Tsien, and T. Pozzan, *J. Cell Biol.* **95,** 189 (1982).
11. R. Y. Tsien, *Methods Cell Biol.* **30,** 127 (1989).

12. J. E. Whitaker, R. P. Haugland, and F. G. Prendergast, *Anal. Biochem.* **194,** 330 (1991).
13. E. K. Hoffmann and L. O. Simonsen, *Physiol. Rev.* **69,** 315 (1989).
14. L. L. Isom, E. J. Cragoe, Jr., and L. E. Limbird, *J. Biol. Chem.* **262,** 6750 (1987); L. L. Isom, E. J. Cragoe, Jr., and L. E. Limbird, *J. Biol. Chem.* **263,** 16513 (1988).
15. S. Grinstein, S. Cohen, and A. Rothstein, *J. Gen. Physiol.* **83,** 341 (1984).
16. S. Wakabayashi, C. Sardet, P. Fafournoux, L. Counillon, S. Meloche, G. Pagés, and J. Pouysségur, *Rev. Physiol. Biochem. Pharmacol.* **119,** 157 (1992).
17. P. S. Aronson, *Annu. Rev. Physiol.* **47,** 545 (1985).
18. A. Roos and W. F. Boron, *Physiol. Rev.* **61,** 296 (1981).

# [10] Brain Cell Acid–Base Changes Measured with Ion-Selective Microelectrodes

Christopher D. Lascola and Richard P. Kraig

## Introduction

Ion-selective microelectrodes (ISMs) provide an unparalleled tool for the simultaneous characterization of electrophysiologic and ionic properties of identified cells within functioning tissue. Simultaneous measurement of ionic activities and cell electrical properties remains a feat unmatched by any single tool for measuring bioelectric events or ionic species. ISMs sensitive to pH have evolved considerably over the past 40 years and their evolution is strongly correlated with progress in our understanding of brain cell acid–base regulation and the relationship of intracellular pH ($pH_i$) to cell membrane physiology. The first use of glass-based pH-sensitive ISMs by Caldwell for study of $pH_i$ helped establish that protons are not passively distributed across crab muscle cell membranes (1). Refinements in glass pH ISMs that included miniaturization in tip size by Thomas (2) facilitated experiments in less robust preparations, demonstrating further that $pH_i$ is actively regulated. These improved glass pH ISMs also permitted investigators to begin to examine plasma membrane mechanisms of $pH_i$ regulation.

A dramatic improvement in pH ISM technology came with the development of liquid membrane pH ISMs based on tridodecylamine (TDDA) (3). Intracellular ISMs based on neutral carriers offer distinct advantages over glass ISMs. Liquid membrane ISMs are considerably easier to fabricate, generally more selective for principal ions, and respond faster than their glass counterparts. Indeed, pH ISMs based on TDDA, a neutral carrier that is highly selective for protons ($H^+$), are the easiest to fabricate among pH ISMs. Furthermore, they are the fastest, finest-tipped, and most selective double-barrel pH-sensitive ISMs fabricated to date. TDDA-based pH ISMs have become the microelectrode of choice for studying $pH_i$ changes from individual brain cells within intact tissues. The theory, construction, and implementation of ISMs have been addressed previously (2, 4). Attention is drawn to the book by Ammann (4), which provides a thorough discussion of all aspects of ISMs based on liquid exchangers and which is required reading for any novice to this field. This review will focus on the fabrication


*Methods in Neurosciences, Volume 27*

and use of pH ISMs based on TDDA for recording $pH_i$ from cells within brain tissue.

## Comparison of Methods for Measuring $pH_i$

Several techniques are now available to measure brain cell $pH_i$. These techniques can be grouped into two distinct categories: those that measure an average $pH_i$ of a cell population and those that measure the $pH_i$ of individual cells. Techniques commonly used to measure the average $pH_i$ of cell groups include measurement of the distribution of weak acids or bases and measurement of $^{31}P$ shifts via nuclear magnetic resonance (NMR) spectroscopy. Methods frequently used to measure the $pH_i$ of individual cells include optical techniques [e.g., fluorescence ratio imaging of 2′,7′-bis(carboxyethyl)-5,6-carboxyfluorescein (BCECF)] and $pH_i$ ISMs. We briefly discuss these techniques below to emphasize the advantages and disadvantages of $pH_i$ ISMs. Other less popular methods for determining $pH_i$ that include use of umbelliferone or neutral red as pH indicators will not be discussed here.

Measurement of the distribution of weak acid or weak base indicator between the intracellular and extracellular compartments is a technically simple, relatively noninvasive approach for measuring $pH_i$ (5). By assuming only the electrically neutral form of a given weak acid or base permeates cell membranes, $pH_i$ is determined once an equilibrium distribution for the uncharged acid or base is reached. By measuring the total (extra- and intracellular) concentrations for the uncharged species in use and the extracellular pH ($pH_o$), $pH_i$ is calculated from combined use of the mass action and Henderson–Hasselbalch equations. 5,5-Dimethyloxazolidine-2,4-dione (DMO) is a weak acid that is often used for determining $pH_i$ via this procedure. Advantages include the ease with which measurements are made and the noninvasive nature of the technique. Disadvantages, however, may outweigh these positive attributes. First, only single measurements of $pH_i$ are attainable, because the method requires destruction of cells to assess indicator content. Second, in practice only steady-state observations of $pH_i$ are possible because of the relatively slow distribution of indicators across cell membranes. Third, measurement errors can be introduced because of inaccuracies in determining either the extracellular volume or indicator ionization equilibrium constants for the intracellular and extracellular spaces. Finally, either metabolic transformation of indicators or permeation of charged species across cell membranes may produce errors in determining accurate values of absolute $pH_i$.

A second, increasingly popular method for measuring average $pH_i$ of brain tissues is the use of $^{31}P$ NMR spectroscopy (6). The technique is based on shifts in the $^{31}P$ peaks of NMR spectra following changes in $pH_i$. The approach is fully discussed in this volume (see Chapter 11 by Prichard). It has the distinct advantage of being completely noninvasive. Furthermore, measurements can be repeated in minutes with a resolution of up to 0.02 pH (6). These attributes have made $^{31}P$ NMR spectroscopy the method of choice for determining average brain $pH_i$ under many circumstances. Perhaps the most important drawback with this approach is the significant investment in equipment, space, and personnel needed to make $pH_i$ measurements. In addition, as with other means for determining regional $pH_i$, resolution of the acid–base status of specific, identified cells cannot be achieved.

Use of fluorescent probes for measuring $pH_i$ is growing in popularity (7). BCECF has become one of the most frequently used probes for such measurements in individual cells (see Chapter 9 by Dixon and Wilson, this volume). Cells loaded with BCECF are excited with ultraviolet light at one wavelength and the pH-dependent absorbance of the dye is then measured at another wavelength. Fluorescent probes have proved to be a useful technique for continuously measuring $pH_i$ from single, very small cells either in suspension or in culture, where they often tend to become flat and difficult to impale with microelectrodes. Many dyes, such as BCECF, are readily taken up into cells and then activated by native intracellular esterases, eliminating the need for cell impalement (7). Dyes can provide excellent temporal resolution, even surpassing that of pH-sensitive ISMs. Compared to potentiometry, however, the use of dyes for continuous $pH_i$ measurement in single cells in tissues (e.g., *in vivo*) remains difficult, being limited by technology available for illuminating and detecting light from isolated cells. In addition, the dye technique yields no parallel information on cell electrical properties unless it is combined with microelectrodes.

A distinct advantage of pH-sensitive microelectrodes, therefore, is that they are the most appropriate technology available for continuous $pH_i$ measurements in single cells of tissue-based preparations. They are, in addition, the most accurate means of attaining absolute values for $pH_i$. As mentioned above, pH ISMs also offer the opportunity for simultaneous measurement of cell electrical properties as well as other ion activities. This attractive feature can be maximized in triple-barrel array ISMs, where one barrel serves as a reference and the other two are selective for pH and another ion of choice (8). Finally, ISMs are capable of measuring pH from localized point sources in the cytoplasm of some cells (9).

The most important disadvantage of pH-sensitive ISMs arises from the need to impale cells. Before the advent of liquid exchanger-based ISMs, measurement of $pH_i$ was only possible in preparations robust enough to withstand penetration from the relatively large tips of microelectrodes based on pH-sensitive glass. The largest of cells had to be used to ensure that the pH-sensitive portion of the glass tip was entirely within the cell. The introduction of the "recessed-tip" design by Thomas (10) helped solve the latter problem by enclosing the pH-sensitive glass tip within a larger micropipette. Currently, electrodes based on liquid ion exchangers can be made with tips less than 1 $\mu$m in diameter. This not only reduces the likelihood of significant injury and subsequent alteration of cellular ionic activities, but also makes more certain the measurement of pH only within the intracellular compartment.

## $pH_i$ ISM Fabrication

The proton-sensitive barrel of a double-barreled microelectrode alone measures a potential related to pH* and the membrane potential of the cell. The reference barrel records the transmembrane potential that is subtracted from the ion barrel response. Several double- and multibarreled configurations have been employed in the construction of $pH_i$ ISMs, including side-by-side, eccentric, theta, and a variety of others with geometrically complex septums yielding multiple capillary compartments. The double-barreled electrodes we describe are based on (1) the eccentric format with delta glass on the inside and (2) Brown–Flaming theta glass. The eccentric format yields microelectrodes that produce stable penetrations, offer excellent pH response, and are easy to fabricate. Theta glass microelectrodes have the advantage of lower impedence reference barrels, facilitating marker injection during recording.

### *Preparing Glass Blanks*

Eccentric electrode blanks (15) are made by gluing longer length (10 cm) delta-shaped borosilicate glass tubes 1.2 mm (outer diameter) inside shorter length (7 cm), 3-mm (outside diameter) borosilicate glass tubes. (All glass in

*pH is traditionally approximated by the reciprocal of the hydrogen ion concentration (11), but this can obscure the complexity of simultaneously acting physiochemical constraints that actually determine pH in biological systems. The traditional approach of describing brain pH in terms of acid–base equilibria originating in the Brønsted–Lowry definitions and Henderson–Hasselbalch formulation (12) is theoretically correct but does not identify clearly those variables that play a dominant role in controlling brain pH. A different treatment has been provided by Stewart (13), in which he describes pH in terms of three independent variables: the strong ion difference (SID), defined as (sum of the strong cations) − (sum of strong anions), the total amount of weak acid present, $[A_{tot}]$, and the partial pressure of $CO_2$, $pCO_2$ (for discussion, see Refs. 13 and 14).

our construction recipes for eccentric ISMs can be attained from A-M Systems (Everett, Washington) or direct from Corning. The delta glass is Model #5910; the 3-mm tube is from Corning (Corning #230169). The inside glass (reference barrel) is made longer (approximately 5–7 mm) than the outside to help electrically isolate the two barrels. For gluing, we recommend using an ultraviolet light-sensitive epoxy glue (such as Crystal Clear; Loctite Corp.). The smallest droplet of glue is placed at both ends of the 3-mm glass tube. Capillary forces draw the glue inside along the point of contact between the tubing for a few millimeters. The blanks are then placed for 10 min under a full-spectrum ultraviolet light, which polymerizes the glue.

Brown–Flaming theta glass is purchased from R&D Scientific Glass Co. (Spencerville, Maryland). The glass has a septum/wall thickness ratio of 1.2 : 1 and outer diameter between 1.8 and 2.7 mm. One tube on each end should be ground down 5–7 mm to isolate the barrels electrically. A fixed, high-speed dremel with an abrasive wheel is appropriate. After pulling each set into two double-barreled microelectrodes, one tube will be longer than the other, and the same format as with the eccentric glass can be followed: the longer tube will serve as the reference, and the shorter will be ion selective.

A note should be made here about the cleaning of glass tubing, which has been the subject of considerable controversy. Some investigators have argued that no cleaning step is required (2). Others have recommended strong acids and organic solvents (16). Empirically, we have seen no differences in either the ease of fabrication or in the final performance whether glass is cleaned or not. This may result from the likelihood that, after pulling, 99.8% of the surface near the tip is freshly exposed glass (17).

## *Micropipette Formation*

Although intracellular micropipette tips in the submicrometer range can be pulled on a number of two-stage pullers, programmable air jet-based pullers, such as the Brown–Flaming models produced by Sutter Instrument Co. (40 Leveroni Court, Novato, California), yield excellent consistency in tip size, and, perhaps more importantly, offer outstanding capability for reducing shank and tip lengths for lower impedance electrodes. Reducing electrode resistance improves both selectivity and speed in the ion-sensitive barrel, lowers electrical noise, and aids ejection of a marker such as horseradish peroxidase or lucifer yellow through the reference barrel. Brown–Flaming pullers cool the glass tubing with an air stream while the tip is being formed, facilitating shorter shank lengths. The timing and strength of the machine's pull can be accurately controlled and specifically programmed, yielding reliable tip sizes in the submicrometer range.

Prior to pulling eccentric glass, we recommend one additional step. Eccentric glass, if glued properly, remains unglued toward the center of both tubes. In the past we have found this potential instability often leads to cracked reference tubing at some point later in the fabrication process. To avoid this frustration, we have started to twist our glass under heat prior to pulling on the Brown–Flaming. The idea is to fuse gently the glass tubes where they are not bonded by cohesive. The eccentric tubing is placed firmly on a PE-2 Narishige verticle puller and its coil is heated sufficiently to melt the glass. The puller's vertical descent is blocked from below after 4 mm. Once the glass has been stretched these few millimeters, we twist one-half of the tubing 180°, then remove the electrode assembly from the heat. Without this prior step, we find more than half our electrodes crack either during pulling or silanization; with it, cracking occurs in less than 5%.

After breaking or beveling the tip (see below), the ion barrel, when filled with electrolyte and no exchanger, should have a resistance not much greater than 20–40 MΩ. The resistance of the reference barrel, on the other hand, should not surpass 150 MΩ, otherwise it is much more difficult to eject intracellular markers. If a reference barrel resistance less than 150 MΩ is unachievable with the eccentric configuration, use theta glass.

## *Silanization*

Silanization of the ion barrel is a prerequisite for loading the ion exchanger and ensuring it is not displaced from the barrel tip throughout the life of the electrode. The basis for silanization arises from the inherent hydrophilic properties of borosilicate glass surfaces, which are endowed with a high density of hydroxyl groups (17). If left untreated, these surface hydroxyl groups prefer the aqueous phases of backfill and sample solutions, and repel organic liquids such as ion exchange membranes. Silanes and their derivatives react with the hydroxyl groups, rendering the glass surface hydrophobic. Thus when loading the hydrophobic exchanger into the ion barrel, one readily determines the success of silanization by observing the ease with which the exchanger flows from the shank to the tip of the electrode. Proper silanization is further evidenced by the stability of the ion exchanger–ion barrel electrolyte interphase when the latter is subsequently loaded. Poor silanization, on the other hand, is manifest in ISM misbehavior. Incomplete silanization produces electrodes that drift, or worse yet, do not respond at all, as the aqueous phase tracks in between the walls of the exchanger and glass, producing leak conductances or dislodging the exchanger all together. Excess silanization can plug the tip of the electrode with silane film and prevent filling of either electrolyte or exchanger.

Techniques designed to render the inner surface of the ion barrel hydrophobic have undergone many modifications and improvements. One approach we have used in the past was a liquid-phase process, in which a diluted siliconizing reagent was simply sucked and expelled from the ion barrel several times (14). Although sufficient for extracellular ISMs with larger diameter tips, the process proved inadequate for finer intracellular microelectrodes. Thus for fine-tipped microelectrodes, we initially used a process in which distilled water was added to the reference barrel and silane reagent to the ion barrel (18–20). The electrode was then heated in a two-step procedure using a hot air gun. The first heating was at 200°C while suction was applied to the ion barrel. This removed the liquid reagent. The second heating was at 700°C with suction applied to the reference barrel. This removed the water. We subsequently discovered the method worked only when fresh silane was diluted in a volatile organic solvent such as ether, suggesting our initial use of concentrated silane was successful only because the silane had degraded and become relatively dilute. Our current silanization method, described below, is a gas-phase process involving the introduction of silane vapor in nitrogen gas to the ion barrel (8).

After pulling the double-barreled tubing into a micropipette, it is mounted in a hood, and both the reference and ion barrel are pressurized with $N_2$ gas passed at 40 psi of pressure through 3-inch-long 28-gauge needles. (The Luer needles may be temporally sealed in the barrels with wax.) Then 1.5 cm of the array's tip is heated with a hot air gun for 2 min at 200°C to evaporate any residual water on the inside surface of the pipette. While the reference barrel remains pressurized with pure $N_2$ gas, a valve is switched, directing the $N_2$ gas from the ion barrel to travel first to a vessel containing *N,N*-dimethyltrimethylsilylamine (Fluka). A second valve is then opened, directing the mixed $N_2$/silane reagent vapor through the Luer needle in the ion barrel. The array is heated for another 2 min at the same temperature. The valves are finally switched back and pure $N_2$ gas is allowed to flow through both barrels for another 2 min with heat. Following this, the array is allowed to cool for 1.5 min, before returning it under a microscope for filling. As silane reagents age, their efficacy may diminish. Moreover, different glass types, wall thicknesses, and tip diameters usually have different optimal silanization requirements. Thus, our times and temperatures should only serve as approximations. The exact requirements for effective silanization ought to be determined empirically for each new series of electrodes if months have passed since the last were fabricated.

## *Preparation of Tridodecylamine-Based Proton Exchanger*

The tridodecylamine-based liquid membrane we describe is made up of the following ingredients in the indicated proportions: 10% w/w TDDA, 0.7%

w/w sodium tetraphenylborate, and 89.3% w/w 2-nitrophenyl octyl ether (3). These can all be obtained from Fluka Chem. Corp. (Ronkonkoma, New York). Although the exchanger can be purchased premixed, we prefer to make our own because it is less expensive and probably fresher. Age of exchanger may be important, because we have measured erroneously high levels of $pH_i$ in mammalian skeletal muscle cells >60 days after mixing the pH-sensitive cocktail (19). Therefore, we avoid using exchanger older than 2 months.

To prepare the exchanger, mix the above components in a total volume of 0.5 ml with constant stirring until dissolved. Next, place the components in 100% $CO_2$ with continued stirring for 12–16 hr. Incubation in $CO_2$ has been shown to reduce considerably the electrical resistance of the electrode. The reasons for this are unknown, but are thought in part to involve salt formation between TDDA and bicarbonate or carbonate (21). Although additional incubation times have previously been thought unnecessary (22), we recommend reexposure of the exchanger nightly to 100% $CO_2$ to help maintain low electrical impedance.

## *Micropipette Tip Beveling and Breaking*

After micropipettes have been silanized, their tips are broken or beveled to facilitate penetration of cells (23). Although beveling has some advantages over breaking—for example, there is more control over tip diameter, and tips may be less prone to obstruction by glass debris—we have found in most cases that breaking is more than sufficient for producing sharp tips that easily penetrate cells and yield stable recordings. We break ISM tips back to approximately 0.5–1.0 $\mu$m in total diameter by slowly advancing a cleaned, glass microscope slide edge toward the ISM using a hydraulic manipulator. The ISM tip is broken while it is viewed with a compound microscope at 400×. Breaking can be accomplished at any stage in the fabrication process after silanization, but breaking immediately after silanization aids filling of both barrels. If beveling is preferred, refer to the text by Brown and Flaming (23) for a thorough discussion. We use a BV-10 micropipette beveler manufactured by Sutter Instruments Co.

## *Filling*

Reference barrels are filled with electrolyte via the introduction of a small glass cannula containing 0.5 *M* KCl. Glass cannulas are made by pulling 1.2-mm outer diameter borosilicate capillary tubes on a verticle PE-2 Narishige puller. Tips long and thin enough to fit past the shank of the electrode are produced by heating the capillary tubes in an 8-mm-long, 7-mm-diameter coil at 15 A for approximately 15–30 sec. Then, as the softened tube elongates, resistance to gravity is tapered off by hand during the first few centimeters of descent. The wisp of the cannula is cut to an appropriate length, and the cannula is attached to a 1-ml syringe using silastic hose.

The concentration of salt in the reference electrolyte can vary somewhat without effecting performance. We have commonly used 0.5 *M* KCl for intracellular ISMs. To test whether different concentrations affect measurement of potential, we have filled individual eccentric electrodes with either 150 m*M* KCl or 0.5 *M* KCl in the outer barrels and 3*M* KCl in the inner barrels, and then used both varieties to measure rat muscle membrane potential. Mean membrane potentials measured with 150 m*M* KCl and 0.5 *M* KCl are, respectively, 0.9 + 0.2 mV and 1.1 + 0.2 mV less than the membrane potential measured with 3 *M* KCl (20). This translates to a pH measurement error of less than 0.02 pH. This error is near the resolution of pH ISMs (e.g., 0.01 pH), which in turn arises from the Ag–AgCl junction potentials of ± 1 mV (24).

To fill the reference barrel, mount an electrolyte-filled syringe and cannula onto a manipulator behind the electrode. Introduce the tip of the cannula and drive it near the shank of the electrode. Again, observation of this with a compound microscope is helpful. The reference barrel should easily fill if no silane reagent entered the tube and the tip of the electrode is not blocked.

The ion barrel is filled next with exchanger to a level that can be reached by each of the glass cannuli. Then a glass cannula filled with phosphate-buffered saline is advanced to just beneath the exchanger–air interface and the barrel is filled with this backfill solution. If this second cannula tip cannot contact the exchanger, a nonconducting air gap may form between the two solutions, because the phosphate buffer can no longer track along the now hydrophobic surface of the glass. Large air gaps within the exchanger column should be avoided. However, we have found that small air spaces do not compromise electrode function. Small air gaps are usually traversed by low-conductance pathways of either exchanger or backfill solution that exist along narrow-angled aspects of the glass barrel.

Occasionally, it may be necessary to produce a shorter ion exchanger column (i.e., <500 $\mu$m in length) because of temperature gradient effects on the millivolt response of the pH barrel (see below, *Characterization*). Shorter columns are produced using a "frontfill" procedure, in which the ion exchanger is drawn into the ion barrel through the electrode tip using suction. Briefly, after breaking the tip of the electrode back to 0.5–1.0 $\mu$m, backfill the reference barrel in the same fashion described above. Next, load the ion barrel with the backfill phosphate-buffered saline. This solution should not spontaneously fill to the tip, however, because the inner wall has been silanized. Insert a 3-inch-long, 28-gauge Luer needle into the ion barrel and seal it with wax. Application of 40–60 psi of positive pressure should now drive the phosphate-buffered saline to the tip of the electrode. If this is insufficient, the tip may be clogged. Briefly immersing the tip in pure $H_2O$ usually corrects this problem. After both the reference and ion barrels are filled to the tip of the electrode with their respective backfill solutions, immerse the tip in ion exchanger held and positioned in a glass capillary tube. Apply suction at 40 mm Hg, and within seconds to minutes an exchanger column should be drawn into the ion barrel.

## *Characterization*

Although corrections for interfering ions are necessary for many ion-selective membranes, the selectivity of the TDDA exchanger for $H^+$ ions is high enough that no significant interference is observable within a wide pH range (5.5–12) (4). Thus the response of the pH-sensitive ISM should be linear with respect to pH and can be described by the Nicolsky–Eisenmann equation, in which the values for interference drop out:

$$\text{mV} = B \log(a_{H^+}) + C, \tag{1}$$

where mV is the electrode potential in millivolts, ($a_{H^+}$ is the activity of hydrogen, $C$ is the reference potential in millivolts, and $B$ is equal to $nRT/F$, where $R$ is the gas constant (8.2 degree$^{-1}$ mol$^{-1}$), $T$ is the temperature (K), $F$ is the Faraday constant, and $n$ is an empirical constant chosen so that $nRT/F$ is the slope of the line when the potential is plotted as a function of log ($a_{H^+}$) with no interfering ions present. Constructing a calibration curve is easily accomplished using solutions made from a common buffer system, such as phosphate-buffered saline solutions. A typical curve relating ISM potential measurements to the negative logarithm of $H^+$ ion activity (i.e., pH) is shown in Fig. 1. ISMs should be calibrated before and after experimental measurements.

The temperature of the calibrating solutions should equal that of the experimental compartment from which measurements are taken. Otherwise, correlations between experimental and calibration potentials cannot be made unless a correction factor is introduced. In general, temperature differences may change activity and selectivity coefficients as well as liquid junction and standard ($RT/zF$) potentials. Differences in temperature have also been shown to affect the sensitivity of the exchanger along the ion-selective column (25). Vaughan-Jones and Kaila demonstrated that ISM responses can vary where there are temperature gradients between an immersed electrode tip and the top of the ion exchanger column. We have tested our TDDA-based

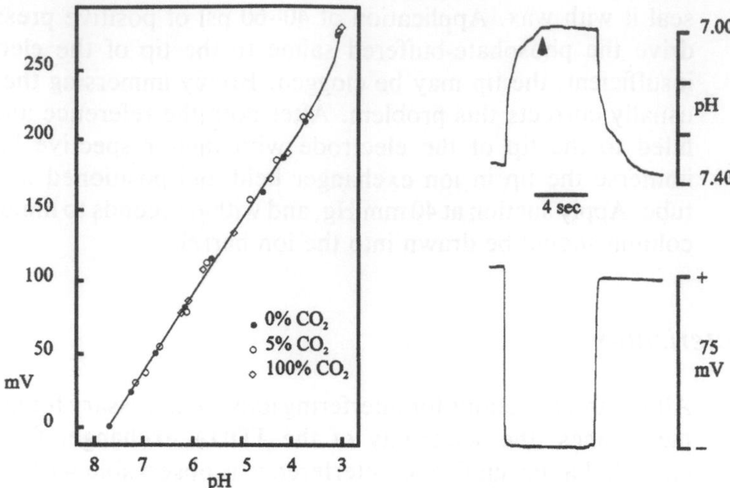


FIG. 1 Characterization of pH-sensitive ISM. On the left is a calibration curve indicating the millivolt response to pH at three separate carbon dioxide tensions. Calibration solutions consisted of 50 m*M* potassium phosphate and were set to different pH values with either 1 *M* NaOH or 1 *M* HCl. Their pH was measured with a benchtop Corning pH meter and rechecked 24 hr later. Three separate groups of calibration solutions are equilibrated at different carbon dioxide tensions (0, 5, and 100%). Here, the electrode response is linear between pH 3.80 and 7.70 with a slope of 53 mV. Electrode response is not influenced by any changes in carbon dioxide tension. On the right, the response time of the ISM is approximated by penetrating skeletal muscle cells of an anesthetized rat. The upper trace demonstrates 95% response time (arrowhead) of 1.9 sec for the pH-sensitive barrel. The lower trace shows the membrane potential change recorded simultaneously by the reference barrel. From Kraig and Chesler, Ref. 20, with permission.

ISMs for this possible error and have observed similar behavior. As an ISM is advanced from room temperature air into a stirred bath warmed to 37°C, the response of the pH barrel can change up to 20 mV, which corresponds to a difference in pH of 0.35. The largest millivolt shift occurs as the surface of the heated bath passes near the shank of the electrode, which in turn corresponds to the interface between the exchanger and backfill solution. Before this depth, only small (i.e., <5mV total) variations in potential are seen. When the bath is at room temperature (approximately 22°C) no such changes in millivolt response are observed at any depth. If temperature gradients are a potential source of error, short exchanger column electrodes can be constructed using the frontfill method. No shift in millivolt response is observed with short (i.e., ≤500 $\mu$m) column electrodes.

The potential interference of $CO_2$ with electrode performance has continued to worry some investigators (26), even though a body of evidence suggests otherwise (reviewed in Ref. 7). We have tested for $CO_2$ interference by testing electrode response in calibrating solutions equilibrated with different carbon dioxide tensions (Fig. 1). No change in slope or offset potential is observed with carbon dioxide tensions ranging from 0 to 100%.

Many different conventions defining response times for ISMs exist in the literature. The investigator should always state the method used. We have defined electrode response time as the amount of time required for electrode potential to change from $E_1$ to 95% of $E_2$. To estimate response time, a double-barreled pH-sensitive electrode tip is quickly advanced into skeletal muscle cells of an anesthetized and artifically ventilated rat. With penetration, the microelectrodes record a 95% response time in $1.9 \pm 0.1$ sec (range 0.9–3.2 sec, $n = 7$) (Fig. 1). This method is convenient because it not only estimates electrode response time but also serves to confirm the ability of an electrode to measure $pH_i$ accurately prior to experiments in brain cells.

## *Reference Junction Potentials*

In general, using more concentrated electrolyte solutions (e.g., 3 *M* KCl vs. 150 m*M* KCl) reduces the junction potential, but higher concentrations increase the likelihood of disrupting normal intracellular concentrations of ions. High electrolyte concentrations may become a significant source for error, because $pH_i$ has often been shown to be interdependent, directly and indirectly, on the distribution of several ionic species (27). Solutions with high osmolarity can also produce cell swelling, which in turn may alter $pH_i$. Using anions other than $Cl^-$, such as acetate, avoids the problem of $Cl^-$ contamination, but usually introduces another potential source of error. Acetate, or any other conjugate base of a weak acid, may produce alkalinization, particularly in smaller cells, because of the subsequent efflux of its protonated form. Care should be taken when choosing electrolyte concentrations; possible errors, for instance, can be ascertained by comparing differences in recorded potentials between two different concentrations of electrolyte.

## *Recording Equipment*

The details of electronics required for microelectrode recordings have been summarized previously (28). Amplifiers, filters, and analog or digital recording hardware and software appropriate for intracellular recording are now readily available from companies such as Axon Instruments, Inc. (1101 Chess Drive, Foster City, California) and often come with manuals and technical support to ease setup and implementation, obviating the need for extensive discussion here. Briefly, amplifier circuitry should have low noise and drift, possess built-in capacitance and resistance compensation, carry an onboard option for injecting controlled current, and have an input impedance capacity at least two orders of magnitude greater than the impedance of the electrode barrel to which it is connected. We have used the Axoprobe A-1 amplifier system from Axon Instruments. The Axoprobe A-1 has the added built-in capacities to properly subtract the reference barrel potential from the ion-selective potential, and match the reference and ion barrel response speeds through capacitance neutralization. Subtraction of electrode potentials can also be accomplished by other means (reviewed in Ref. 29). Our data is frequently filtered at 2 Hz, displayed on a strip chart recorder, and stored on videotape (DR-484; Neurodata Instruments, New York, New York), as well as digitized (Digidata, Axon Instruments) and stored (Axotape 2.02, Axon Instruments) on computer for subsequent analysis. Although chart recorders are now being replaced by analog/digital converters and computers loaded with data acquisition software such as Axotape, which speed data analysis, chart recorders are not an unwise investment, because they provide an on-line hard copy that can aid the investigator during experimentation. An entire experiment can be seen with one sweeping glance; on a computer screen, one must switch back and forth to assess the entire record. Subtle electrode shifts occurring over a long period, for instance, are more quickly realized on the chart.

Proper grounding and shielding of the recording setup is essential for preventing noise interference. High-impedance intracellular ISMs pick up 60-Hz electrical noise from a variety of sources as well as higher frequency noise from heaters, pumps, microscopes, oscilloscopes, and computers. In addition, ISMs are particularly sensitive to nearby motion because of resultant electrostatic noise. Shielding from outside sources can be accomplished by enclosing the recording rig inside a Faraday cage. The 60-Hz hum from light sources, electronic manipulators, heating devices, etc. within the cage is avoided by using only DC current power supply. Internal noise from recording devices is reduced through proper grounding. Effective grounding techniques have been covered by Geddes (24).

## Measurement of Astrocytic $pH_i$ *in Vivo*

The utility of pH-sensitive ISMs for measuring $pH_i$ in brain has been demonstrated in a series of experiments in which glial $pH_i$ was measured under both physiological and pathological conditions. Following electrical stimulation of the brain, ISM measurements in astrocytes record marked alkaline shifts, whereas parallel measurements in neurons and the interstitial space indicate a predominant acid shift. The astrocytic change in $pH_i$ appears to correlate generally with the extent of depolarization; thus, with maximum depolarization during spreading depression (SD), astrocytic $pH_i$ rises from 7.0 to 7.6–7.8 before returning to normal as cells repolarize. During complete ischemia with hyperglycemia, however, astrocytes exhibit an extreme acidosis, with a measured $pH_i$ more acid than surrounding neurons and interstitial space. A peak astrocytic $pH_i$ of <4.0 has been observed, with more typical values approximating 5.30.

Cortical glia are identified, after penetration, from electrophysiological criteria consisting of a lack of injury discharge, absence of spontaneous or synaptic potentials, and a high membrane potential (>65 mV). Cells with these electrical characteristics are consistently identified as protoplasmic astrocytes following injection of horseradish peroxidase (18). Neurons can also be impaled, as indicated by injury and spontaneous discharges and lower membrane potentials. Stable glial impalements can last as long as 80 min, but are typically maintained for approximately 10 min. With penetration, resting $pH_i$ rapidly settles to steady state (Fig. 2). In all glial cells measured to date in our laboratory, the mean resting $pH_i$ was 7.04 ± 0.02 (range 6.73–7.38; Fig. 2) (Refs. 18, 19, and 30).

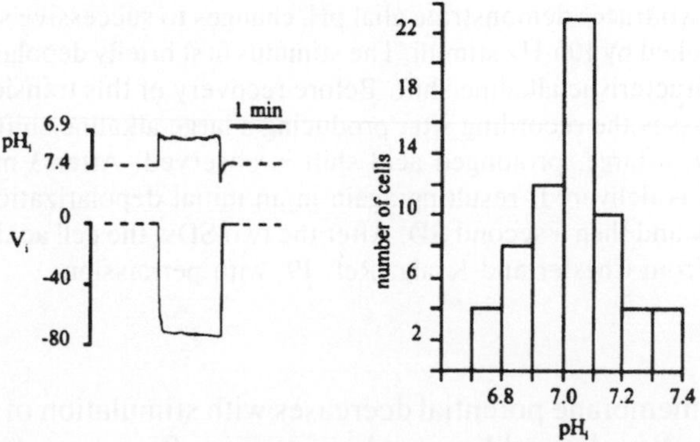


FIG. 2 Characterization of resting $pH_i$ in glial cells. The left-hand figure demonstrates the pH response (upper trace) rapidly settling to a new steady state following penetration of a neocortical glial cell. The lower trace shows a slower recovery of membrane potential to a final maximum. In this study (19), average membrane potential was 75 mV in 51 cells. More recent measurements demonstrate an average membrane potential of 81 mV in 12 cells (30). On the right is a histogram representing the $pH_i$ of all glial cells measured to date in this laboratory. Average $pH_i$ is 7.04, with a range of 6.73–7.38. Of note is the observation that cells closer to the pial surface are typically more alkaline. Modified from Chesler and Kraig, Ref. 19, with permission.

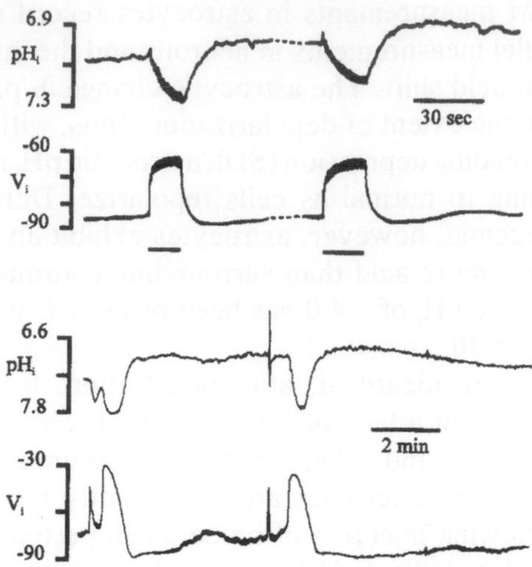


FIG. 3 Glial cell response to surface stimulation. The two upper traces show a recording from a glial cell near the pial surface at a resting $pH_i$ of 7.08 and membrane potential of 86 mV. With stimulation (20 Hz, indicated by the two horizontal bars), $pH_i$ shifts alkaline to 7.25 as membrane potential falls to a new steady state of 62 mV. Note that after stimulation, membrane potential hyperpolarizes before returning to baseline. After 4 min, a second 20-Hz stimulation is delivered, and glial pH shows a stereotypic alkaline, then acid, response, followed by an acidic plateau of 6.85. Thus recurrent stimuli, sufficiently close in time, can progressively acidify glia. The bottom two traces demonstrate glial $pH_i$ changes to successive spreading depressions (SD), evoked by 100-Hz stimuli. The stimulus first briefly depolarizes the cell, producing a characteristic alkaline shift. Before recovery of this transient, however, a wave of SD passes the recording site, producing a large alkaline shift of nearly 7.80. With recovery, a large, prolonged acid shift is observed. After 3 min, a second 100-Hz stimulus is delivered, resulting again in an initial depolarization, followed by a 40-sec delay and then a second SD. After the two SDs, the cell acidifies to a new plateau of 6.6. From Chesler and Kraig, Ref. 19, with permission.

Glial membrane potential decreases with stimulation of the cortical surface (Fig. 3). A prolonged hyperpolarization is often seen after stimulation. The depolarizing response evoked by stimulation ranges from 4 to 38 mV, and depends on both the frequency and the proximity of the stimulation. This electrical behavior matches previous electrophysiological observations in glial cells (31). The membrane depolarization is always accompanied by an intracellular alkaline shift (Fig. 3). $pH_i$ recovers in tens of seconds after stimulation and sometimes continues into an acid rebound (19 of 50 responses; Fig. 3).

Parallel observations suggest these $pH_i$ responses reflect genuine biological phenomena. First, the alkaline shift is still seen in poorly penetrated glial cells, despite membrane potentials of only 30–40 mV. In comparable neuronal penetrations, however, stimulation never evokes an alkaline shift. Only a slow acidification is observed within neurons, in agreement with observations from frog spinal cord (32). Moreover, only an acid shift is measured when pH ISMs record from the interstitial space (Fig. 4). Finally, older ISMs with exceedingly slow pH response times record normal membrane depolarizations, but little or no pH response. Thus only pH ISMs with functional ion barrels recorded the glial alkaline response, suggesting the alkaline $pH_i$ shifts are not due to any subtraction artifacts.

A greater depolarization, as occurs during cortical SD, produces an even larger intracellular alkaline shift. During SD, $K_o^+$ can reach levels in excess of 40 m*M* (33). Indeed, SD evoked by two brief trains of stimuli at 100 Hz produce glial alkaline transients that are larger than those initiated by electrical stimulation (Fig. 3). For 26 observations of SD, the peak alkaline shift was 0.32 ± 0.04 (range 0.11–0.78). During repolarization, $pH_i$ returns and goes into acid rebound concomitant with hyperpolarization (Fig. 3). The final acid shift above baseline was 0.19 ± 0.02 (range 0.03–0.39) in these experiments.

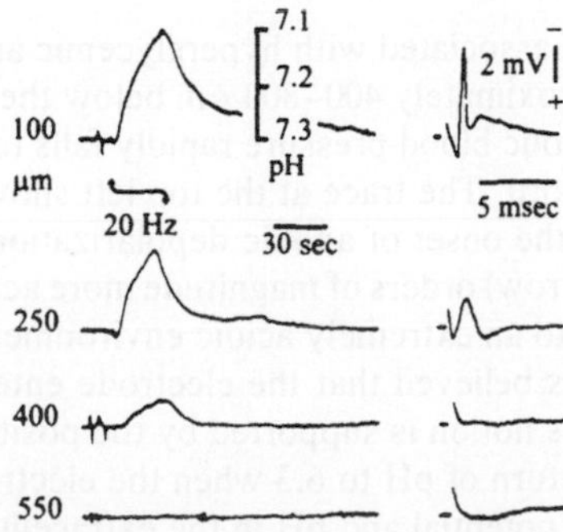


FIG. 4 Extracellular pH changes in response to surface stimulation at 20 Hz; recordings made from rat cerebellar cortex. $pH_o$ changes in the cerebral cortex are analogous to changes in the cerebellar cortex. The traces on the left demonstrate a depth profile of $pH_o$ changes associated with recurrent surface stimulation. $pH_o$ changes in response to surface stimulation for 30 sec at 20 Hz (beginning with the upward arrow and ending with the downward arrow) are recorded along a parallel fiber beam, with the stimulus being repeated as the pH ISM is advanced at increments of 150 $\mu$m. pH in the extracellular microenvironment swings principally acid in both cerebral and cerebellar tissue. Traces on the right indicate the simultaneously evoked field potentials at these depths. From Kraig *et al.*, Ref. 14, with permission.

During hyperglycemic and complete ischemia, measurements in presumed astrocytes suggest extreme acidosis. Following cardiac arrest, two astrocytic loci more acidic than the interstitial space (6.17–6.20 pH) are found. The more acidic locus (4.30 ± 0.19, $n$ = 5; range 3.82–4.89) is observed closer to the onset of anoxic depolarization; the less acidic locus (5.30 ± 0.07, $n$ = 53; range 4.46–5.93) is measured generally 5–46 min after cardiac arrest. The most extreme acid peak measured (pH 3.82) is shown in Fig. 5 on the left; this observation illustrates the considerable range of TDDA-based intracellular ISMs. More typical $pH_i$ changes are shown on the right in Fig. 5.

The measurement of glial $pH_i$ in these experiments showcase the use of intracellular pH-sensitive ISMs in identified cells of nervous tissue. Intracellular pH ISMs are capable of simultaneously recording changes in the electri-

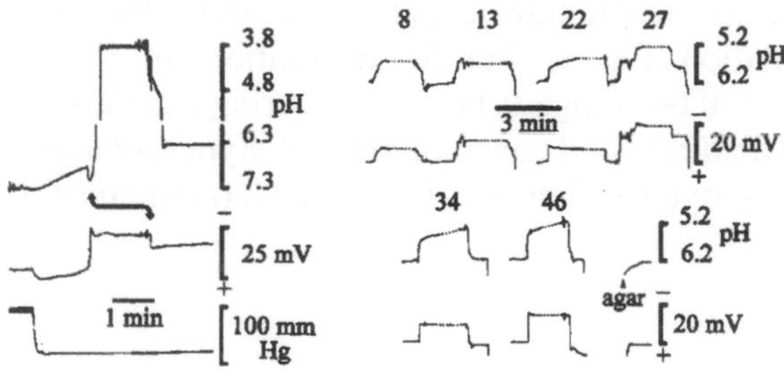


FIG. 5 Acid shifts associated with hyperglycemic and complete ischemia. pH ISMs are positioned approximately 400–800 $\mu$m below the parietal pial surface. Following cardiac arrest, systolic blood pressure rapidly falls to a minimum, as indicated in the lower left-hand record. The trace at the top left shows first a slow extracellular acid shift, and then, at the onset of anoxic depolarization (left middle trace), indicates a pH shift (upward arrow) orders of magnitude more acidic than the extracellular space, suggesting entry into an extremely acidic environment. In this example, $pH_i$ reaches a peak of 3.82. It is believed that the electrode enters glial cells as they swell with depolarization. This notion is supported by the positive shift in DC potential (downward arrow) and return of pH to 6.3 when the electrode is raised 200 $\mu$m, which are expected values of potential and pH in the extracellular space of hyperglycemic and completely ischemic brain. The record of pH is interrupted to allow visualization of lower and uppermost portions of the record. On the right, the top and third traces show pH changes and the second and fourth panel show simultaneous DC changes. The numbers above each pH change indicate the number of minutes after ischemia. Breaks in pH and DC traces (e.g., at 34 and 46 min) occur when the electrode was withdrawn from the brain. After a final measurement, the electrode was removed from the brain and advanced into a phosphate-buffered (6.20 pH) agar gel (3.5%) containing 150 m$M$ NaCl. The dotted lines in each record indicate a 3-min period, during which horseradish peroxidase was electrophoretically injected for later morphological identification. From Kraig and Chesler, Ref. 20, with permission.

cal properties while accurately capturing the dynamics of $pH_i$ changes in both the acid and alkaline direction. Compared to other techniques for measuring $pH_i$ in intact preparations, intracellular pH ISMs offer superior temporal and spatial resolution, are relatively inexpensive, and are easy to construct and implement. Because of these attributes, pH ISMs are a superior technology for capturing intracellular acid–base dynamics in brain tissues.

## Acknowledgments

R. P. K. was supported by The National Institute of Neurological Disorders and Stroke Grant NS-19108, an Established Investigator Award from the American Heart Association, the American Heart Association of Metropolitan Chicago, and the Brain Research Foundation of the University of Chicago. C. L. was supported by the Training Grant HD-07009 from the USPHS.

## References

1. P. C. Caldwell, *J. Physiol.* (*London*) **126,** 169 (1954).
2. R. C. Thomas, "Ion-Selective Intracellular Microelectrodes: How to Make and Use Them," Academic Press, London, 1978.
3. D. Ammann *et al.*, *Anal. Chem.* **53,** 2267 (1981).
4. D. Ammann, "Ion-Selective Microelectrodes." Springer-Verlag, Berlin, 1986.
5. A. Roos and W. F. Boron, *Physiol. Rev.* **61**(2), 296 (1981).
6. R. A. Kaupinen *et al.*, *Trends Neurosci.* **16,** 88 (1993).
7. I. Kurtz, *in* "Fluorescent and Luminescent Probes for Biological Activity" (W. T. Mason, ed.), Academic Press, London, 1993.
8. K. Wietasch and R. P. Kraig, *Am. J. Physiol.* **261,** R760 (1990).
9. B. C. Gibbon and D. L. Kropf, *Science* **263,** 1419 (1994).
10. R. C. Thomas, *J. Physiol.* (*London*) **238,** 159 (1974).
11. A. K. Covington, *et al., Pure Appl. Chem.* **57,** 531 (1985).
12. B. K. Siesjo, "Brain Energy Metabolism." Wiley, New York, 1978.
13. P. A. Stewart, "How to Understand Acid-Base." Elsevier, New York, 1981.
14. R. P. Kraig *et al.*, *J. Neurophys.* **49,** 831 (1983).
15. R. C. Thomas, *J. Physiol.* (*London*) **371,** 24P (1985).
16. J. F. Garcia-Diaz and W. M. Armstrong, *J. Membr. Biol.* **55,** 213 (1980).
17. F. Deyhimi and J. A. Coles, *Helv. Chim. Acta* **65,** 1752 (1982).
18. M. Chesler and R. P. Kraig, *Am. J. Physiol.* **253,** R666 (1987).
19. M. Chesler and R. P. Kraig, *J. Neurosci.* **9,** 2011 (1989).
20. R. P. Kraig and M. Chesler, *J. Cereb. Blood Flow Metab.* **10,** 104 (1990).
21. M. Muhammed, *Acta Chem. Scand.* **26,** 412 (1972).
22. A. C. Kurkdjian and H. Barbier-Brygoo, *Anal. Biochem.* **132,** 96 (1983).

23. K. T. Brown and D. G. Flaming, "Advanced Micropipette Techniques for Cell Physiology." Wiley, Chichester, 1986.
24. L. A. Geddes, "Electrodes and the Measurement of Biolelectric Events." Wiley, New York, 1972.
25. R. D. Vaughan-Jones and K. Kaila, *Pfluegers Arch.* **406,** 641 (1986).
26. J. A. Coles, *in* "Practical Electrophysiological Methods" (H. Kettenmann and R. Grantyn, eds.), Wiley, New York, 1992.
27. M. Chesler, *Prog. Neurobiol.* **34,** 401 (1990).
28. R. D. Purves, "Microlelectrode Methods for Intracellular Recording and Ionophoresis." Academic Press, New York, 1981.
29. W. G. Carlini and B. R. Ransom, *in* "Neurophysiological Techniques" (A. A. Boulton, G. B. Baker, and C. H. Vanderwolf, eds.), p. 227. Humana, New Jersey, 1990.
30. M. S. Swain, A. T. Blei, R. F. Butterworth, and R. P. Kraig, *Am. J. Physiol.* **261,** R1491 (1991).
31. R. K. Orkand, J. G. Nicholls, and S. W. Kuffler, *J. Neurophysiol.* **29,** 788 (1966).
32. W. K. Endres, K. Bellanyi, G. Serve, and P. Grafe, *Neurosci. Lett.* **72,** 54 (1986).
33. F. Vyskocil, N. Kriz, and J. Bures, *Brain Res.* **39,** 255 (1972).

# [11] Measurement of Cytosolic pH by Nuclear Magnetic Resonance Spectroscopy

Ognen A. C. Petroff and James W. Prichard

## Introduction

Hydrogen ion concentration is basic to the function of all biological fluids. Free protons are hydrated in aqueous solution to form hydronium ions, the concentration of which is usually expressed as its negative logarithm, or pH. This important quantity is difficult to measure accurately in living tissues, because any kind of tissue-disruptive sampling procedure changes it, and exogenous indicators may be unevenly distributed in tissue, be toxic, or both.

In 1973, Moon and Richards (1) reported nondestructive measurement of cytosolic pH in red blood cells by nuclear magnetic resonance (NMR) spectroscopy. Much subsequent work has shown that NMR methods are especially well suited for pH measurements in living tissue. Being noninvasive, they can be used repeatedly on the same sample without disturbing it and on human subjects without fear of harm; their versatility ranges from cell suspensions to intact, functioning mammalian organs; for the most part, they use endogenous indicators, but exogenous ones are available for special tasks. The widespread and growing availability of NMR machines has, or soon will have, placed the capability within reach of most investigators whose work can benefit from it.

This chapter describes the basis of NMR pH measurements in biological tissues. Because much work, including that of the authors, has been done on measurement of brain pH by $^{31}P$ spectroscopy, most of the specific examples given are from that area. Examples from other tissues and other nuclei are given to emphasize the versatility of NMR methods and provide the interested reader access to branches of the literature that can only be mentioned briefly here.

## Physical Basis of Cytosolic pH Measurement by NMR

### *NMR History*

The NMR phenomenon was described in 1946 (2, 3). It has been productive in a wide range of scientific applications since then. When NMR instruments large enough to use on humans became available in the 1980s,

*Methods in Neurosciences, Volume 27*

magnetic resonance imaging (MRI) emerged as the premier imaging technique for medical diagnosis (4), replacing X-ray methods for many purposes.

The utility of NMR methods for noninvasive measurement of many biological variables derives from the fact that resonant frequencies of nuclear precession can be detected from $^{1}H$, $^{31}P$, $^{13}C$, $^{23}Na$, and several other magnetic nuclei that are natural constituents of biological tissue, and from nonbiological nuclei such as $^{19}F$, which is administered as exogenous tracers. Slight differences in resonant frequency from a single nucleus often identify and allow measurement of various specific compounds containing the nucleus. Measurements are made by placing the subject in a strong, static magnetic field that aligns magnetic nuclei in an orderly way, allowing radiofrequency signals from large populations of the nuclei to be detected after controlled perturbation by small injections of radiofrequency energy. Many different variables can be measured, depending on which nuclei and which compounds containing them are chosen for study. The very strong $^{1}H$ signal from tissue water is the basis of MRI. Much weaker signals from nonwater $^{1}H$ and other nuclei allow measurement of specific compounds such as adenosine triphosphate (ATP), lactate, and some amino acids. For convenience, study of biological compounds other than water is often designated magnetic resonance spectroscopy (MRS) to distinguish it from MRI.

## *Inorganic Phosphate: The Best NMR Indicator of* $pH_i$

The $^{31}P$ spectrum of most biological tissues includes a signal from inorganic phosphate ($P_i$) that can be used to measure mean cytosolic pH in volumes of tissue containing millions of cells. Figure 1 shows such a spectrum from the brain of a living rabbit. The resonant frequency of the $P_i$ signal changes with pH. Because most NMR-observable $P_i$ is in cytosol, the signal principally reflects the pH of the intracellular compartment ($pH_i$).

NMR measurements of pH depend on resonant frequency differences ("chemical shift" in NMR language) between the conjugate acid and base forms of the compound used as a pH indicator. If the rate of proton exchange between the acid and base forms is slow relative to the duration of the NMR measurement, separate signals from each can be detected, and the ratio of their intensities—corrected if necessary for differences in relaxation properties—can be substituted for the base/acid concentration ratio in the Henderson–Hasselbalch equation

$$\mathrm{pH} = \mathrm{p}K + \log([\mathrm{base}]/[\mathrm{acid}]).$$

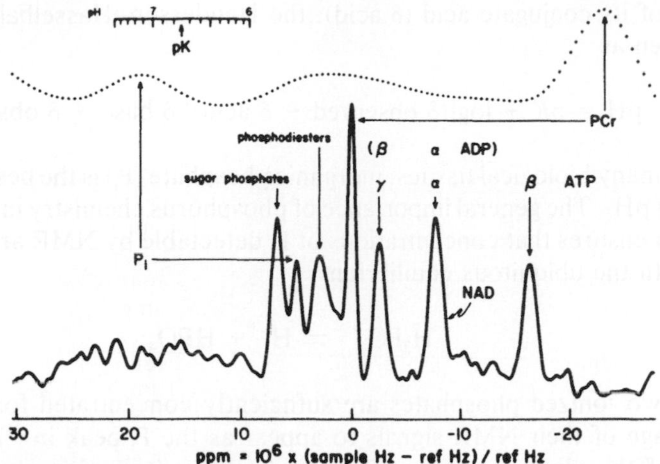


FIG. 1 The $^{31}P$ spectrum of rabbit cerebrum. The seven resonances labeled from left to right are assigned to phosphomonoesters (sugar phosphates), phosphate ($P_i$), phosphodiesters, phosphocreatine (PCr), and the three resonances from the nucleoside phosphates ($\gamma$-, $\alpha$-, and $\beta$-ATP, $\beta$- and $\alpha$-ADP, and NAD). The dotted line at the top is an expansion of the spectrum highlighting the chemical shift difference between $P_i$ and PCr, which is the basis for the pH measurement. A scale showing how the chemical shift of $P_i$ is related to pH is located above the $P_i$ signal. The p$K_2$ of phosphate under conditions approximating those of cerebral cytosol (see Table I) is marked with an arrow. The formula on the chemical shift axis shows how its units (parts per million) are calculated from the frequency of the signal of interest (sample Hz) and the reference frequency (ref Hz). The figure is reproduced with permission from *Neurology* **35,** 781–788 (1985).

In this case, the precision of pH estimation depends on precision of the intensity measurements of the two conjugates and determination or assumption of a p$K$ value appropriate for the conditions.

If the rate of proton exchange between the base and acid forms is fast relative to the duration of the NMR measurement, a single signal is observed. Its chemical shift is the concentration weighted average of the chemical shifts of the acid and base forms of the indicator. Here, precision of the pH estimation depends on (1) the precision with which the chemical shift of the exchange-averaged signal can be measured and (2) determination or assumption of the chemical shifts of both the acid and base forms of the indicator, as well as p$K$. For an observed chemical shift (δ observed) composed of a weighted average between the shift of a pure base (δ base) and that of its conjugate acid (δ acid), the Henderson–Hasselbalch equation is written as

$$pH = pK + \log[(\delta\ \text{observed} - \delta\ \text{acid})/(\delta\ \text{base} - \delta\ \text{observed})].$$

In many biological tissues, inorganic phosphate ($P_i$) is the best NMR indicator of $pH_i$. The general importance of phosphorus chemistry in energy metabolism ensures that concentrations of $P_i$ detectable by NMR are usually present. In the ubiquitous equilibrium

$$H_2PO_4^{1-} \rightleftharpoons H^+ + HPO_4^{2-},$$

the two ionized phosphates are sufficiently concentrated for the weighted average of their NMR signals to appear as the $P_i$ peak in $^{31}P$ spectra. The p$K$ of this dissociation is 6.77, which places its region most sensitive to titration well within the range of biological interest. (The p$K$ values of the first and third dissociations of phosphoric acid are so far from the mammalian pH range that $H_3PO_4$ and $PO_4^{3-}$ are too dilute to contribute to the $P_i$ signal.) Hence, for NMR measurements of pH based on dissociation of phosphoric acid, the appropriate form of the Henderson–Hasselbalch equation is

$$pH = pK + \log[(\delta P_i - \delta H_2PO_4^{1-})/(\delta HPO_4^{2-} - \delta P_i)],$$

where $\delta P_i$ is the observed chemical shift of the weighted average signal from $H_2PO_4^{1-}$ and $HPO_4^{2-}$, and $\delta H_2PO_4^{1-}$ and $\delta HPO_4^{2-}$ are the chemical shifts of the pure compounds known from separate measurements.

Anything that affects the ratio $[H_2PO_4^{1-}]/[HPO_4^{2-}]$ changes $\delta P_i$ and therefore changes the pH inferred from it. Factors known to affect $P_i$ titration include temperature, ionic strength, and the concentrations of magnesium, calcium, protein, sodium, potassium, and the inorganic phosphates (5). Kushmerick and colleagues have compiled a number of pertinent association constants (6). Rising temperature shifts the p$K$ values of phosphoric acid in the alkaline direction (7, 8). Because monobasic and dibasic phosphates are both charged, ionic strength and cations can affect both the chemical shift of the end points and the p$K$ of the $P_i$ titration curve (8), but these factors have been shown to be minor ones. Changes in ionic strength, which can occur under pathological conditions, have little effect (9, 10). Even the drastic change of substituting 150 m$M$ potassium chloride for 150 m$M$ sodium chloride shifts the p$K$ by only 0.1, and its effect on chemical shift of the titration end points is small (7, 11). Changes in the concentrations of calcium and magnesium over the range expected in brain have little influence on the p$K$ of the $P_i$ titration curve (5, 7, 11–13).

Table I lists Henderson–Hasselbalch constants for titration of $P_i$ in various tissues. The differences are due to factors such as those discussed above, some of which vary across tissues. They are large enough to affect NMR $pH_i$ measurements in living tissue at its usual level of precision of about 0.1 pH unit (see below).

## *Chemical Shift References for NMR Measurement of $pH_i$*

The chemical shift of any NMR indicator of $pH_i$ must be measured relative to some signal of known frequency. Many workers have used 85% phosphoric acid as an external chemical shift reference for measurement of $pH_i$ by $^{31}P$ spectroscopy (1, 10, 13). However, several problems afflict this procedure (10, 14). The main one is that concentrated phosphoric acid has a different magnetic susceptibility from the aqueous solutions that comprise all animal tissues. Chemical shifts measured relative to it vary with respect to orienta-

TABLE I Henderson–Hasselbalch Constants for $P_i$ Titration in Various Tissues[a]

| Tissue | p*K* | Acid (ppm) | Base (ppm) | Ref. |
|---|---|---|---|---|
| Brain | 6.77 | 3.29 | 5.68 | (5) |
| Heart | 6.72 | 3.27 | 5.69 | [b] |
| Liver | 6.75 | 3.36 | 5.76 | [c] |
| Kidney | 6.80 | 3.19 | 5.76 | [d] |
| Smooth muscle | 6.77 | 3.39 | 5.69 | (6) |
| Striated muscle | 6.54 | 3.38 | 5.70 | [e] |
| Erythrocytes | 6.85 | 4.05 | 6.18 | [f] |
| Mitochondria | 6.85 | 3.19 | 5.62 | [g] |
| Bacteria | 6.83 | 3.15 | 5.65 | [h] |

[a] All normalized to the resonant frequency of brain phosphocreatine as 0.

[b] T. A. Bailey, S. R. Williams, G. K. Radda, and D. G. Gadian, *Biochem. J.* **196,** 171 (1981).

[c] C. R. Malloy, C. C. Cunningham, and G. K. Radda, *Biochim. Biophys. Acta* **885,** 1 (1986).

[d] J. J. H. Ackerman, M. Lowry, G. K. Radda, B. D. Ross, and G. G. Wong, *J. Physiol.* (*London*) **319,** 65 (1981).

[e] D. J. Taylor, P. J. Bore, P. Styles, D. G. Gadian, and G. K. Radda, *Mol. Biol. Med.* **1,** 77 (1983).

[f] I. M. Stewart, B. E. Chapman, K. Kirk, P. W. Kuchel, V. A. Lovric, and J. E. Raftos, *Biochim. Biophys. Acta* **885,** 23 (1986).

[g] S. Ogawa, C. C. Boens, and T. Lee, *Arch. Biochem. Biophys.* **210,** 740 (1981).

[h] J. L. Slonczewski, B. P. Rosen, J. R. Alger, and R. M. Macnab, *Proc. Natl. Acad. Sci. U.S.A.* **78,** 6271 (1981).

tion of the samples to the magnetic field; a difference of 0.85 ppm has been reported (10). Signals from concentrated phosphoric acid overlap with signals from phosphorus metabolites present in biological tissues, thus not providing the freedom from signal interaction that can be one of the advantages of external shift references.

An internal chemical shift reference is desirable for measurement of $pH_i$ *in vivo* because it is subject to the same magnetic fields as the indicator signal. In $^{31}P$ spectra of brain and muscle, the phosphocreatine (PCr) signal is useful for this purpose; it is large, it has a narrow linewidth (Fig. 1), and its frequency is not sensitive to pH over most of the range of physiological interest. Figure 2 shows the relationship between pH and the chemical shift of $P_i$ with PCr as an internal chemical shift reference. Most $pH_i$ changes of biomedical interest occur along the steep portion of the curve where the $P_i$

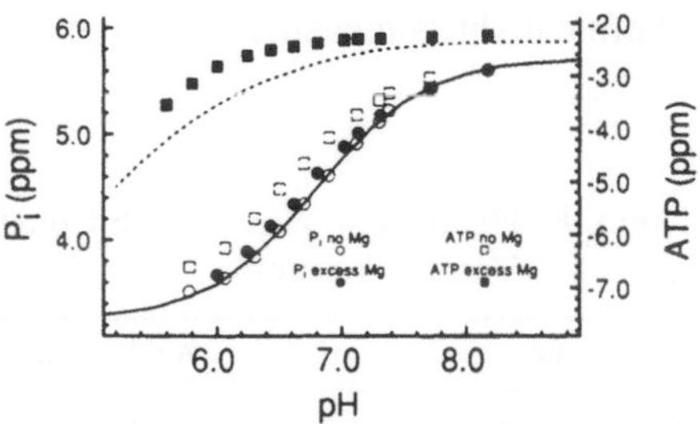


FIG. 2 Titration curves for phosphate ($P_i$) and γ-ATP in solutions modeling brain cytosol. The chemical shift of phosphate in model solutions was measured as the pH was changed in the absence and presence of excess (6 m*M* Mg) magnesium. Phosphocreatine was assigned a chemical shift of 0 ppm. Its stability was checked by comparison to signals from an external reference solution. The solid line shows the Henderson–Hasselbalch equation using the constants listed for brain in Table I. The chemical shift of γ-ATP with respect to phosphocreatine was measured simultaneously with the phosphate shift in the absence and presence of excess (6 m*M* Mg) magnesium. The dashed line shows the Henderson–Hasselbalch equation for γ-ATP when 80% of the total is complexed with magnesium. This percentage was based on unpublished measurements of the percentage MgATP of rabbit cerebrum under basal conditions. The $^{31}P$ spectra of model solutions were obtained at 37°C with a 10-mm broadband probe operating at 145.78 MHz. A fixed concentric inner tube, oriented parallel to the long axis of the main tube, contained 50 m*M* methyl phosphonate and 18 m*M* glycerophosphorylcholine in 20% (v/v) methanol in deuterium oxide reference solution and the sensor of a fiber-optic thermometer. Model solutions intended to approximate conditions in cerebral cytosol were made from a stock solution containing 150 m*M* KCl, 3 m*M* $K_2HPO_4$, 5 m*M* phosphocreatine, and 3.1 m*M* ATP. Serial pH titration measurements were made as the magnesium concentration was increased in 0.5 m*M* increments up to 10 m*M*.

shift has its greatest sensitivity to them. Measurements of $pH_i$ from the $P_i$–PCr chemical shift difference have often been used in work on intact brain, heart, and muscle (5, 15–19).

Other internal chemical shift references have been proposed. Examples include signals from glycerophosphorylcholine (6, 10, 20, 21) and $\alpha$-ATP (13, 22). Both are less suitable than the PCr signal (23). Both have intrinsically broader linewidths because of their short $T_2$ values. Both overlap the resonances from other compounds with different chemical shifts (24–26). The $\alpha$-ATP signal from biological tissues is actually a mixture of resonances from ATP, other nucleoside triphosphates, nucleoside diphosphates, and the nicotinamide adenine dinucleotide phosphates. The glycerophosphorylcholine signal from intact tissue is a component of the phosphodiester (PDE) resonance (Fig. 1), which also includes signals from glycerophosphorylethanolamine and other phospholipids in proportions that vary with tissue and age, among other things. In living brain, glycerophosphorylcholine and glycerophosphorylethanolamine together account for less than 25% of the PDE signal. At fields strengths less than 2 T, head groups of membrane phospholipids are responsible for most of this signal (26), including its many very broad components (26–30). Only under unusual circumstances could such signals be useful chemical shift references.

The $^1$H signal from water, which is large and has an intrinsically narrow linewidth, has been suggested as an internal chemical shift reference when phosphocreatine is not visible in the $^{31}$P spectrum (23, 31), despite the fact that its own shift is temperature dependent. [The shift of the water signal in the $^1$H spectrum can be used to measure sample temperatures (32).] Derivatives of phosphonic acid have been used both as external shift references and as exogenous internal references (9, 14, 33).

## *Other NMR Indicators of $pH_i$*

Several intracellular metabolites in addition to $P_i$ have resonant frequencies sensitive to pH and produce $^{31}$P NMR signals large enough to detect under some experimental conditions. These include 2,3-diphosphoglycerate (1, 34), fructose phosphates (9, 35), glucose phosphate (36), and phosphomonoesters (PME) such as phosphoethanolamine (37, 38). The PME resonance is prominent in $^{31}$P spectra of many tissues, but it can be used to determine $pH_i$ accurately only if it consists principally of a single resonance that titrates around a favorable p$K$. This condition is met partially in neonatal brain, where the PME signal comes mainly from phosphoethanolamine (24, 37–39), although its p$K$ of 5.63 (19) suits it best for observation of very acidic $pH_i$. In adult brain, phosphocholine and other esters contribute a large portion of the PME signal (39–42). Because the p$K$ values of the various phosphomonoesters are not the same (10, 13, 19), accurate $pH_i$ measurements cannot be made from a composite PME signal.

The $^{31}P$ resonance from $\gamma$-ATP has been suggested as an indicator of $pH_i$ (13). However, most intracellular ATP is complexed with magnesium (10, 22, 30, 43, 44), which has a large effect on the p$K$ of the $\gamma$-ATP resonance. At 4.85, the p$K$ of MgATP is too acidic to allow accurate measurements within the pH range of greatest biological interest (10). Both of these features are illustrated in Fig. 2.

Exogenous NMR indicators of $pH_i$ are available. Examples include methyl phosphonate (9, 20, 34, 45) and 2-deoxyglucose 6-phosphate (19, 35, 46). Various phosphonate derivatives have been used to measure both $pH_i$ and extracellular pH (34, 45, 47). Because *in vivo* NMR methods usually cannot detect signals from compounds less concentrated than about 1 m$M$, exogenous indicators must be used in quantities that can intoxicate biological tissues and alter their acid/base balance. To wit, deoxyglucose competes with glucose for transport into the brain, resulting in electroencephalogram slowing and ATP depletion when deoxyglucose is present at the concentrations necessary for detection by NMR (O. A. C. Petroff and J. W. Prichard, unpublished observations). Phosphorylation of deoxyglucose requires hydrolysis of ATP to ADP, which releases $H^+$ (46). Deoxyglucose 6-phosphate is a base that contributes to the passive buffering capacity of the cell.

Spectroscopy of nuclei other than $^{31}P$ can yield $pH_i$ measurements. Some tissues contain compounds with strong $^{1}H$ signals whose resonant frequencies are sensitive to pH. In muscle, the histidine signal from the small dipeptides carnosine and anserine have been used for this purpose (48–52). Citrate has been suggested as a $pH_i$ indicator in the prostate (53). Signals from carnosine, anserine, citrate, and malate measured by $^{13}C$ spectroscopy have been used to measure $pH_i$ in muscle, prostate, and plants (51, 54).

Exogenous indicator molecules can sometimes be used for NMR measurements of pH. To be useful, they must be distributed predictably within the tissue of interest in sufficient concentration to produce a detectable NMR signal, without affecting the function of the tissue or the buffering capacity. These requirements can be met in tissue cultures, and early efforts to meet them in whole animals have been reported. Exogenous fluorinated compounds measured by $^{19}F$ spectroscopy have been used to measure pH in a variety of tissues in cell culture (55, 56) and in rat brain *in vivo* (57).

Table II lists Henderson–Hasselbalch constants for several $pH_i$ indicators in spectra of various nuclei, together with examples of systems in which they have been used. Most of them have been only briefly investigated. The range and versatility they represent suggest that many aspects of $pH_i$ measurement by NMR have yet to be explored.

TABLE II Henderson–Hasselbalch Constants for Various Indicators of $pH_i$

| Indicator | p$K$ | Acid (ppm) | Base (ppm) | Tissue | Ref. |
|---|---|---|---|---|---|
| **Endogenous** | | | | | |
| Diphosphoglycerate ($^{31}P$) | 7.32 | 6.28 | 7.32 | Erythrocytes | (34) |
| α-Fructose diphosphate ($^{31}P$) | 5.70 | 3.65 | 7.85 | Liver | (35) |
| Glucose phosphate ($^{31}P$) | 6.20 | 3.82 | 7.49 | Plants | (10) |
| Phosphoethanolamine ($^{31}P$) | 5.63 | 3.19 | 6.95 | Neocortex | (38) |
| Anserine ($^{1}H$) | 7.09 | 7.25 | 6.77 | Muscle | [a] |
| Carnosine ($^{1}H$) | 6.73 | 8.57 | 7.67 | Muscle | (52) |
| Citrate p$K_1$ ($^{1}H$) | 3.13 | 0.19 | 0.11 | Prostate | (53) |
| Citrate p$K_3$ ($^{1}H$) | 5.65 | 0.11 | 0.14 | Prostate | (53) |
| **Exogenous** | | | | | |
| Methyl phosphonate ($^{31}P$) | 7.52 | 27.6 | 23.6 | Erythrocytes | (34) |
| Difluoromethylalanine ($^{19}F$) | 7.35 | 4.85 | 2.85 | Lymphocytes | (56) |
| | 6.82 | 4.85 | 2.85 | Brain (37°C) | (57) |

[a] C. Arus and M. Barany, *Biochim. Biophys. Acta* **886**, 411 (1986).

## *Sources of Error in NMR Measurements of $pH_i$*

In living tissues, $pH_i$ can often be determined with precision of about 0.1 pH unit (10, 58–61). Many factors that influence the limit of precision under particular circumstances are common to all NMR methods for measurement of $pH_i$ and are illustrated by the case of $P_i$ discussed below.

At neutral pH, the $P_i$ titration curve is nearly linear (Fig. 2), affording maximum sensitivity for monitoring of pH. The spectrum shown in Fig. 1 was obtained with digital resolution of 0.06 ppm, which translates to 0.07 pH unit if the $P_i$ and PCr signals are assumed to have lorentzian lineshapes. In fact, NMR signals from living tissue never have pure lorentzian shapes. The precision of $pH_i$ measurement from $P_i$ is reduced by low spectral signal-to-noise ratio and greater linewidths as well as by coarser digital resolution. It is less below pH 6 and above pH 7.6, where $\delta P_i/\Delta pH$ drops significantly (Fig. 2). Nonlinear line fitting to lorentzian or gaussian models in the frequency domain does not improve precision; uncertainties are 0.04–0.17 ppm, corresponding to errors of up to 0.14 pH unit (23, 62–64). More recently, signal analysis in the time domain has been reported to improve the precision, with standard deviations of 0.02–0.06 pH unit (62, 65–67).

The effects of chemical exchange, compartmentation, and overlapping resonances complicate $pH_i$ measurement from $P_i$. At 1.5–2 T, the natural linewidth of $P_i$ is 4 Hz (0.11–0.16 ppm), equivalent to a $T_2$ of 80 msec (10). Static field heterogeneity and chemical exchange broadening increase the linewidth observable *in vivo* (16), and the error of chemical shift measurement increases in proportion to observed linewidth.

## *Compartmentation of pH*

Living tissues have a number of compartments that maintain different values of pH, and nearly all of them are of dimensions well below the spatial resolution of NMR methods. Consequently, NMR measurements of pH must always be interpreted in light of what is known about the compartmental distribution of the nuclei that generate the pH-sensitive signal. Because such knowledge is rarely complete, tissue compartmentation is responsible for a special kind of uncertainty affecting interpretation of $pH_i$ measurements. Although not a source of error in the same sense as the factors discussed above, it imposes limits that must be considered carefully in reaching biological conclusions based on NMR data from intact tissue. Again, the case of $P_i$ provides an enlightening example.

When $P_i$ exchanges slowly between two chemical environments, two $P_i$ signals may be observed if the environments have different pH (16). As the rate of exchange ($k$) becomes faster, the two signals coalesce. If $P_i$ yields signals of equal strength from the two environments and resonates at frequencies $f_A$ and $f_B$, the separation of the two signals is given by

$$\Delta f = [(f_A - f_B)^2 - 2k^2/\pi^2]^{1/2}.$$

At the frequency of the spectrum shown in Fig. 1 (32.5 MHz), signals from two $P_i$ pools at pH 7.1 and 7.0 would coalesce as the exchange rate approached 10.6 $sec^{-1}$. If the $P_i$ signals from the two regions are of different intensity, reflecting different numbers of nuclei, calculations are more complicated, because the signal from the smaller pool will be relatively broadened.

When exchange of $P_i$ between two compartments with differing pH is negligible, the $P_i$ signals in $^{31}P$ spectra are approximated by the sum of lorentzian functions plus noise. The intensity of the $P_i$ signal from each compartment is proportional to the phosphate concentration in the compartment and the fraction that each compartment is of the whole volume. Total intracellular $P_i$ has been estimated at 2–3.5 m$M$ (68–71) [also see Helpern (Chapter 16), this volume]. Cerebrospinal fluid $P_i$ is 0.4–0.6 m$M$ and its pH is about 7.3 (72). Extracellular fluid (ECF) has similar pH, 7.25–7.3 (73), and it is thought to have a similar $P_i$ concentration as well. Therefore, if ECF occupies 20% of an observed volume of brain, and intracellular fluid (ICF) occupies 80%, the ICF/ECF ratio of $P_i$ is 20 to 1. Unless the signal-to-noise ratio in $^{31}P$ spectra is considerably greater, the ECF $P_i$ signal cannot be detected.

Heterogeneity of $pH_i$ measured from $P_i$ has been demonstrated in a number of systems, including perfused adrenal gland, platelets, plant vacuoles, and unicellular organisms (36, 45, 54, 74). Intramitochondrial pH has been measured from the $P_i$ signal of isolated mitochondria (58), isolated liver cells (75), and perfused heart (76). In brain, the mitochondrial matrix space occupies only about 5% of cell volume. A $P_i$ signal plausibly attributable to mitochondria has not been detected in brain slices or cultured brain cells (19, 21, 77) or in intact brain (30, 71, 78–80). However, heterogeneity of $pH_i$ across populations of different cells in the same tissue has been reported in brain, heart, muscle, and kidney (15, 61, 74, 81–84). In ischemic tissue, $P_i$ concentration rises severalfold, and $pH_i$ moves in the acidic direction, along the most sensitive portion of the $P_i$ titration curve (Fig. 2). During and after a hypoxic challenge, both factors weight the $pH_i$ reported by $P_i$ heavily toward the most hypoxic regions in the observed volume and thus obscure the contribution of less affected regions (30).

Although signal from phosphate in the extracellular space can be easily seen in cell suspensions (9, 20, 21, 35, 77), it has only been reported infrequently in $^{31}P$ spectra of perfused organs (15, 85, 86). The extracellular phosphate signal has not been reported in spectra from intact vertebrates (30, 59, 71, 74, 78). Neoplastic tissues often have increased ECF/ICF ratios; nevertheless, pH measured by $^{31}P$ spectroscopy in tumors is probably a measure principally of $pH_i$ (87).

As a whole, the findings discussed above support the view that the pH reported by the $P_i$ signal can be considered to be principally that of the intracellular compartment. However, the measurement often cannot distinguish among multiple intracellular compartments that may be present in the observed volume; all are represented, in proportion to their sizes, $P_i$ concentrations, and their position on the $P_i$ titration curve. Only large differences can be detected *in vivo*.

## *Validation of $pH_i$ Measurements by NMR*

Values of $pH_i$ derived from the $P_i$ signal compare well with values based on other methods (5, 10). A number of laboratories have also made direct comparisons, with the same result (74, 88–91). Experimental manipulation of $pH_i$ is reliably detected by NMR methods. The buffering capacity of the living brain—defined as the $pH_i$ response to administration of carbon dioxide and acetazolamide—has been studied in a number of animal models (5, 38, 91–94) and in humans (95–97), with results that are in general agreement with estimates made by other techniques. The response of NMR-monitored $pH_i$ to hypoxic/ischemic injury, seizures, and ammonia challenge is as expected from earlier work (30, 98–101). Similarly good agreement was achieved in studies of the relationship of $pH_i$ to contractile force and fatigue in striated muscle (79, 84, 102) and studies of acid challenge to liver and heart (78, 103). When more than one NMR indicator has been used to measure $pH_i$ in the same tissue, the numbers have agreed well (34, 52).

## Brain $pH_i$ Measurements by NMR

Measurement of $pH_i$ in the brain offers a number of challenges due to the protected location of the brain and its complexity. For many years, the most useful estimates were based on analysis of the distribution of carbon dioxide and bicarbonate (12, 73, 104–106), a weak acid/base pair whose concntration ratio is related to pH by the Henderson–Hasselbalch equation in the form

$$pH = pK + \log([HCO_3^{1-}]/[CO_2]),$$

where p$K$ is the first dissociation constant of $H_2CO_3$ and $[CO_2]$ is the tissue partial pressure of $CO_2$ multiplied by a solubility coefficient, which is 0.0314 $\mu$mol/g/mm Hg under mammalian biological conditions (107).

Useful though it has been, this approach has several well-recognized theoretical and practical limitations. It assumes that the carbon dioxide concentration of venous blood reflects the carbon dioxide concentration of brain quantitatively. The p$K$ of carbonic acid (6.12) is at the lower extreme of even pathological pH excursions in the brain, so that the method is relatively insensitive to pH changes in the range of greatest biological interest. It is invasive, requiring removal of tissue for measurement of bicarbonate concentration. It is vulnerable to technical variations in the rapid freezing of the brain (or a portion of it) necessary to prevent agonal changes in bicarbonate concentration. Obviously, it cannot be used in humans or for serial measurements in a single experimental animal.

Microelectrode techniques have added a great deal to modern understanding of pH behavior in the brain (73), and methods based on optical detection of pH-sensitive substances have an important role in some kinds of preparations (108). Positron emission tomographic techniques that can be used to estimate brain pH *in vivo* have been developed (109). However, NMR methods for pH measurement eclipse all others in versatility and noninvasiveness. They can be used with tissues ranging from cell cultures to the living human brain, they can employ a variety of endogenous indicators, and several other variables of biological importance can be measured from the same preparation in the same session.

Following the report of Moon and Richards (1) that $^{31}$P NMR spectroscopy could be used to measure $pH_i$ in red blood cells, Hoult and colleagues (110) used the method to study $pH_i$ changes in ischemic muscle. Within the next decade, measurements of $pH_i$ in the brains of living animals (43, 111) and humans (112–114) had been reported. Concomitant with these developments, numerous $^{31}$P spectroscopic studies were done on pH phenomena in a variety of organisms, tissues, and organelles (1, 7, 10, 11, 15, 35, 46, 58, 110). This early literature has been well reviewed (16, 115).

The considerable potential of $pH_i$ measurement by NMR indicated by the first decade of work led several investigators interested in the brain to evaluate various NMR methods for use in that organ (5, 13, 18, 19, 24, 37, 44, 88, 116). $P_i$ emerged clearly as the best indicator under most conditions, but others are available and may be useful for special purposes. For measurement of $pH_i$ from $P_i$ in mammalian brain, an appropriate numerical form of the Henderson–Hasselbalch equation is

$$pH = 6.77 + \log[(\delta P_i - 3.29)/(5.68 - \delta P_i)].$$

The constants were determined by *in vitro* titration under conditions intended to mimic the brain environment (5). Actual measurements in a living animal are illustrated in Fig. 3, which shows the $pH_i$ of rabbit brain plotted against arterial $pCO_2$. Data from several animals are included. The scatter of the data points along the $pH_i$ axis at each $pCO_2$ represents inter- as well as

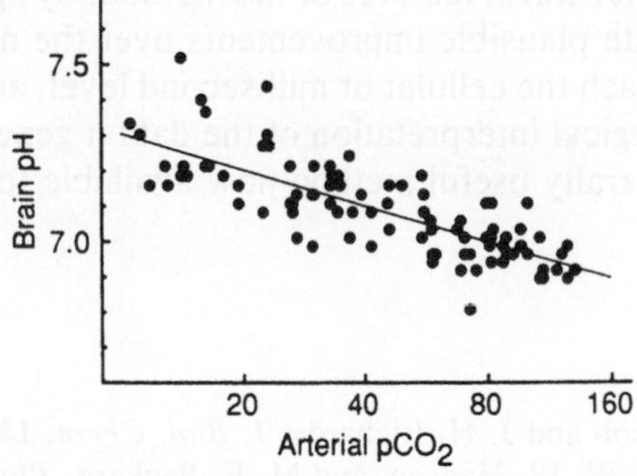


FIG. 3 Cerebral cytosolic pH ($pH_i$) as a function of arterial $pCO_2$. The solid line is fitted to the $pH_i$ serially measured in four rabbits as the arterial carbon dioxide tension ($pCO_2$) was varied by adjusting the inhaled carbon dioxide concentration and minute ventilation. Data replotted from *Neurology* **35,** 1682 (1985).

intraanimal variation and serves as an empirical demonstration of the reproducibility of the method. Superimposed on the scatter is a clear relationship between the two variables, which can be used to estimate the total buffering capacity of the living brain (117).

Numerous studies of $pH_i$ measured from $P_i$ have been carried out in animals and humans. Most of these have verified what was already suspected from less direct measurements, but a few have provided new findings with potential to increase understanding of disease. Dissociation of $pH_i$ from lactate was demonstrated during experimental status epilepticus (118). In human cerebral infarcts, a change from acidosis to alkalosis a few days after onset may indicate the outer limit of effectiveness for some kinds of stroke therapy (119). Acidosis was observed in the temporal lobes of patients with intractable complex partial seizures by one group of investigators (120), but not by another (121).

## Conclusion

Use of NMR methods for study of $pH_i$ in the living brain is in its infancy. Technological improvements will refine the measurement considerably in the next few years. Yet even at the level of precision described in this chapter, a level that is coarse by the standard of test-tube chemistry, NMR measurement of $pH_i$ from the $P_i$ signal of brain and a number of other tissues is adequate for many research purposes. Its anatomical resolution ranges from a few cubic millimeters in small animals studied in high-field research spectrometers to a few cubic centimeters in the human brain at the usual medical imaging field of 1.5 T, and its time resolution was a few minutes in late 1993. It is always limited by the requirement that it be made in a magnet and by other technical features of the method, by space and time resolutions that, even with plausible improvements over the numbers given above, are unlikely to reach the cellular or millisecond level, and by assumptions necessary for biological interpretation of the data it generates. Nevertheless, it is the most generally useful method now available for measurement of $pH_i$ in living tissue.

## References

1. R. B. Moon and J. H. Richards, *J. Biol. Chem.* **148,** 7276 (1973).
2. F. Bloch, W. W. Hansen, and M. E. Packard, *Phys. Rev.* **69,** 127 (1946).
3. E. M. Purcell, H. C. Torrey, and R. V. Pound, *Phys. Rev.* **69,** 37 (1946).
4. D. D. Stark and W. G. Bradley, "Magnetic Resonance Imaging," p. 2520. Mosby Yearbook Inc., St. Louis, 1992.
5. O. A. C. Petroff, J. W. Prichard, K. L. Behar, J. R. Alger, J. A. den Hollander, and R. G. Shulman, *Neurology* **35,** 781 (1985).

6. M. J. Kushmerick, P. F. Dillon, R. A. Meyer, T. R. Brown, J. M. Krisanda, and H. L. Sweeney, *J. Biol. Chem.* **262,** 14420 (1986).
7. Y. Seo, M. Murakami, H. Watari, Y. Imai, K. Yoshizaki, H. Nishikawa, and T. Morimoto, *J. Biochem.* (*Tokyo*) **94,** 729 (1983).
8. R. M. C. Dawson, D. C. Elliott, W. H. Elliott, and K. M. Jones, "Data for Biochemical Research," 3rd Ed., p. 580. Oxford Univ. Press (Clarendon), Oxford, 1987.
9. R. J. Gillies, J. R. Alger, J. A. den Hollander, and R. G. Shulman, *in* "Intracellular pH: Its Measurement, Regulation, and Utilization in Cellular Functions" (R. Nucitelli and D. Deamer, eds.), p. 79. Alan R. Liss, New York, 1982.
10. D. G. Gadian, G. K. Radda, R. E. Richards, and P. J. Seeley, *in* "Biological Applications of Magnetic Resonance" (R. G. Shulman, ed.), p. 463. Academic Press, New York, 1979.
11. J. K. M. Roberts, N. Wade-Jardetzky, and O. Jardetzky, *Biochemistry* **20,** 3389 (1981).
12. J. W. R. Lawson and R. L. Veech, *J. Biol. Chem.* **254,** 6528 (1979).
13. J. Pettegrew, G. Withers, K. Panchalingam, and J. F. M. Post, *Magn. Reson. Imaging* **6,** 135 (1988).
14. J. K. Gard and J. J. H. Ackerman, *J. Magn. Reson.* **51,** 124 (1983).
15. P. B. Garlick, G. K. Radda, and P. J. Seeley, *Biochem. J.* **184,** 547 (1979).
16. D. G. Gadian, "Nuclear Magnetic Resonance and Its Applications to Living Systems." Oxford Univ. Press (Clarendon), Oxford, 1982.
17. W. E. Jacobus, I. H. Pores, S. K. Lucas, C. H. Kallman, M. L. Weisfeldt, and J. T. Flaherty, "Intracellular pH: Its Measurement, Regulation, and Utilization in Cellular Functions" (R. Nucitelli and D. Deamer, eds.), p. 79. Alan R. Liss, New York, 1982.
18. D. G. Gadian, R. S. J. Frackowiak, H. A. Crockard, E. Proctor, K. Allen, S. R. Williams, and R. W. Ross Russell, *J. Cereb. Blood Flow Metab.* **7,** 199 (1987).
19. K. J. Brooks, R. Porteous, and H. S. Bachelard, *J. Neurochem.* **52,** 604 (1989).
20. J. L. Slonczewski, B. P. Rosen, J. R. Alger, and R. M. Macnab, *Proc. Natl. Acad. Sci. U.S.A.* **78,** 6271 (1981).
21. Y. H. H. Lien, H. Z. Zhou, C. Job, J. A. Barry, and R. J. Gillies, *Biochimie* **74,** 931 (1992).
22. C. R. Malloy, C. C. Cunningham, and G. K. Radda, *Biochim. Biophys. Acta* **885,** 1 (1986).
23. A. Madden, M. O. Leach, J. C. Sharp, D. J. Collins, and D. Easton, *NMR Biomed.* **4,** 1 (1991).
24. L. Gyulai, L. Bolinger, J. S. Leigh, C. Barlow, and B. Chance, *FEBS Lett.* **178,** 137 (1984).
25. J. W. Pettegrew, S. J. Kopp, N. J. Minshew, T. Glonek, J. M. Feliksik, J. P. Tow, and M. M. Cohen, *J. Neuropathol. Exp. Neurol.* **46,** 419 (1987).
26. E. J. Murphy, B. Rajagopalan, K. M. Brindle, and G. K. Radda, *Magn. Reson. Med.* **12,** 282 (1989).
27. J. J. H. Ackerman, J. L. Evelhoch, B. A. Berkowitz, G. M. Kichura, R. K. Deuel, and K. S. Lown, *J. Magn. Reson.* **56,** 318 (1984).

28. R. Gonzales-Mendez, L. Litt, A. P. Koretsky, J. von Colditz, M. W. Weiner, and T. L. James, *J. Magn. Reson.* **57,** 526 (1984).
29. P. M. Kilby, N. M. Bolas, and G. K. Radda, *Biochim. Biophys. Acta* **1085,** 257 (1991).
30. D. G. Gadian, S. R. Williams, T. E. Bates, and R. A. Kauppinen, *Acta Neurochir.* **57,** 1 (1993).
31. J. J. H. Ackerman, M. Lowry, G. K. Radda, B. D. Ross, and G. G. Wong, *J. Physiol. (London)* **319,** 65 (1981).
32. N. W. Lutz, A. C. Kuesel, and W. E. Hull, *Magn. Reson. Med.* **29,** 113 (1993).
33. J. A. Barry, K. A. McGovern, Y.-H. H. Lien, B. Ashmore, and R. J. Gillies, *Biochemistry* **32,** 4665 (1993).
34. I. M. Stewart, B. E. Chapman, K. Kirk, P. W. Kuchel, V. A. Lovric, and J. E. Raftos, *Biochim. Biophys. Acta* **885,** 23 (1986).
35. G. Navon, S. Ogawa, R. G. Shulman, and T. Yamane, *Biochemistry* **74,** 87 (1977).
36. J. K. M. Roberts, P. M. Ray, N. Wade-Jardetzky, and O. Jardetzky, *Nature (London)* **283,** 870 (1980).
37. R. J. T. Corbett, A. R. Laptook, and R. L. Nunnally, *Neurology* **37,** 1771 (1987).
38. R. J. T. Corbett, A. R. Laptook, A. Hassan, and R. L. Nunnally, *Magn. Reson. Med.* **6,** 99 (1988).
39. R. Burri, F. Lazeyras, W. P. Aue, P. Straehl, P. Bigler, U. Althaus, and N. Herschkowitz, *Dev. Neurosci.* **10,** 213 (1988).
40. J. W. Pettegrew, S. J. Kopp, J. Dadok, N. J. Minshew, J. M. Feliksik, T. Glonek, and M. M. Cohen, *J. Magn. Reson.* **67,** 443 (1986).
41. I. Kwee and T. Nakada, *Magn. Reson. Med.* **6,** 296 (1988).
42. R. Gruetter, *Magn. Reson. Med.* **29,** 804 (1993).
43. J. J. H. Ackerman, T. H. Grove, G. G. Wong, D. G. Gadian, and G. K. Radda, *Nature (London)* **283,** 167 (1980).
44. H. R. Halvorson, A. M. Q. van Linde, J. A. Helpern, and K. M. A. Welch, *NMR Biomed.* **5,** 53 (1992).
45. F. Brenot, L. Aubry, J. B. Martin, M. Satre, and G. Klein, *Biochimie* **74,** 883 (1992).
46. I. A. Bailey, S. R. Williams, G. K. Radda, and D. G. Gadian, *Biochem. J.* **196,** 171 (1981).
47. K. Clarke, L. C. Stewart, S. Neubauer, J. A. Balschi, T. W. Smith, and J. S. Ingwall, *NMR Biomed.* **6,** 278 (1993).
48. K. Yoshizaki, Y. Seo, and H. Nishikawa, *Biochim. Biophys. Acta* **678,** 283 (1981).
49. S. R. Williams, D. G. Gadian, E. Proctor, D. B. Sprague, D. F. Talbot, I. R. Young, and F. F. Brown, *J. Magn. Reson.* **63,** 406 (1985).
50. C. Arus, W. M. Westler, M. Barany, and J. L. Markley, *Biochemistry* **25,** 3346 (1986).
51. P. Lundberg and H. J. Vogel, *Ann. N.Y. Acad. Sci.* **508,** 516 (1987).
52. J. W. Pan, J. R. Hamm, D. L. Rothman, and R. G. Shulman, *Biochemistry* **85,** 7836 (1988).
53. G. J. Moore and L. O. Sillerud, *Proc. Soc. Magn. Reson. Med.* **2,** 958 (1993).

54. L. Walter, R. Callies, and R. Altenburger, *in* "Magnetic Resonance Spectroscopy in Biology and Medicine" (J. D. de Certaines, W. M. M. J. Bovee, and F. Podo, eds.), p. 573. Pergamon, New York, 1992.
55. J. S. Taylor and C. Deutsch, *Biophys. J.* **43,** 261 (1983).
56. C. J. Deutsch and J. S. Taylor, *in* "NMR Spectroscopy of Cells and Organisms" (R. K. Gupta, ed.), p. 55. CRC Press, Boca Raton, Florida, 1987.
57. C. Chiles, Y. Sachar-Hill, and K. L. Behar, *Soc. Magn. Reson. Med. Abstracts* **9,** 1239 (1990).
58. S. Ogawa, C. C. Boens, and T. Lee, *Arch. Biochem. Biophys.* **210,** 740 (1981).
59. D. J. Taylor, P. J. Bore, P. Styles, D. G. Gadian, and G. K. Radda, *Mol. Biol. Med.* **1,** 77 (1983).
60. J. R. Alger, A. Brunetti, G. Nagashima, and K.-A. Hossmann, *J. Cereb. Blood Flow Metab.* **9,** 506 (1989).
61. K. L. Behar, D. L. Rothman, and K.-A. Hossmann, *J. Cereb. Blood Flow Metab.* **9,** 655 (1989).
62. M. Joliot, B. M. Mazoyer, and R. H. Huesman, *Magn. Reson. Med.* **18,** 358 (1991).
63. J. Haselgrove and M. Elliott, *Magn. Reson. Med.* **17,** 496 (1991).
64. S. Morikawa, T. Inubushi, K. Kito, and C. Kido, *Magn. Reson. Med.* **29,** 249 (1993).
65. J. W. van der Veen, R. de Beer, P. R. Luyten, and D. van Ormondt, *Magn. Reson. Med.* **6,** 92 (1988).
66. A. R. Mazzeo and G. C. Levy, *Magn. Reson. Med.* **17,** 483 (1991).
67. Y. Zaim-Wadghiri, A. Diop, D. Graveron-Demilly, and A. Briguet, *Biochimie* **74,** 769 (1992).
68. R. L. Veech, J. W. Lawson, N. W. Cornell, and H. A. Krebs, *J. Biol. Chem.* **254,** 6538 (1979).
69. H. McIlwain and H. S. Bachelard, "Biochemistry and the Central Nervous System," 5th Ed., p. 660. Churchill Livingstone, New York, 1985.
70. R. J. Connett, *Am. J. Physiol.* **254,** R949 (1988).
71. P. R. Luyten, J. P. Groen, J. W. Vermeulen, and J. A. den Hollander, *Magn. Reson. Med.* **11,** 1 (1989).
72. R. A. Fishman, "Cerebrospinal Fluid in Diseases of the Nervous System," 2nd Ed., Saunders, Philadelphia, Pennsylvania, 1992.
73. M. Chesler, *Prog. Neurobiol.* **34,** 401 (1990).
74. D. G. Gadian and G. K. Radda, *Annu. Rev. Biochem.* **50,** 69 (1981).
75. S. M. Cohen, S. Ogawa, H. Rottenberg, P. Glynn, T. Yamane, T. R. Brown, R. G. Shulman, and J. R. Williamson, *Nature (London)* **273,** 554 (1978).
76. P. B. Garlick, T. R. Brown, R. H. Sullivan, and K. Ugurbill, *J. Mol. Cell. Cardiol.* **15,** 855 (1983).
77. M. Merle, I. Pianet, P. Canioni, and J. Labouesse, *Biochimie* **74,** 919 (1992).
78. J. L. Gallis and P. Canioni, *in* "Magnetic Resonance Spectroscopy in Biology and Medicine" (J. D. de Certaines, W. M. M. J. Bovee, and F. Podo, eds.), p. 345. Pergamon, New York, 1992.
79. B. Barbiroli, *in* "Magnetic Resonance Spectroscopy in Biology and Medicine"

(J. D. de Certaines, W. M. M. J. Bovee, and F. Podo, eds.), p. 369. Pergamon, New York, 1992.

80. J. Murphy-Boesch, R. Stoyanova, R. Srinivasan, T. Willard, D. Vigneron, S. Nelson, J. S. Taylor, and T. R. Brown, *NMR Biomed.* **6,** 173.
81. B. M. Massie, M. Conway, R. Yonge, S. Frostick, P. Sleight, J. Ledingham, G. Radda, and B. Rajagopalan, *Am. J. Cardiol.* **60,** 309 (1987).
82. B. Chance, J. S. Leigh, S. Nioka, T. Sinwell, D. Younkin, and D. S. Smith, *Ann. N.Y. Acad. Sci.* **508,** 309 (1987).
83. N. M. Bolas, B. Rajagopalan, F. Mitsumori, and G. K. Radda, *Stroke* **19,** 608 (1988).
84. M. W. Weiner, R. S. Moussavi, A. J. Baker, M. D. Boska, and R. G. Miller, *Neurology* **40,** 1888 (1990).
85. J. M. Salhany, G. M. Pieper, S. Wu, G. L. Todd, F. C. Clayton, and R. S. Eliot, *J. Mol. Cell. Cardiol.* **11,** 601 (1979).
86. D. D. Gilboe, D. Kintner, M. E. Anderson, J. H. Fitzpatrick, S. E. Emoto, and J. L. Markley, *J. Neurochem.* **60,** 2192 (1993).
87. M. Stubbs, Z. M. Bhujwalla, G. M. Tozer, L. M. Rodrigues, R. J. Maxwell, R. Morgan, F. A. Howe, and J. R. Griffiths, *NMR Biomed.* **5,** 351 (1992).
88. M. Hoehn-Berlage, Y. Okada, O. Kloiber, and K. A. Hossman, *NMR Biomed.* **2,** 240 (1989).
89. R. Jayasundar, L. D. Hall, and N. M. Bleehen, *NMR Biomed.* **5,** 360 (1992).
90. R. Gonzalez-Mendez, G. M. Hahn, N. G. Wade-Jardetzky, and O. Jardetzky, *Magn. Reson. Med.* **6,** 373 (1988).
91. S. Adler, V. Simplaceanu, and C. Ho, *J. Appl. Physiol.* **64,** 1829 (1988).
92. L. Litt, R. Gonzalez-Mendez, J. W. Severinghaus, W. K. Hamilton, J. Shuleshko, J. Murphy-Boesch, and T. L. James, *J. Cereb. Blood Flow Metab.* **5,** 537 (1985).
93. K. L. Behar, D. L. Rothman, S. M. Fitzpatrick, H. P. Hetherington, and R. G. Shulman, *Ann. N.Y. Acad. Sci.* **508,** 81 (1987).
94. P. E. Bickler, L. Litt, D. L. Banville, and J. W. Severinghaus, *J. Appl. Physiol.* **65,** 422 (1988).
95. P. C. van Rijen, P. R. Luyten, J. W. Berkelbach van der Sprenkel, V. Kraaier, A. C. van Huffelen, C. A. Tulleken, and J. A. den Hollander, *Magn. Reson. Med.* **10,** 182 (1989).
96. S. Vorstrup, K. E. Jensen, C. Thomsen, O. Henriksen, N. A. Lassen, and O. B. Paulson, *J. Cereb. Blood Flow Metab.* **9,** 417 (1989).
97. P. G. Bain, M. D. O'Brien, S. F. Keevil, and D. A. Porter, *Ann. Neurol.* **31,** 147 (1992).
98. D. Sessler, P. Mills, G. Gregory, L. Litt, and T. James, *J. Pediatr.* **111,** 817 (1987).
99. B. J. Andersen, A. W. Unterberg, G. D. Clarke, and A. Marmarou, *J. Neurosurg.* **68,** 601 (1988).
100. J. M. Rosenberg, G. B. Martin, N. A. Paradis, R. M. Nowak, D. Walton, T. J. Appleton, and K. M. A. Welch, *Ann. Emerg. Med.* **18,** 341 (1989).
101. K. L. Behar, *Acta Neurochir* **57,** 9 (1993).
102. Z. Argov and W. J. Bank, *Ann. Neurol.* **30,** 90 (1991).

103. M. Bernard and P. J. Cozzone, *in* "Magnetic Resonance Spectroscopy in Biology and Medicine" (J. D. de Certaines, W. M. M. J. Bovee, and F. Podo, eds.), p. 387. Pergamon, New York, 1992.
104. B. K. Siesjo, *Acta Neurol. Scand.* **38,** 121 (1962).
105. B. K. Siesjo and U. Ponten, *Ann. N.Y. Acad. Sci.* **133,** 180 (1966).
106. B. K. Siesjo and K. Messeter, *in* "Ion Homeostasis of the Brain" (B. Siesjo and S. Sorensen, eds.), p. 244. Munksgaard, Copenhagen, 1971.
107. B. K. Siesjo, *Kidney Int.* **1,** 360 (1972).
108. J. C. LaManna, J. K. Griffith, B. R. Cordisco, C.-W. Lin, and W. D. Lust, *Can. J. Physiol. Pharmacol.* **70,** S269 (1992).
109. D. A. Rottenberg, J. Z. Ginos, K. J. Kearfort, L. Junck, and D. D. Bigner, *Ann. Neurol.* **15**(Suppl.), S98 (1984).
110. D. I. Hoult, S. J. W. Busby, D. G. Gadian, G. K. Radda, R. E. Richards, and P. J. Seeley, *Nature (London)* **252,** 285 (1974).
111. B. Chance, S. Eleff, and J. S. Leigh, *Proc. Natl. Acad. Sci. U.S.A.* **77,** 7430 (1980).
112. E. B. Cady, A. M. Costello, M. J. Dawson, D. T. Delpy, P. L. Hope, E. O. R. Reynolds, P. S. Tofts, and D. R. Wilkie, *Lancet* **2,** 1059 (1983).
113. P. A. Bottomley, H. R. Hart, W. A. Edelstein, J. F. Schenk, L. S. Smith, W. M. Leue, O. M. Mueller, and R. W. Reddington, *Radiology* **150,** 441 (1984).
114. D. P. Younkin, M. Delivoria-Papadopoulos, J. C. Leonard, V. H. Subramanian, S. Eleff, J. S. Leigh, and B. Chance, *Ann. Neurol.* **16,** 581 (1984).
115. R. Nucitelli and D. Deamer (ed.), "Intracellular pH: Its Measurement, Regulation, and Utilization in Cellular Functions." Alan R. Liss, New York, 1982.
116. S. Naruse, S. Takada, I. Koizuka, and H. Watari, *Jpn. J. Physiol.* **33,** 19 (1983).
117. O. A. C. Petroff, J. W. Prichard, K. L. Behar, D. Rothman, J. R. Alger, and R. G. Shulman, *Neurology* **35,** 1681 (1985).
118. O. A. C. Petroff, J. W. Prichard, T. Ogino, M. J. Avison, J. R. Alger, and R. G. Shulman, *Ann. Neurol.* **20,** 185 (1986).
119. K. M. A. Welch, S. R. Levine, G. Martin, R. Ordidge, A. M. Vande Linde, and J. A. Helpern, *Neurol. Clin.* **10,** 1 (1992).
120. J. W. Hugg, K. D. Laxer, G. B. Matson, A. A. Maudsley, C. A. Husted, and M. W. Weiner, *Neurology* **42,** 2011 (1992).
121. R. Kuzniecky, G. A. Elgavish, H. P. Hetherington, W. T. Evanochko, and G. M. Pohost, *Neurology* **42,** 1586 (1992).

# [12] Manipulation and Regulation of Cytosolic pH

Mark O. Bevensee and Walter F. Boron

## Introduction

Many cellular processes are sensitive to pH (1). For instance, decreasing *in vitro* pH by ~0.1 decreases the activity of phosphofructokinase, which catalyzes the rate-limiting step in glycolysis, more than 90% of maximal (2). pH can also influence cell structure. For example, late endosomes of cultured rat hippocampal neurons display microtubule-dependent retrograde movement under acidic conditions (3). The pH sensitivity of some processes, such as the activity of ion channels (4, 5) as well as neurotransmitter release, binding to receptors, and reuptake into cells, is especially relevant to the nervous system. As far as channels are concerned, some are sensitive to changes in intracellular pH ($pH_i$), and others to changes in extracellular pH ($pH_o$). An example of a $pH_i$-sensitive channel is the ATP-sensitive, $Ca^{2+}$-dependent $K^+$ channel in leech Retzius neurons; this channel becomes more active (greater open-state probability) when the solution bathing the cytoplasmic face of inside-out patches is more alkaline (6). An example of a channel sensitive to $pH_o$ in the physiological range is the NMDA-stimulated channel, which is inhibited by decreases in $pH_o$, with an apparent p*K* between ~6.9 (7) and ~7.3 (8). Although changes in $pH_i$ and $pH_o$ can affect channel activity independently, the distinction between $pH_i$ and $pH_o$ changes in the central nervous system (CNS) may be somewhat artificial. For example, the extrusion of acid from an astrocyte during intense neuronal activity not only increases $pH_i$ in the astrocyte, but also decreases $pH_o$ in the extracellular space. Moreover, because the distances between neighboring cells in the CNS are extremely small (e.g., 0.02 $\mu$m; see Ref. 9), this $pH_o$ decrease would be expected to lead rapidly to a $pH_i$ decrease in nearby neurons.

In the preceding three chapters, the authors discussed powerful methods for measuring $pH_i$ in neural tissue. In this chapter, we discuss techniques used to manipulate $pH_i$. Such $pH_i$ alterations can have two important practical applications. First, by clamping $pH_i$ to various values, one can study the $pH_i$ dependence of a process. Second, by acutely introducing acid or alkali into a cell, one can study the compensatory acid–base transport processes that return $pH_i$ to normal. In the remainder of the chapter we will consider factors that determine steady-state $pH_i$, namely, the balance between pro-



*Methods in Neurosciences, Volume 27*

cesses that load the cell with either acid or alkali. Next we will discuss how one can exploit this understanding of steady-state $pH_i$ regulation to modulate $pH_i$. Finally, we will examine how one can use the recovery of $pH_i$ from an acid or alkali load to study the activity of a single acid–base transporter.

## Factors That Affect Steady-State $pH_i$

When $pH_i$ is constant, the sum of acid extruded from the cell ($J_E$) by processes such as $Na^+/H^+$ exchange must balance the sum of acid loaded into the cell ($J_L$) by processes such as cellular metabolism or $Cl^-/HCO_3^-$ exchange. $J_E$ and $J_L$ are fluxes* and have the units of moles per unit membrane area, per unit time (typically pmol $cm^{-2}$ $sec^{-1}$). Because it is difficult to estimate membrane area in most cells, and because both pH and buffering power are intensive properties, it has become customary to express acid–base "pseudofluxes" in terms of moles per unit volume of cytoplasm, per unit time (typically $\mu M$ $sec^{-1}$). Here, we introduce the notation $\varphi$ to refer to these pseudofluxes.

A common misconception is that, when $pH_i$ is in a steady state, $J_E$ and $J_L$ must both be zero. Rather, $J_E$ must be equal to $J_L$ in the steady state. $pH_i$ can change only if $J_E$ and $J_L$ become unequal. Imagine, for example, a situation in which a cell is initially in a steady state (i.e., $J_E = J_L$) at a $pH_i$ of 7.2 (point a in Fig. 1A). We now apply an agent that stimulates an acid-extruding mechanism, but does not affect $J_L$. This change reflects a permanent shift in the kinetics of the acid extruder ($J_{E1} \rightarrow J_{E2}$ in Fig. 1A). Because $J_E$ now exceeds $J_L$ at a $pH_i$ of 7.2 (point a′ in Fig. 1A and B), $pH_i$ will initially begin to increase (point a in Fig. 1C). The speed of the $pH_i$ increase obviously depends both on the magnitude of the difference ($J_E - J_L$) as well as on the ability of the cell to buffer the excess alkali being introduced into the cell. The latter is quantitated by a parameter termed intracellular buffering power ($\beta$), which will be described in more detail later in this chapter. The rate of $pH_i$ increase will be greater the larger the difference, $J_E - J_L$, and the smaller the buffering power. This statement can be put into mathematical form [Eq. (1)]:

$$dpH_i/dt = (J_E - J_L)/\beta, \quad (1)$$

which can be regarded as the fundamental law of $pH_i$ regulation.

Although the agent we applied initially causes $pH_i$ to increase in our

*Although the production/consumption of acid via cellular metabolism is not strictly a "flux" across the cell membrane, the rate of $H^+$ production/consumption can be expressed in the same units as a flux.

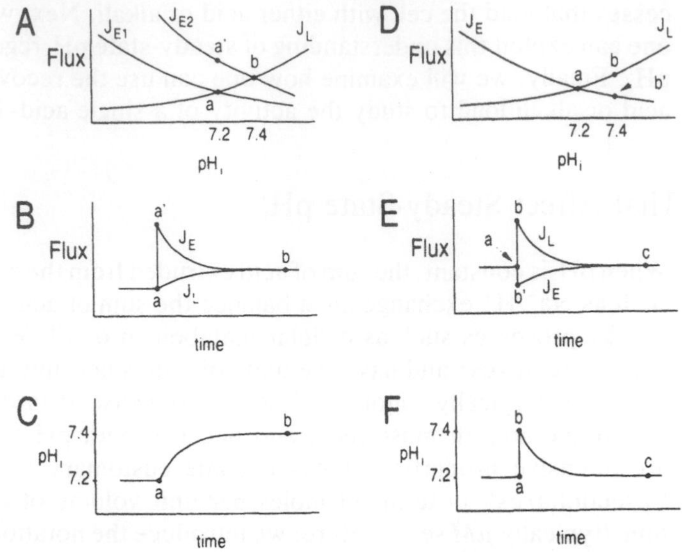


FIG. 1 Relationship between acid-extrusion rate ($J_E$), acid-loading rate ($J_L$), and $pH_i$ when acid extrusion is stimulated in a hypothetical cell (A–C) or when a cell is injected with a bolus of KOH (D–F). (A) $pH_i$ dependence of $J_E$ and $J_L$. At initial steady-state $pH_i$ of 7.2 (point a), acid-extrusion rate ($J_{E1}$) equals acid-loading rate ($J_L$). $J_{E2}$ represents the new acid-extrusion rate after stimulation of an acid-extrusion mechanism. Immediately after stimulation ($pH_i$ = 7.2), $J_{E2}$ (a') exceeds $J_L$ (a). However, $J_{E2}$ and $J_L$ eventually balance at a new steady-state $pH_i$ of 7.4 (b). (B) Time dependence of $J_E$ and $J_L$. Initially $J_E$ equals $J_L$ at a. Stimulating acid extrusion causes $J_E$ to increase instantaneously (a'). However, because this increase in $J_E$ causes $pH_i$ to increase as well (see panel C), $J_E$ gradually decreases to a value (b) that is above the initial $J_E$. This $J_E$ decrease parallels the $pH_i$ increase (panel C). Also paralleling the $pH_i$ increase is a $J_L$ increase. $J_L$ increases gradually and stabilizes at a value (b) greater than the initial $J_L$. Although equal at b, both $J_E$ and $J_L$ are larger than at a. (C) Time dependence of $pH_i$. Stimulating acid extrusion causes $pH_i$ to increase from 7.2 (a) to 7.4 (b). (D) $pH_i$ dependence of $J_E$ and $J_L$. The cell is initially in a steady state ($J_E = J_L$), with a $pH_i$ of 7.2 (a). Injecting KOH causes $pH_i$ to increase to 7.4 (b) without affecting the kinetics of any acid-base transporter. However, the unchanged kinetics dictate that, at $pH_i$ 7.4, acid extrusion is inhibited (b') and acid loading is stimulated (b). Thus, $pH_i$ decreases and causes $J_E$ to increase and $J_L$ to decrease until the two balance again at the initial $pH_i$ of 7.2 (a). (E) Time dependence of $J_E$ and $J_L$. Initially $J_E$ equals $J_L$ at a. Injecting KOH causes an instantaneous increase in $J_L$ (b) and decrease in $J_E$ (b'). However, the slowly developing $pH_i$ decrease (see panel F) causes $J_L$ to decrease and $J_E$ to increase to values (c) that equal the initial fluxes. (F) Time dependence of $pH_i$. Injecting KOH causes $pH_i$ to increase instantaneously from 7.2 at a to 7.4 at b. However, the changes in $J_E$ and $J_L$ discussed in panel E cause $pH_i$ to return to the initial $pH_i$ of 7.2 (c).

hypothetical cell, $pH_i$ will not continue to rise indefinitely. Generally, the activity of acid-extruding transporters such as $Na^+/H^+$ exchangers will decrease as $pH_i$ increases, whereas the activity of acid-loading transporters such as $Cl^-/HCO_3^-$ exchangers will tend to increase. Thus, as $pH_i$ increases, $J_E$ and $J_L$ will gradually come into balance (Fig. 1A and B), creating a new steady state at a $pH_i$ (e.g., 7.4) that is higher than the initial one (point b in Fig. 1C). It is important to note that although the rate of the $pH_i$ change during segment ab in Fig. 1C is inversely proportional to $\beta$, the magnitude of the $pH_i$ change is independent of $\beta$. For example, if $\beta$ were very large, the application of the agent would cause $pH_i$ to increase very slowly, although the final $pH_i$ would still be 7.4.

One could also imagine performing the opposite experiment, in which we apply an agent that stimulates an acid-loading process, but has no effect on acid extrusion. Thus, at the initial $pH_i$ of 7.2, $J_L$ will exceed $J_E$ and, according to Eq. (1), $pH_i$ will decrease. However, as $pH_i$ decreases, we would expect $J_E$ to increase gradually and $J_L$ to decrease gradually, so that the two will eventually balance at a new steady-state $pH_i$ (e.g., 7.0) that is lower than the initial one.

It should be clear from the preceding discussion that a mere shift in steady-state $pH_i$ can provide little information about the activities of specific acid–base transporters. For example, an increase in steady-state $pH_i$ could reflect the stimulation of one or more acid-extruding transporters or the inhibition of one or more acid-loading transporters. However, a useful approach for gaining insight into the acid–base transporters that regulate $pH_i$ is to introduce acutely a bolus of acid or alkali into the cell and monitor the subsequent recovery of $pH_i$ to its initial value. For example, we could microinject KOH into the cytoplasm of a cell initially at $pH_i$ 7.2, and shift $pH_i$ to 7.4. We will make the reasonable assumption that this maneuver has no effect on the overall kinetics of the various acid–base transporters; it merely causes $pH_i$ to increase (e.g., point a → point b in Fig. 1D). However, after $pH_i$ has increased, the cell is no longer in a steady state: given reasonable $pH_i$ dependencies for $J_E$ and $J_L$, $J_L$ will now exceed $J_E$ (points b and b′ in Fig. 1D and E) and $pH_i$ will begin to decrease as required by Eq. (1) (bc in Fig. 1F). As $pH_i$ decreases, $J_E$ will gradually increase to its initial value, and $J_L$ will decrease toward its initial value (Fig. 1E). Eventually, $pH_i$ will stabilize at the original value of 7.2. At this point (c in Fig. 1F), all of the KOH previously injected will have been extruded from the cell.

At the highest $pH_i$ achieved after the injection of KOH (point b in Fig. 1F), the rate of $pH_i$ change tends to be dominated by acid-loading processes. However, because acid-extruding processes may not be totally inactive at such high $pH_i$ values, sorting out acid-loading rates from acid-extrusion rates requires inhibiting specific transporters by removing individual ions (e.g.,

$Na^+$ in the case of the $Na^+/H^+$ exchanger) or applying specific blockers (e.g., amiloride analogs in the case of the $Na^+/H^+$ exchanger), as discussed later in this chapter.

## Approaches for Manipulating $pH_i$

Although many techniques are available for manipulating $pH_i$, some are more practical than others for a particular preparation. The following brief description covers some of these techniques, providing a more detailed discussion of the more common ones.

### *Changes in $pH_o$*

Perhaps the simplest way to alter $pH_i$ is to change $pH_o$. Work on numerous cell types (see Ref. 10) has shown that imposed changes in $pH_o$ evoke changes in $pH_i$ of the same direction. The magnitude of the evoked $pH_i$ change is typically about one-third of the $pH_o$ change. However, in some cells the sensitivity of $pH_i$ to changes in $pH_o$ is much larger. For example, the $\Delta pH_i/\Delta pH_o$ ratio in mesenteric vascular smooth muscle is ~0.7 (11). The mechanism of the $pH_i$ change is believed to be either a change in the passive flux of protons (or proton buffers) across the cell membrane, or a change in the activity of individual acid–base transporters. Advantages of this approach for changing $pH_i$ are that a shift in $pH_o$ generally will produce a stable change in $pH_i$, and that it can be applied to even the smallest of cells. A disadvantage is that highly nonphysiological $pH_o$ changes may be required to achieve a desired $pH_i$ change. Furthermore, because acid–base transporters typically are $pH_o$ sensitive, transport activity at an altered $pH_o$ value may differ substantially from that at a more physiological $pH_o$.

### *Internal Dialysis*

A powerful approach for altering the composition of large cylindrical cells, such as the squid giant axon (12) or barnacle muscle fiber (13), is to dialyze them using a permeable dialysis capillary. For example, dialyzing the cell with an acidic fluid can rapidly decrease $pH_i$. The chief advantage of this technique is that many intracellular parameters (e.g., $[Na^+]$ and $[Cl^-]$) as well as $pH_i$ can be controlled. Furthermore, drugs and nucleotides (e.g., γ-ATP and γ-GTP) can be easily introduced into the cell. Of course, $pH_i$ can be altered without changing $pH_o$. Unfortunately, the technique requires that the cell be large and tubular to accommodate the dialysis tubing. This chief disadvantage limits the usefulness of dialysis to large invertebrate cells.

## *Microinjection of Acid or Base*

$pH_i$ can also be manipulated by directly injecting acid or base into a cell with a micropipette, using either iontophoresis (14) or pressure (15) to achieve the injection. These approaches have been successfully used in snail neurons, but in principle could be applied to any cell large enough to tolerate impalement by one or more microelectrodes. In practice, however, this requirement would limit the technique to large invertebrate cells as well as to certain vertebrate cells, such as oocytes. Although theoretically simple, altering $pH_i$ via microinjection is technically demanding.

## *Ionophores*

A fourth approach for altering $pH_i$ involves exposing the cell to an $H^+$ ionophore, the most commonly used of which is nigericin. Nigericin is an ion exchanger that equilibrates $K^+$ (and to a lesser extent, $Na^+$) and protons across cell membranes, and is used routinely at the end of experiments to calibrate intracellularly trapped pH-sensitive dyes (16) such as carboxy seminaphthofluorescein-1 (carboxy SNAFL-1) and 2′,7′-bis(carboxyethyl)-5,6-carboxyfluorescein (BCECF). In such dye calibration experiments, cells are exposed to a solution containing nigericin and $K^+$ at an external concentration ($[K^+]_o$) equal to the internal $K^+$ ($[K^+]_i$). If it is assumed that $[K^+]_i$ equals $[K^+]_o$, then one can conclude that the ionophore has set $[H^+]_i$ equal to $[H^+]_o$. Grinstein *et al.* have extended this nigericin approach to acid load cells during the course of an experiment (17). For a cell bathed in a typical physiological solution, where $[K^+]_i$ is larger than $[K^+]_o$, nigericin will cause exchange of internal $K^+$ and external $H^+$ across the cell membrane, and thus cause $pH_i$ to decrease. After $pH_i$ has reached a stable value, the nigericin is removed, and nigericin attached to the cell is scavenged with albumin. This method for altering $pH_i$ has the advantage that it is simple and can be used to set $pH_i$ to a wide range of values. Moreover, if the nigericin is maintained in the extracellular solution, the ionophore can be used to clamp $pH_i$ to a desired value. A disadvantage of using the nigericin approach for manipulating $pH_i$ is that the nigericin must be completely removed from the cell membrane if one is to study acid–base transport accurately. Because nigericin can function as an exogenous $K^+/H^+$ and $Na^+/H^+$ exchanger, the presence of residual ionophore could complicate the assessment of endogenous cellular acid–base transporters.

## Weak Bases

### *General Effects of Weak Acids and Bases*

The most common and perhaps the easiest way to manipulate $pH_i$ consists of exposing a cell to a permeant weak acid or base. Although a wide variety of weak acids (e.g., $CO_2$, 5,5-dimethyl-2,4-oxazolidinedione, acetic acid, and propionic acid) and bases (e.g., $NH_3$ and trimethylamine) have been used to manipulate $pH_i$, they all exert their effects on $pH_i$ by similar mechanisms. Therefore, we will focus on the two most commonly used and best described of the permeant weak acids and bases, the weak base $NH_3$ and the weak acid $CO_2$ (18).

### *Intracellular Alkalinization Caused by Weak Bases*

The application of the weak base $NH_3$ can be used to alkali load cells, whereas the removal of this weak base can be used to acid load cells. In solution, $NH_3$ is in a chemical equilibrium with its conjugate weak acid, $NH_4^+$:

$$NH_3 + H^+ \rightleftharpoons NH_4^+. \tag{2}$$

The pK (i.e., the pH at which $[NH_3]$ is the same as $[NH_4^+]$) is ~9.2 at 25°C and ~8.9 at 37°C. As shown in the model experiment in Fig. 2A, when a cell is exposed to an $NH_3/NH_4^+$ solution, the $NH_3$ rapidly enters the cell because it is uncharged and can thus freely cross the cell membrane. To date, only three membrane surfaces seem to be impermeable to $NH_3$: (1) the apical membrane of the thick ascending limb of Henle's loop in the kidney (19, 20), (2) the cell membrane of *Xenopus* oocytes (21), and (3) the apical barrier of gastric glands (22). After $NH_3$ enters a cell, it associates with a proton to form $NH_4^+$. As shown in Fig. 2A, the resulting decrease in free proton concentration causes an increase in $pH_i$ and thus alkali loads the cell (ab). The rapid increase in $pH_i$ in response to the $NH_3$ can also be termed the alkalinization phase of the experiment, during which time $[NH_3]_i$ approaches $[NH_3]_o$ (i.e., $NH_3$ approaches an equilibrium state). The magnitude and rate of the $NH_3$-induced $pH_i$ increase depends on five factors. First, the magnitude of the $pH_i$ increase in response to the $NH_3/NH_4^+$ is inversely related to the buffering power ($\beta$) of the cell. Moreover, the rate of the $pH_i$ increase is inversely proportional to $\beta$. Second, at lower initial $pH_i$ values (i.e., when $pH_i$ is further away from the $pK$ of $NH_3/NH_4^+$), more protons

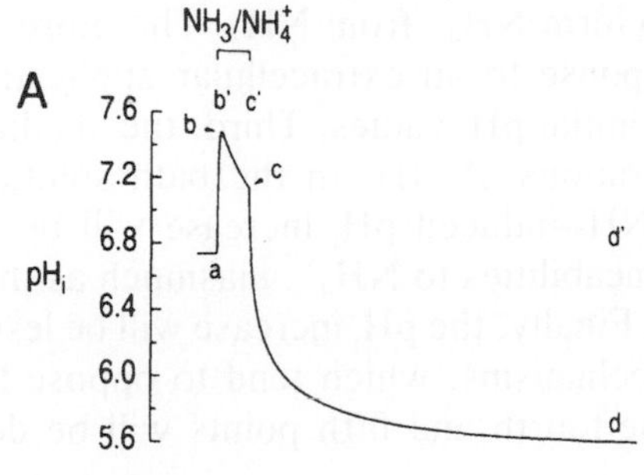


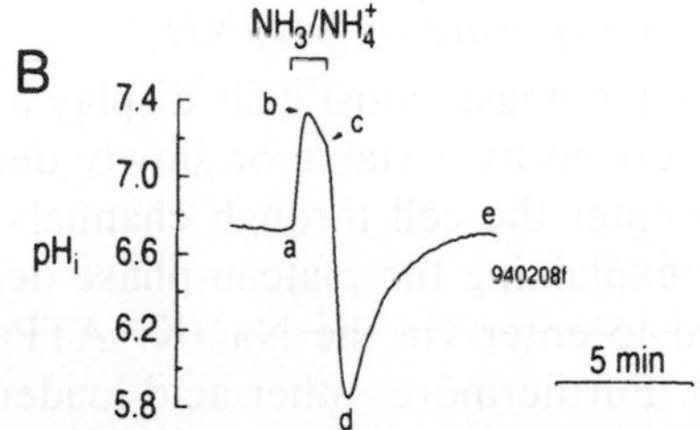


FIG. 2 Effect on $pH_i$ of exposing a cell to a solution containing $NH_3$ and $NH_4^+$. (A) Results of theoretical calculations for cells exposed to $NH_3/NH_4^+$, assuming that only $NH_3/NH_4^+$ transport can influence $pH_i$. Mathematical model assumes that $NH_3$ transport is governed by Fick's law, and that $NH_4^+$ transport is governed by the Goldman-Hodgkin-Katz equation (18). The light and dark records represent expected $pH_i$ changes for a cell in which no $NH_4^+$ transport occurs ($NH_4^+/NH_3$ permeability ratio of zero, thin tracing), and a cell in which the cell membrane is assumed to be permeable to $NH_4^+$ ($NH_4^+/NH_3$ permeability ratio of 0.0014, thick tracing). When the membrane is permeable to $NH_4^+$, applying a solution containing $NH_3/NH_4^+$ is predicted to cause $pH_i$ to increase rapidly (ab) and then to decrease more slowly during the plateau phase (bc). Removing $NH_3/NH_4^+$ at point c causes $pH_i$ to fall rapidly to a value well below the initial $pH_i$ (cd). When the membrane is not permeable to $NH_4^+$, applying $NH_3/NH_4^+$ causes a rapid increase in $pH_i$ (ab') to a value slightly greater than the maximal $pH_i$ achieved in the first example (dark record). There is no plateau-phase acidification in the continued presence of $NH_3/NH_4^+$ (b'c'). When $NH_3/NH_4^+$ is removed, $pH_i$ decreases to a value that is the same as the initial one (c'd'). (B) Results of an actual experiment on a single cultured hippocampal astrocyte. The cell was initially bathed in a HEPES-buffered solution. When the extracellular solution was switched to one that contained 20 mM $NH_3/NH_4^+$ $pH_i$ rapidly increased (ab) and then began to decrease more slowly (bc). When $NH_3/NH_4^+$ was removed from the bath, $pH_i$ abruptly decreased (cd) but then recovered to a value similar to the initial value (de). Segment-de $pH_i$ recovery is due to the activity of an amiloride-sensitive Na-H exchanger.

are available to form $NH_4^+$ from $NH_3$. Therefore, the magnitude of the $pH_i$ increase in response to an extracellular application of $NH_3/NH_4^+$ will be larger at lower initial $pH_i$ values. Third, the alkalinization will be greater at higher concentrations of $NH_3$ in the bath solution. Fourth, the rate and degree of the $NH_3$-induced $pH_i$ increase will be less in cells with greater membrane permeabilities to $NH_4^+$, inasmuch as the influx of $NH_4^+$ tends to acidify the cell. Finally, the $pH_i$ increase will be less in cells with more active acid-loading mechanisms, which tend to oppose the alkalinizing effects of $NH_3$ influx. The fourth and fifth points will be described in greater detail below.

*Plateau Phase of an Exposure to $NH_3/NH_4^+$*

After the alkalinization phase, most cells display a plateau phase (bc in Fig. 2A) that is characterized by a stable or slowly decreasing $pH_i$. With time, $NH_4^+$ may slowly enter the cell through channels and dissociate into $NH_3$ and protons, thus explaining the plateau-phase decrease in $pH_i$. $NH_4^+$ has also been reported to enter via the $Na^+/K^+$-ATPase (23) and $Na^+/K^+/Cl^-$ cotransporter (24). Furthermore, other acid loaders may also contribute to the decrease in $pH_i$ during the plateau phase. For example, the alkali load imposed by the entry of $NH_3$ into the cell may enhance $Cl^-/HCO_3^-$ exchange activity. This acid loader would transport $Cl^-$ into the cell and $HCO_3^-$ out, the net movement of base equivalents out of the cell causing $pH_i$ to decline. Therefore, one could examine a declining $pH_i$ during the plateau phase for evidence of compensatory acid–base transport activity (e.g., $Cl^-/HCO_3^-$ exchange).

If a cell membrane were completely impermeable to $NH_4^+$, and no acid–base transport were to occur other than passive $NH_3$ fluxes, there would be no acidification during the plateau phase (b′c′ in Fig. 2A). Also note, as discussed above, that the rate and magnitude of the alkalinization phase would be greater in a cell lacking $NH_4^+$ permeability and acid loading mechanisms (ab′).

*Effect of Removing Extracellular $NH_3/NH_4^+$*

Regardless of the mechanism(s) by which the $pH_i$ decreases during the plateau phase, the net result is an accumulation of intracellular $NH_4^+$ (i.e., an increase in $[NH_4^+]_i$). The removal of $NH_3/NH_4^+$ from the solution bathing the cell results in the exit of both $NH_3$ and $NH_4^+$; however, $NH_3$ efflux is by far predominant. As the $NH_3$ exits, intracellular $NH_4^+$ dissociates into protons and more $NH_3$. Because excess $NH_4^+$ had accumulated in the cell during the plateau phase, excess protons are released to the cell on the removal of the $NH_3/NH_4^+$ solution. As a result, $pH_i$ falls well below the initial $pH_i$ that prevailed before the $NH_3/NH_4^+$ was added (compare points a and d in Fig. 2A). Therefore, the extent to which $pH_i$ falls during this acidification phase depends on the amount of $NH_4^+$ entry during the plateau phase. In a cell that does not display a decrease in $pH_i$ during the plateau phase (i.e., has no net accumulation of $[NH_4^+]_i$), the removal of the $NH_3/NH_4^+$ solution would cause $pH_i$ to decline only to the initial value and not generate an acid load (c'd').

It should be clear from the above discussion that, in cells with a plateau-phase acidification, the removal of extracellular $NH_3/NH_4^+$ will lead to a $pH_i$ undershoot. This undershoot represents an intracellular acid load, much as would be produced by the microinjection of HCl into the cell. The approach of using a brief exposure to $NH_3/NH_4^+$ for acid loading cells has been termed the $NH_4^+$-prepulse technique.

### *Recovery of $pH_i$ from an Acid Load*

In cells that have been acid loaded and possess active acid-extrusion mechanisms, $pH_i$ typically returns to its initial value. The acid–base transporters responsible for this recovery phase can be determined by reducing or blocking the $pH_i$ recovery with various ion substitutions and/or pharmacological agents.

Figure 2B is a record of an experiment in which a hippocampal astrocyte was briefly exposed to a 20 m*M* $NH_4Cl$ solution. The addition of the $NH_4Cl$ solution caused a rapid increase in $pH_i$ (ab), followed by a distinct acidification during the plateau phase (bc). The rapid decline in the $pH_i$ during the plateau phase was responsible for the robust acid load on removal of the $NH_3/NH_4^+$ solution. The removal of the $NH_3/NH_4^+$ solution caused the $pH_i$ to decrease rapidly and well below the initial $pH_i$ prior to the addition of the $NH_3/NH_4^+$ (cd). Following the acid load, the $pH_i$ recovered to a value nearly identical to the initial one (de). This recovery from the acid load is dependent on external $Na^+$ and is inhibited by the $Na^+/H^+$ exchange inhibitor amiloride (25, M. O. Bevensee and W. F. Boron, 1993, unpublished observations; 26). Therefore, hippocampal astrocytes possess an amiloride-sensitive $Na^+/H^+$ exchanger. Similar transporters, which function as acid-extrusion mechanisms, are found in virtually all animal cells.

## *Weak Acids*

### *Chemical Reactions Involved in the $CO_2/HCO_3^-$ Equilibrium*

Just as applying the weak base $NH_3$ can be used to increase $pH_i$, whereas removing the weak base can be used to acid load the cell, applying a weak acid such as $CO_2$ can be used to decrease $pH_i$, whereas removing the weak acid can be used to alkali load the cell. The $CO_2$ in a $CO_2/HCO_3^-$-buffered solution is in chemical equilibrium with its conjugate weak base, $HCO_3^-$. In the presence of the enzyme carbonic anhydrase, this equilibrium can be mediated by three separate mechanisms. The first is actually two sequential reactions, the first of which involves the formation of carbonic acid and the second of which leads to the formation of bicarbonate:

$$CO_2 + H_2O \rightleftharpoons H_2CO_3 \rightleftharpoons HCO_3^- + H^+. \tag{3}$$

These two reactions can be combined into a single pseudoreaction,

$$CO_2 + H_2O \rightleftharpoons HCO_3^- + H^+, \tag{4}$$

which has a p$K$ of 6.1 at 25°C.

In the second mechanism, the formation of $HCO_3^-$ occurs without $H_2CO_3$ as an intermediary:

$$CO_2 + OH^- \rightleftharpoons HCO_3^-. \tag{5}$$

It is physiologically important only at relatively high pH values (e.g., >8). Because the $OH^-$ consumed in Eq. (5) is formed from $H_2O$,

$$H_2O \rightleftharpoons H^+ + OH^-, \tag{6}$$

the net effect of the second mechanism, seen by summing Eqs. (5) and (6), is the same as Eq. (3). Thus, mechanisms one and two are equivalent thermodynamically, and in terms of their effect on $CO_2$, $H^+$, and pH.

The third mechanism is similar to the second, but is catalyzed by the enzyme carbonic anhydrase:

$$CO_2 + OH^- \overset{CA}{\rightleftharpoons} HCO_3^-. \tag{7}$$

This mechanism is orders of magnitude faster than the other two.

*Acidifying Effect of Exposing Cells to $CO_2/HCO_3^-$*

As shown for the model experiment in Fig. 3A, when a cell is exposed to a solution containing $CO_2/HCO_3^-$, the highly lipophilic $CO_2$ rapidly enters the cell and continues to enter until $[CO_2]_i$ reaches $[CO_2]_o$. To date, the only cell barriers that appear to be impermeable to $CO_2$ are the apical surfaces of gastric gland cells (22). Once inside the cell, the $CO_2$ leads to the production of $H^+$ and $HCO_3^-$ via the three reaction mechanisms outlined above. The resulting increase in $[H^+]_i$ (i.e., decrease in $pH_i$) constitutes the acidification phase (ab, ab′, ab″, and ab‴ in Fig. 3A) of the cell's response to $CO_2/HCO_3^-$. This intracellular acidification represents an acute intracellular acid load, formally equivalent (from the perspective of acid–base chemistry) to the microinjection of HCl. The rate and magnitude of the $CO_2$-induced acidification of the cell are dependent on factors analogous to those discussed above in connection with the cell's response to $NH_3$ (see Fig. 2A). These include $\beta$, initial $pH_i$, concentration of the neutral weak acid (i.e., $[CO_2]_o$), permeability of the membrane to the conjugate weak base ($HCO_3^-$), and the activity of various acid–base transporters. In response to $CO_2$, $pH_i$ will decrease more slowly, and by a smaller amount, when $\beta$ for the cell is higher. In contrast, the $pH_i$ will decrease more rapidly, and by a greater amount when the initial $pH_i$ is more alkaline (i.e., the further away $pH_i$ is from the p$K$). The reason is that under more alkaline conditions (i.e., with fewer free protons), there is a greater tendency for the newly entering $CO_2$ to form $HCO_3^-$ and $H^+$. The $CO_2$-induced acidification also will be greater for higher values of $[CO_2]_o$, inasmuch as this will lead to the influx of more $CO_2$. In the first moments of the exposure to $CO_2/HCO_3^-$, the rate and degree of the $CO_2$-induced acidification is expected to be reduced when the cell membrane has an increased permeability to $HCO_3^-$. The reason is that the influx of $HCO_3^-$ will tend to alkalinize the cell. However, several seconds into the $CO_2/HCO_3^-$ exposure, after $[HCO_3^-]_i$ has increased, the electrochemical gradient for $HCO_3^-$ would favor an efflux of this ion and thus lead to an augmented acidification. Finally, the rate and degree of the acidification will be reduced to the extent that the cell possesses transporters that respond to the decrease in $pH_i$ by extruding acid from the cell.

*The Plateau Phase of $CO_2/HCO_3^-$ Exposure*

The period after the rapid, $CO_2$-induced intracellular acidification is known as the plateau phase. During this plateau phase, $pH_i$ may be stable, drift down, or drift up. In the absence of any permeability to $HCO_3^-$, or of any acid–base transport, $pH_i$ would be stable during the plateau phase (b′c′ in Fig. 3A), reflecting the equilibration of $CO_2$ across the cell membrane. On the other hand, a cell membrane permeable to $HCO_3^-$ would allow the passive efflux of $HCO_3^-$ and thus a slow intracellular acidification during the plateau

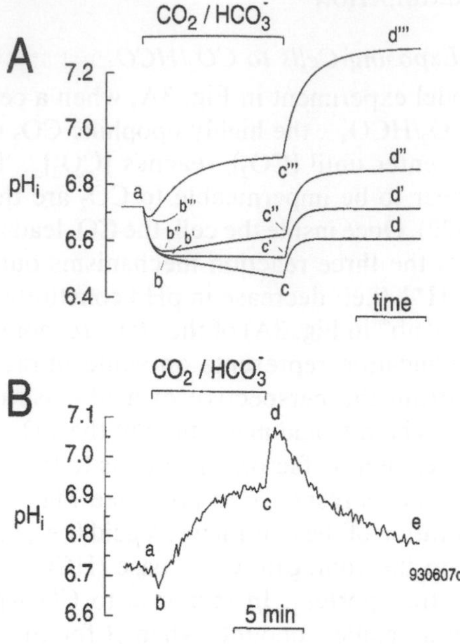


FIG. 3 Effect on $pH_i$ of exposing a cell to a solution containing $CO_2$ and $HCO_3^-$. (A) Results of theoretical calculations for cells exposed to $CO_2/HCO_3^-$. Mathematical model assumes that $CO_2$ transport is governed by Fick's law, that $HCO_3^-$ transport is governed by the Goldman-Hodgkin-Katz equation, and that acid-extrusion rate increases linearly with $[H^+]_i$ (18). We consider four effects of $CO_2/HCO_3^-$ on $pH_i$. First, in a cell permeable to $HCO_3^-$ ($HCO_3^-/CO_2$ permeability ratio of 0.005) but with no $H^+$ transport, $pH_i$ decreases rapidly in response to $CO_2/HCO_3^-$ (ab), due to $CO_2$ influx, and then continues to fall (bc) as a result of passive $HCO_3^-$ efflux, for example. Removing $CO_2/HCO_3^-$ causes $pH_i$ to increase to a value below the initial value (cd). Second, in a cell with neither $HCO_3^-$ permeability ($HCO_3^-/CO_2$ permeability ratio of zero) nor $H^+$ transport, $pH_i$ stabilizes after the initial $CO_2$-induced acidification (b'c'). $CO_2/HCO_3^-$ removal causes $pH_i$ to return to its initial level (c'd'). Third, in a cell with no $HCO_3^-$ permeability and modest acid extrusion, applying $CO_2/HCO_3^-$ causes less of an initial acidification (ab''), and a modest alkalinization during the plateau phase (b''c''). When the $CO_2/HCO_3^-$ is subsequently removed, $pH_i$ overshoots the initial value by a modest amount (c''d''). Fourth, in a cell with no $HCO_3^-$ permeability and substantial acid extrusion, the $CO_2$-induced acidification is even less than in the previous case (ab'''), and the plateau-phase alkalinization drives $pH_i$ to a value that is even higher than the initial value (b'''c'''). Washing away $CO_2/HCO_3^-$ leads to an exaggerated $pH_i$ overshoot (c'''d'''). (B) Results of an experiment on a single cultured hippocampal astrocyte. The cell was initially bathed in a HEPES-buffered solution. When the cell was switched from a nominally $CO_2/HCO_3^-$-free solution to one containing 5% $CO_2$/17 mM $HCO_3^-$, the $pH_i$ briefly decreased (ab), and then subsequently increased to a value well above the initial $pH_i$ (bc). When the cell was returned to the HEPES-buffered solution, the $pH_i$ rapidly rose (cd) and then relaxed to the initial $pH_i$ (de).

phase (bc). Finally, a cell membrane possessing a mechanism for accumulating $HCO_3^-$ and/or extruding $H^+$ would lead to a plateau-phase alkalinization (b″c″). This $pH_i$ recovery (i.e., alkalinization) represents an attempt by the cell to compensate for the acute intracellular acid load produced by the influx of $CO_2$. By far, the most common plateau-phase response is a $pH_i$ recovery. In many cases, this $pH_i$ recovery is incomplete (i.e., $pH_i$ never reaches the initial, pre-$CO_2$ value). On the other hand, in cells with powerful $HCO_3^-$-dependent acid extrusion mechanisms, it is possible for the final $pH_i$ during the plateau to be substantially greater than the initial $pH_i$ (b‴c‴). Cells in which the plateau-phase $pH_i$ is higher than the pre-$CO_2$ $pH_i$ include mesangial cells (27), sheep cardiac Purkinje fibers (28), invertebrate glial cells (29, 30), mammalian astrocytes (31), and mammalian hippocampal neurons (32, 33).

### *Effect of Removing Extracellular $CO_2/HCO_3^-$*

Regardless of the course of $pH_i$ during the plateau phase, the removal of extracellular $CO_2/HCO_3^-$ will lead to a rapid increase in $pH_i$ as the reactions that led to the $CO_2$-induced acidification reverse: intracellular $HCO_3^-$ combines with $H^+$ and ultimately forms $CO_2$, which rapidly leaves the cell. In the case that $pH_i$ was stable during the plateau phase, $pH_i$ returns to exactly the pre-$CO_2$ value during the alkalinization phase (c′d′ in Fig. 3A). The reason is that $[HCO_3^-]_i$ at the end of the plateau phase exactly represents the $HCO_3^-$ formed from the entry of $CO_2$ during the acidification phase. In the case that $pH_i$ fell during the plateau phase, $HCO_3^-$ was depleted from the cell. Thus, the removal of $CO_2/HCO_3^-$ causes $pH_i$ to increase rapidly, but to a value that is less than the pre-$CO_2$ value (cd). In the cases in which $pH_i$ increased during the plateau phase, $HCO_3^-$ accumulated inside the cell. Thus, $pH_i$ during the alkalinization phase overshoots the pre-$CO_2$ $pH_i$ (c″d″ and c‴d‴). The degree of the overshoot reflects the degree of accumulation of intracellular $HCO_3^-$ during the plateau phase. Hence, the $pH_i$ at point d‴ is greater than that at point d″.

### *Recovery of $pH_i$ from an Alkali Load*

The $pH_i$ overshoot that follows the removal of $CO_2/HCO_3^-$ in an experiment with a rising plateau phase represents an acute intracellular alkaline load. An important issue is whether the cell will recover from such an alkaline load as it recovered from an acid load in Fig. 2B. Figure 3B shows an experiment in which a single cultured astrocyte from the rat hippocampus was exposed for 7.5 min to a solution containing 5% $CO_2$/17 m*M* $HCO_3^-$ (pH 7.3). The application of $CO_2/HCO_3^-$ caused a transient $pH_i$ decrease (ab), due to $CO_2$ influx, followed by a sustained $pH_i$ increase (bc), due to a $Na^+$-dependent, $HCO_3^-$ uptake mechanism that is inhibited by the stilbene derivative 4,4′-diisothiocyanatostilbene-2,2′-disulfonic acid (DIDS) (34). When the cell then was returned to a nominally $CO_2/HCO_3^-$-free bath solution buffered with HEPES, the $pH_i$ rapidly increased (cd), reflecting the efflux of $CO_2$, and then recovered more slowly (de) to a value comparable to the initial $pH_i$ in the HEPES-buffered solution. This slow $pH_i$ decrease can be termed the recovery phase. Presently, the acid-loading mechanisms responsible for the $pH_i$ recovery in the nominal absence of $CO_2/HCO_3^-$ have not been determined. It has been suggested that $Cl^-/HCO_3^-$ exchange may be involved in the $pH_i$ decrease. Although this may indeed be the case, we note that because most of the $HCO_3^-$ was washed out of the cell by point d in Fig. 3B, $Cl^-/HCO_3^-$ exchange would have to occur at very low levels of intracellular $HCO_3^-$. Alternative explanations include the generation of $H^+$ from cellular metabolism, the passive influx of $H^+$, $K^+/H^+$ exchange (35), and the $K^+/HCO_3^-$ cotransporter identified in squid axons (36–38).

### *An Approach for Studying $pH_i$ Recovery from an Alkali Load in the Presence of Physiological Levels of $CO_2/HCO_3^-$*

A disadvantage of alkali loading the cell, as shown in Fig. 3B, in which the cell was switched from a HEPES buffer to a 5% $CO_2$ buffer and back to a HEPES buffer, is that the $pH_i$ recovery from the alkali load (de) occurs in the nominal absence of $CO_2/HCO_3^-$. An approach that has been employed to avoid this problem is to expose the cell sequentially to a 5% $CO_2$ buffer, followed by a 10% $CO_2$ buffer (at the same $pH_o$), and then a 5% $CO_2$ buffer (39). This technique has the advantage that it allows one to evaluate the contribution of $HCO_3^-$-dependent transporters to the $pH_i$ recovery from an alkali load. Figure 4 (39) illustrates an experiment on mesangial cells, which are related to vascular smooth muscle. The cells, initially bathed in a 5% $CO_2$/25 m*M* $HCO_3^-$ solution (pH 7.4), were then exposed to a solution buffered with 10% $CO_2$/50 m*M* $HCO_3^-$, which caused a sustained $pH_i$ increase. In this experiment, this $HCO_3^-$-induced alkalinization was so robust that the expected, initial $CO_2$-induced acidification was not apparent. When the cells were returned to a bath solution buffered with 5% $CO_2$/25 m*M* $HCO_3^-$, $pH_i$ rapidly increased, reflecting the efflux of $CO_2$, and then recovered more slowly toward the initial $pH_i$. Figure 4 also shows that this $pH_i$ recovery from the alkali load is blocked by the stilbene derivative DIDS, and requires external $Cl^-$. These properties are hallmarks of the $Cl^-/HCO_3^-$ exchanger. Thus, in mesangial cells studied in the presence of physiological levels of $CO_2/HCO_3^-$, the $pH_i$ recovery from an alkali load is mediated almost exclusively by a $Cl^-/HCO_3^-$ exchanger.

### *Acid or Alkali Loading Cells in the Presence of $CO_2/HCO_3^-$*

A cell whose $pH_i$ has stabilized in the presence of $CO_2/HCO_3^-$ (e.g., point c in Fig. 3B) can also be acid or alkali loaded by means other than changing

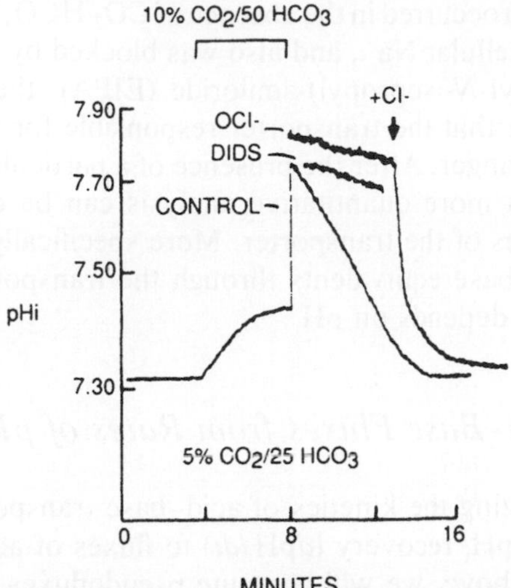


FIG. 4 Contribution of a $Cl^-/HCO_3^-$ exchanger to the recovery of $pH_i$ from an alkali load in mesangial cells, in the presence of physiological $CO_2/HCO_3^-$. When mesangial cells, initially bathed in a 5% $CO_2$/25 m*M* $HCO_3^-$ solution, were exposed to 10% $CO_2$/50 m*M* $HCO_3^-$, $pH_i$ increased by ~0.12 pH units. Later, returning the cells to the initial 5% $CO_2$/25 m*M* $HCO_3^-$ solution caused $pH_i$ first to increase rapidly and then to recover to the initial $pH_i$. The $pH_i$ recovery was inhibited ~90% by the application of DIDS or by the removal of external $Cl^-$. Reprinted with permission from *Nature*, Ganz *et al.* (39). Copyright 1989, Macmillan Magazines Ltd.

levels of $CO_2/HCO_3^-$. For instance, a cell bathed in a $CO_2/HCO_3^-$ solution can be acid loaded using the $NH_4^+$-prepulse approach discussed above. This approach would allow one to evaluate both $HCO_3^-$-dependent and -independent acid-extrusion mechanisms that contribute to the $pH_i$ recovery.

## Use of $pH_i$ Recoveries from Acid or Alkali Loads to Determine Transporter Activity

### *Identification of Acid–Base Transport Systems*

The initial step in understanding the $pH_i$-regulating mechanisms of a cell is to identify the acid–base transporters present. As discussed above, transporters that regulate $pH_i$ can be determined by acid or alkali loading cells and evaluating the characteristics of the $pH_i$ recovery from the acid or alkali load. For example, if the recovery of $pH_i$ from an acid load (e.g., imposed by an $NH_4^+$

prepulse) occurred in the absence of $CO_2/HCO_3^-$, was blocked by the removal of extracellular $Na^+$, and also was blocked by amiloride derivatives such as 5-(*N*-ethyl-*N*-isopropyl)-amiloride (EIPA), then it would be reasonable to conclude that the transporter responsible for the $pH_i$ recovery was a $Na^+/H^+$ exchanger. After the presence of a particular transporter has been determined, a more quantitative analysis can be done to elucidate the kinetic properties of the transporter. More specifically, one can determine the flux of acid–base equivalents through the transporter, and also determine how this flux depends on $pH_i$.

## *Computing Acid–Base Fluxes from Rates of pH$_i$ Recovery*

In analyzing the kinetics of acid–base transporters, one must first convert rates of $pH_i$ recovery ($dpH_i/dt$) to fluxes of acid–base equivalents. As discussed above, we will compute pseudofluxes, designated by the symbol $\varphi$ and expressed in the units $\mu M$ $sec^{-1}$. $\varphi$ is simply the product of $dpH_i/dt$ (pH units $sec^{-1}$) and total intracellular buffering power ($\beta_T$) expressed in the units $\mu M$ (pH units)$^{-1}$. Because, as shown in the idealized example in Fig. 5, both $dpH_i/dt$ (Fig. 5A) and $\beta_T$ (Fig. 5B) may vary with $pH_i$, $\varphi$ may also be $pH_i$ dependent (see Fig. 5C). Cellular buffering power (discussed in detail in Ref. 10) is a measure of the ability of a cell to buffer $pH_i$ changes by either consuming or releasing protons. $\beta$ is defined as $\Delta[base]_i/\Delta pH_i$, where $\Delta[base]_i$ is the number of micromoles of strong base added to a volume of cytoplasm and $\Delta pH_i$ is the resultant increase in $pH_i$. $\beta_T$ is the sum of the buffering power due to intrinsic intracellular buffers ($\beta_I$) (i.e., buffers that cannot traverse the cell membrane) and the buffering power due to extrinsic buffers such as $CO_2/HCO_3^-$ ($\beta_{HCO_3^-}$) (which can cross the membrane) . Thus, in a typical physiological solution that contains $CO_2/HCO_3^-$, $\beta_T = \beta_I + \beta_{HCO_3^-}$ (see Fig. 5B). $\beta_I$ can be measured experimentally by recording changes in $pH_i$ (i.e., $\Delta pH_i$ in the above definition of $\beta$) in response to the addition or removal of weak acids or bases. In the case of an experiment in which $[NH_3]$ is increased, $\Delta[base]_i$ represents the amount of $H^+$ consumed as $NH_3$ entering the cell goes on to form $NH_4^+$. In such experiments, acid–base transporters should be blocked so that they cannot affect the measured $\Delta pH_i$. If $\beta_I$ is to be measured, the experiment must be done in the absence of extrinsic buffers such as $CO_2/HCO_3^-$. $\beta_{HCO_3^-}$ can be computed from theoretical considerations by differentiating the Henderson–Hasselbalch equation for intracellular $CO_2/HCO_3^-$: $\beta_{HCO_3^-} = \delta[HCO_3^-]_i/\delta pH_i = 2.3[CO_2]_i 10^{pH_i - pK} = 2.3[HCO_3^-]_i$. Our laboratory has recently confirmed the validity of this theoretical calculation (38). Both $\beta_I$ and $\beta_{HCO_3^-}$ should be determined or computed for a range of $pH_i$ values, inasmuch as both parameters can vary substantially with $pH_i$.

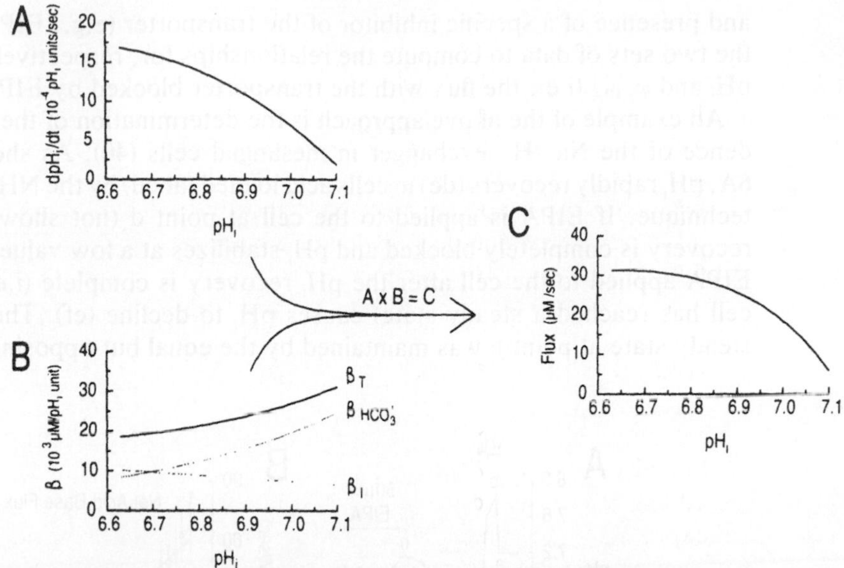


FIG. 5 Computing the $pH_i$ dependence of net acid extrusion from the $pH_i$ dependency of the rate of $pH_i$ recovery from an acid load and the $pH_i$ dependency of buffering power. (A) A $dpH_i/dt$ vs. $pH_i$ relationship. The curve illustrates how the rate of $pH_i$ recovery from an acid load might vary with $pH_i$. This curve was taken from an experiment on a single pyramidal neuron acutely isolated from the rat hippocampus. (B) $pH_i$ dependence of intrinsic buffering power ($\beta_I$), open-system $CO_2/HCO_3^-$ buffering power ($\beta_{HCO_3^-}$), and total buffering ($\beta_T$). $\beta_{HCO_3^-}$ was computed from $pH_i$, assuming a constant $CO_2$ level of 5%. $\beta_T$ is the sum of $\beta_I$ and $\beta_{HCO_3^-}$ at individual $pH_i$ values. The $\beta_I$ vs. $pH_i$ relationship was taken from data on mesangial cells (27). (C) Hypothetical $pH_i$ dependence of net acid-extruding flux. The coordinates of the curve were computed, point by point, by multiplying $dpH_i/dt$ at a particular $pH_i$ (taken from panel A) by the $\beta_T$ at the same $pH_i$ (taken from panel B).

## *Determining the Flux due to Individual Acid–Base Transporters*

After computing acid–base fluxes from $pH_i$ recovery rates and buffering powers, fluxes due to individual transporters can be determined from data generated with specific inhibitors. The flux due to a $Na^+/H^+$ exchanger, for example, is not necessarily the same as the total acid–base flux ($\varphi_{Total}$) computed from the recovery of $pH_i$ from an acid load. $\varphi_{Total}$ is the sum of all acid-loading ($\varphi_L$) and acid-extruding ($\varphi_E$) fluxes. The flux due to the $Na^+/H^+$ exchanger ($\varphi_{Na/H}$) is simply one component of $\varphi_E$. Therefore, if the goal is to obtain $\varphi_{Na/H}$ as a function of $pH_i$, this can only be done by stripping the $\varphi_{Na/H}$ vs. $pH_i$ relationship out of the $\varphi_{Total}$ vs. $pH_i$ relationship. The most straightforward approach is to monitor the recovery of $pH_i$ in the absence and presence of a specific inhibitor of the transporter (e.g., EIPA), and use the two sets of data to compute the relationships for, respectively, $\varphi_{Total}$ vs. $pH_i$ and $\varphi_{EIPA}$ (i.e., the flux with the transporter blocked by EIPA) vs. $pH_i$.

An example of the above approach is the determination of the $pH_i$ dependence of the $Na^+/H^+$ exchanger in mesangial cells (40). As shown in Fig. 6A, $pH_i$ rapidly recovers (de) in cells acid loaded (abcd) by the $NH_4^+$-prepulse technique. If EIPA is applied to the cell at point d (not shown), the $pH_i$ recovery is completely blocked and $pH_i$ stabilizes at a low value. However, EIPA applied to the cell after the $pH_i$ recovery is complete (i.e., after the cell has reached a steady state) causes $pH_i$ to decline (ef). Therefore, the steady state at point e was maintained by the equal but opposing actions of

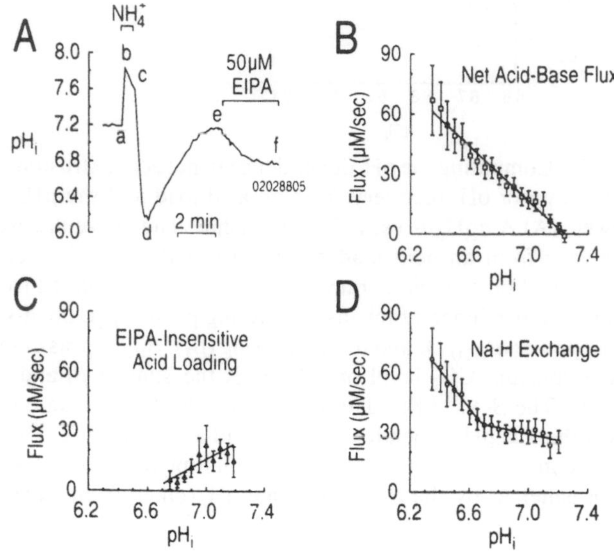


FIG. 6 Determination of $pH_i$ dependence of the $Na^+/H^+$ exchanger in single mesangial cells. Reproduced from Boyarsky *et al.* (40), with permission of G. Boyarsky. (A) Recovery of $pH_i$ from an acid load. A mesangial cell, in a HEPES-buffered solution, recovered from an acid load (de) that was imposed using the $NH_4^+$-prepulse technique (abcd). After the recovery was complete, the cell was exposed to 50 $\mu M$ EIPA and the $pH_i$ decreased (ef). (B) $pH_i$ dependence of the net acid–base flux. Fluxes were computed from $dpH_i/dt$ data obtained in experiments like that shown in panel A (segment de) and from the known $pH_i$ dependence of $\beta_I$ (see Fig. 5B). (C) $pH_i$ dependence of the EIPA-insensitive acid-loading flux. Fluxes were computed from $dpH_i/dt$ data obtained in experiments like that shown in panel A (segment ef) and from the known $pH_i$ dependence of $\beta_I$ (see Fig. 5B). (D) $pH_i$ dependence of $Na^+/H^+$ exchange. Fluxes were computed as the sum of the net acid–base flux (panel B) and the EIPA-insensitive acid-loading flux (panel C).

$Na^+/H^+$ exchange (supplemented, perhaps, by other acid extruders) and background acid-loading mechanisms. Applying EIPA blocks $Na^+/H^+$ exchange and thereby unmasks the effects of all non-$Na^+/H^+$ exchange processes, which have the net effect of acid loading the cell. Figure 6B is a plot of the $pH_i$ dependence of average $\varphi_{Total}$ values, computed from data such as that in segment de of Fig. 6A. Figure 6C is a plot of the $pH_i$ dependence of average $\varphi_{EIPA}$ values, the EIPA-insensitive fluxes computed from data such as that in segment ef of Fig. 6A. Positive $\varphi_{EIPA}$ values represent net acid loading. Finally, Fig. 6D is the result of adding the $\varphi_{EIPA}$ vs. $pH_i$ relationship in Fig. 6C to the $\varphi_{Total}$ vs. $pH_i$ relationship in Fig. 6B. If EIPA blocks 100% of $Na^+/H^+$ exchange, and has no other actions, the plot in Fig. 6D represents the $pH_i$ dependence of the $Na^+/H^+$ exchanger.

As shown in Fig. 6D, the $Na^+/H^+$ exchanger in mesangial cells is most active at acidic $pH_i$ values. However, even though the activity of the exchanger decreases with increasing $pH_i$, the transporter remains functional even at the steady-state $pH_i$ that prevails in the nominal absence of $CO_2/HCO_3^-$. Therefore, the $Na^+/H^+$ exchanger is not quiescent at the steady-state $pH_i$. Similar observations have also been made in the IEC-6 colonic cell line (41), UMR-106 osteoblast-like cells (42), and *ras*-transformed NIH-3T3 fibroblasts (43).

## Concluding Remarks

Evaluating how a cell responds to acid or alkali loads is a powerful technique for elucidating the mechanisms, kinetics, and regulation of acid–base transporters responsible for $pH_i$ regulation. From intracellular buffering power, as well as the rate at which $pH_i$ recovers from acute acid or alkali loads, one can compute the equivalent acid–base flux responsible for the $pH_i$ recovery. With the results from additional experiments exploiting specific inhibitors of transporters, it is possible to determine the $pH_i$ dependence of these transporters. This is a necessary step for elucidating the physiological control of these transporters. For example, by comparing the $pH_i$ dependencies of the $Na^+/H^+$ exchanger under various conditions, it is possible to determine the effect of second messengers, changes in cell volume, growth factors, and oncogenes on the $pH_i$ dependence of the $Na^+/H^+$ exchanger. In the nervous system, regulating the $pH_i$ dependence of the acid–base transporters that regulate $pH_i$ could play a role in neuromodulation, and also could be a mechanism for correcting for acid–base imbalances that arise during electrical activity.

## Acknowledgments

We thank Duncan Wong for computer assistance and help in preparing the manuscript. MOB was supported by an NIH Training Grant 5-T32-6M07527-181716. This work was supported by NIH Grant NS18400.

## References

1. W. B. Busa and R. Nuccitelli, *Am. J. Physiol.* **246,** R409 (1984).
2. B. Trivedi and W. H. Danforth, *J. Biol. Chem.* **241,** 4110 (1966).
3. R. G. Parton, C. G. Dotti, R. Bacallao, I. Kurtz, K. Simons, and K. Prydz, *J. Cell Biol.* **113,** 261 (1991).
4. W. Moody, Jr., *Annu. Rev. Neurosci.* **7,** 257 (1984).
5. M. Chesler, *Prog. Neurobiol.* **34,** 401 (1990).
6. G. Frey, W. Hanke, and W.-R. Schlue, *J. Membr. Biol.* **134,** 131 (1993).
7. C.-M. Tang, M. Dichter, and M. Morad, *Proc. Natl. Acad. Sci. U.S.A.* **87,** 6445 (1990).
8. S. F. Traynelis and S. G. Cull-Candy, *Nature (London)* **345,** 347 (1990).
9. B. R. Ransom, *Prog. Brain Res.* **94,** 37 (1992).
10. A. Roos and W. F. Boron, *Physiol. Rev.* **61,** 296 (1981).
11. C. Austin and S. Wray, *J. Physiol. (London)* **466,** 1 (1993).
12. J. M. Russell and W. F. Boron, *Nature (London)* **264,** 73 (1976).
13. W. F. Boron, J. M. Russell, M. S. Brodwick, D. W. Keifer, and A. Roos, *Nature (London)* **276,** 511 (1978).
14. R. C. Thomas, *J. Physiol. (London)* **255,** 715 (1976).
15. R. W. Meech and R. C. Thomas, *J. Physiol. (London)* **298,** 111 (1980).
16. J. A. Thomas, R. N. Buchsbaum, A. Zimniak, and E. Racker, *Biochemistry* **81,** 2210 (1979).
17. S. Grinstein, S. Cohen, and A. Rothstein, *J. Gen. Physiol.* **83,** 341 (1984).
18. W. F. Boron and P. De Weer, *J. Gen. Physiol.* **67,** 91 (1976).
19. J. L. Garvin, M. B. Burg, and M. A. Knepper, *Am. J. Physiol.* **255,** F57 (1988).
20. D. Kikeri, A. Sun, M. L. Zeidel, and S. C. Hebert, *Nature (London)* **339,** 478 (1989).
21. B.-C. Burckhardt and E. Fromter, *Pfluegers Arch.* **420,** 83 (1992).
22. S. J. Waisbren, J. Geibel, I. M. Modlin, and W. F. Boron, *Nature (London)* **368,** 332 (1994).
23. F. V. Bielen, H. G. Glitsch, and F. Verdonck, *J. Physiol. (London)* **442,** 169 (1991).
24. D. Kikeri, A. Sun, M. L. Zeidel, and S. C. Hebert, *J. Gen. Physiol.* **99,** 435 (1992).
25. C. A. Pappas and B. R. Ransom, *Glia* **9,** 280 (1993).
26. M. O. Bevensee and W. F. Boron, unpublished observations (1993).
27. G. Boyarsky, M. B. Ganz, B. Sterzel, and W. F. Boron, *Am. J. Physiol.* **255,** C844 (1988).
28. C. Dart and R. D. Vaughan-Jones, *J. Physiol. (London)* **451,** 365 (1992).

29. E. A. Newman and M. L. Astion, *Glia* **4,** 424 (1991).
30. J. W. Deitmer and W.-R. Schlue, *J. Physiol.* (*London*) **411,** 179 (1989).
31. G. Boyarsky, B. Ransom, W.-R. Schlue, M. B. E. Davis, and W. F. Boron, *Glia* **8,** 241 (1993).
32. K. M. Raley-Susman, R. M. Sapolsky, and R. R. Kopito, *J. Biol. Chem.* **266,** 2739 (1991).
33. C. J. Schwiening and W. F. Boron, *J. Physiol.* (*London*) **475,** 59 (1994).
34. M. O. Bevensee, R. A. Weed, and W. F. Boron, *FASEB J.* **7,** A186 (1993).
35. A. M. Hofer and T. E. Machen, *J. Membr. Biol.* **126,** 245 (1992).
36. E. M. Hogan, M. A. Cohen, and W. F. Boron, *J. Gen. Physiol.*, in press (1995a).
37. E. M. Hogan, M. A. Cohen, and W. F. Boron, *J. Gen. Physiol.*, in press (1995b).
38. J. Zhao, E. M. Hogan, M. O. Bevensee, and W. F. Boron, *Nature* (*London*) **374,** 636 (1995).
39. M. B. Ganz, G. Boyarsky, R. B. Sterzel, and W. F. Boron, *Nature* (*London*) **337,** 648 (1989).
40. G. Boyarsky, M. B. Ganz, E. J. Cragoe, and W. F. Boron, *Proc. Natl. Acad. Sci. U.S.A.* **87,** 5921 (1990).
41. M. D. Sjaastad, E. Wenzl, and T. E. Machen, *Am. J. Physiol.* **262,** C164 (1992).
42. A. Gupta, C. J. Schwiening, and W. F. Boron, *Am. J. Physiol.* **266,** C1083 (1994).
43. D. Kaplan and W. F. Boron, *J. Biol. Chem.* **269,** 4116 (1994).

# [13] Fluorescence Measurements of Cytosolic Sodium Concentration

Marli A. Robertson and J. Kevin Foskett

## Introduction

The activity of sodium in the cytoplasm ($[Na^+]_i$) is highly regulated in most cells. Whereas cells are bathed in extracellular fluids that contain high $[Na^+]$ (100–500 m$M$), $[Na^+]_i$ is generally maintained at much lower levels, ranging from 5 to 30 m$M$. The chemical driving force for $Na^+$ entry into cells is complemented by the negative inside electrical potential across the plasma membrane. Thus, the electrochemical driving force for $Na^+$ entry across the plasma membrane is usually considerable. The implications for cellular physiology are twofold. First, cells must expend energy to maintain the $[Na^+]$ gradient. Second, the $[Na^+]$ gradient can be used by cells to perform work. Thus, many cells utilize the energy in the $[Na^+]$ gradient to couple unfavorable transmembrane solute flows to $Na^+$ transport. Examples include nutrient (e.g., sugars, amino acids) uptake, neurotransmitter uptake, and ion uptake or exchange, processes that contribute to transepithelial transport and intracellular ion regulation, including $[Ca^{2+}]_i$ and $[H^+]_i$ homeostasis, and are therefore important in signal transduction in cells. Of critical importance in electrically excitable cells, the $[Na^+]$ gradient provides the basis for most action potentials and synaptic depolarization. In addition to the $[Na^+]$ gradient, there is evidence that the absolute $[Na^+]_i$ in cells is also important. Cytoplasmic $[Na^+]$ plays an important role in mitochondrial $[Na^+]$ and $[Ca^{2+}]$ homeostasis in most cells, as a substrate for mitochondrial $Na^+/H^+$ and $Na^+/Ca^{2+}$ exchangers (1). Furthermore, the activities of intracellular enzymes, including adenylyl cyclase (2, 3), heterotrimeric G proteins (2, 4), and $K^+$ channels (5, 6) have been reported to be directly modulated by changes in $[Na^+]_i$.

The numerous transport processes that utilize the energy of the $Na^+$ gradient to couple solute transport to $Na^+$ transport contribute a $Na^+$ load to cells, which tends to minimize this gradient. Cells maintain $[Na^+]_i$ homeostasis by counteracting this $Na^+$ leak by active (ATP-consuming) transport of $Na^+$ out of the cell by the ubiquitous $Na^+$ pump, the $Na^+/K^+$-ATPase. Nevertheless, the activities of many $Na^+$-coupled solute transporters can be rapidly modulated, by acute changes in substrate concentrations or by allosteric modification (e.g., phosphorylation). Furthermore, the activity of the $Na^+/$


*Methods in Neurosciences*, Volume 27

$K^+$-ATPase can be acutely regulated (7). Thus, the pump–leak balance can be rapidly perturbed and consequently $[Na^+]_i$ might be expected to change under a variety of physiological as well as pathophysiological conditions. Because of the critical role of the $[Na^+]$ gradient, as outlined above, such changes would be likely to have diverse physiological consequences.

## Measurement of $[Na^+]_i$ Using Fluorescent Indicator Dyes

The advantages of using fluorescence techniques for measuring $[Na^+]_i$ are similar to those described for other cell parameters. Fluorescence techniques allow the determination of $[Na^+]_i$ in a noninvasive manner even in small intact cells. As described below, many cells can be loaded with dyes by passive permeation of the dye across the plasma membrane, and low-light-level detection techniques can minimize possible toxic effects of illuminating the dye, e.g., generation of free radicals. Fluorescence techniques allow $[Na^+]_i$ to be determined in populations of cells, as well as at the single-cell level, thereby permitting population responses to be understood in terms of intercellular variability. Furthermore, low-light-level imaging techniques provide the capability to spatially resolve $[Na^+]_i$ within single cells. Although it is unlikely that significant $[Na^+]_i$ gradients can be generated in the cytoplasm of most cells, gradients of $[Na^+]_i$ might be generated in cells of complicated geometry, which impose obstacles to free diffusion. For example, $[Na^+]_i$ gradients have been observed in hippocampal neurons (8). Imaging provides the ability to resolve organellar $[Na^+]_i$, it allows determination in more than one cell simultaneously, and it permits correlation of changes in $[Na^+]_i$ with specific changes in cell physiology (9, 10).

## $Na^+$-Binding Benzofuran Isophthalate

### *Structure*

To date, all reported fluorometric determinations of $[Na^+]_i$ have employed the $Na^+$-sensitive indicator, $Na^+$-binding benzofuran isophthalate (SBFI) (11). Our discussion will therefore be limited to this one indicator dye. The reader should note, however, that Molecular Probes, Inc. (Eugene, Oregon) has recently announced the availability of newer $Na^+$-sensitive fluorescent indicator dyes (12), and Smith *et al.* (13) have also described a fluorescent version of a fluorine-substituted cryptand.

SBFI, synthesized by Minta and Tsien in 1987 (11), consists of a crown ether with additional ether oxygens capping both poles (Fig. 1). The macrocyclic ring forms an equatorial belt around the sodium cation. The size of the ring appears to largely determine the cation selectivity of the dye (discussed below) (11). The fluorophores are benzofurans linked to isophthalate groups.

## *Spectral Properties*

The attached fluorophores of SBFI are similar to those in the calcium indicator dye fura-2; the dyes therefore exhibit similar spectral properties. The emission maximum is ~525 nm for the $Na^+$-bound form of the dye. *In vitro*, the $Na^+$-bound form of the dye absorbs maximally at 336 nm, whereas the unbound form absorbs maximally at 345 nm (Fig. 2). Therefore, $Na^+$ binding to SBFI shifts its excitation spectrum slightly to shorter wavelengths. In addition, $Na^+$ binding causes the longer wavelength side of the excitation spectrum to fall off faster, resulting in an isosbestic ($Na^+$ insensitive) wavelength (~375 nm) and an inverted response of excitation efficiencies at wavelengths longer than 375 nm. These changes in the excitation spectrum are therefore similar to the changes in the fura-2 spectrum caused by $Ca^{2+}$,

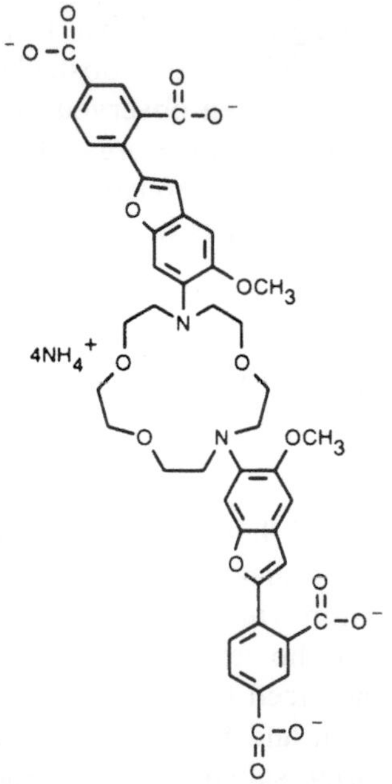

FIG. 1 Structure of SBFI (reprinted with permission from Molecular Probes, Inc.).

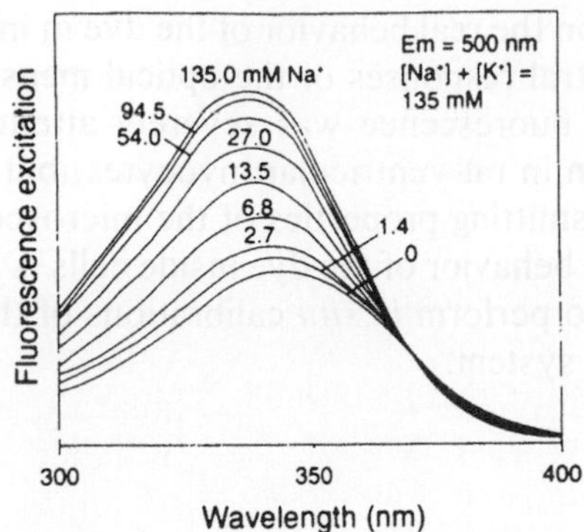


FIG. 2 Fluorescence emission response of SBFI to increasing concentrations of $Na^+$ in solutions containing both $Na^+$ and $K^+$; the combined concentrations of both ions equaled 135 m*M* (reprinted with permission from Molecular Probes, Inc.).

although their magnitudes are smaller. As with fura-2, therefore, the ratio of excitation efficiencies at 335–340 nm to that at 380–390 nm increases with increasing [$Na^+$]. *In vitro,* this ratio increases by approximately threefold (11), or approximately an order of magnitude less than fura-2 in response to $Ca^{2+}$. The fluorescence of an indicator dye is determined specifically by the level of the sensitive parameter (i.e., $Na^+$ for SBFI) and nonspecifically by other factors such as dye concentration, excitation intensity, and optical pathlength. The spectral shift that results from interaction with $Na^+$ allows the nonspecific parameters to be normalized by ratio techniques. Therefore, the preferred mode of use of SBFI is by excitation ratioing.

It is important to note that SBFI in the cytoplasm of intact cells may exhibit substantially altered spectral properties compared with its properties *in vitro* (14–20). For example, the excitation maximum was shifted to slightly longer wavelengths, the quantum efficiency of the dye was enhanced, and the difference in quantum efficiencies of the $Na^+$-bound compared with unbound dye was minimized in the cytoplasm of Jurkat lymphocytes compared with those properties of the dye after it had been released from the cells into the extracellular buffer (15). Titration of lysates showed that dye released from the cells had the same properties as the free acid, indicating that the alteration of its spectral characteristics was not due to incomplete deesterification of the dye within the cells. In rat ventricular myocytes (16–18) and in rabbit gastric parietal cells (19, 21) an apparent blue shift of the isosbestic wavelength is observed, with the result that at wavelengths approximately ≥340 nm, fluorescence emission intensity *decreased* with increasing $[Na^+]_i$. It is worthwhile noting that the apparent behavior of SBFI inside cells depends not only on the real behavior of the dye in intracellular compartments, but on the spectral responses of the optical measuring system as well. For example, SBFI fluorescence was severely attenuated at excitation wavelengths ≤340 nm in rat ventricular myocytes, but this effect was likely due to the light-transmitting properties of the microscope system (16). As such, and because the behavior of the dye inside cells is variable among cell types, it is necessary to perform *in situ* calibrations of the dye in each cell type in each measuring system.

## *Cation Selectivity*

*In vitro*, in a medium in which the sum of the concentrations of $Na^+$ and $K^+$ is maintained constant at 135 m$M$, a situation that reflects the physiological condition of the cytoplasm in most cells, the dissociation constant ($K_d$) of SBFI for $Na^+$ is ~18 m$M$ (11) at pH 7.05. Nevertheless, SBFI fluorescence is sensitive to $K^+$ as well as to $Na^+$. In 100 m$M$ tetramethylammonium chloride, the $K_d$ for $K^+$ was 166 m$M$, compared with 7.4 m$M$ for $Na^+$ (11). This selectivity appears to be sufficient to allow interference by $K^+$ to be ignored in most cases. First, $[K^+]_i$ in most cells is normally less than the $K_d$ for $K^+$, and $[K^+]_i$ will unlikely ever exceed resting levels under physiological conditions. Because electroneutrality and osmotic balance require that the sum of $Na^+$ and $K^+$ concentrations will remain relatively constant in cells, any decrease in $[K^+]_i$ will be accompanied by equimolar increase in $[Na^+]_i$. The much lower $K_d$ for $Na^+$ ensures that SBFI will report nearly exclusively the $[Na^+]_i$ during such changes in cytoplasmic cation composition. Second, experimental manipulations in intact cells have verified the lack of significant $[K^+]_i$ sensitivity of SBFI (15, 18, 21, 22). This was most convincingly demonstrated in SBFI-loaded rabbit parietal cells exposed to gramicidin D to create nonselective cation channels in the plasma membrane (21). By exploiting the permeability of the gramicidin channels to $K^+$ as well as $Cs^+$ and the insensitivity of SBFI to $Cs^+$, it could be demonstrated that manipulations of $[K^+]_i$, at constant $[Na^+]_i$ by replacement with $Cs^+$, had only minor effects on the SBFI signal.

SBFI fluorescence is sensitive to other cations as well. However, the sensitivities to $Mg^{2+}$ and $Ca^{2+}$ are so low that these ions will not affect the fluorescence *in vivo*. The highest p$K_a$ of SBFI in solutions containing physiological cation composition is 6.1 (11), suggesting that cytoplasmic SBFI will be relatively unaffected by changes in intracellular pH ($pH_i$). Furthermore, *in vitro* acidification depressed proportionately the fluorescence at 340 and 380 nm excitations, with the resultant ratio being little affected. In agreement, $pH_i$ ranging from 6.4 to 8.0 had no effect on excitation ratios determined for SBFI in a pituitary cell line (23), $pH_i$ over the range

6.6 to 7.6 was without effect on SBFI fluorescence in rat ventricular myocytes (24), and we have observed that acidification of salivary acinar cells to $pH_i$ 6.9 had no effect on the SBFI fluorescence ratio (10). In contrast, raising $pH_i$ from 7.0 to 7.5 shifted the SBFI ratio the equivalent of 4 m$M$ $Na^+$ in isolated rabbit parietal cells (19). Over the range 6.5 to 7.5, the shift was equivalent to that induced by 8 m$M$ $Na^+$. Because, under physiological conditions, $pH_i$ in intact cells is highly regulated and is unlikely to ever vary more than 0.3 units, in most instances any errors in determination of $[Na^+]_i$ during such $pH_i$ changes will likely be small at most. Some compartments in cells, for example, lysosomes, endosomes, and golgi, are much more acidic than the cytoplasm. Minta and Tsien (11) determined that over the range of pH values found in these organelles (5.0 to 6.5), decreasing pH diminished the fluorescence over the entire excitation spectrum. At the lowest pH values (5.0–5.5) the dye was relatively $[Na^+]$ insensitive. Because SBFI has been observed to accumulate in intracellular organelles in some cells (discussed below), such $Na^+$-insensitive fluorescence may contribute, in addition to the cytoplasmic dye, to the total signal measured from cells unless appropriate precautions are taken (discussed below).

### *Other Influences on SBFI*

Minta and Tsien (11) reported that large increases in ionic strength have relatively minor effects on SBFI fluorescence, probably because the binding site for $Na^+$ in the crown ether is uncharged. Increasing solution viscosity by addition of 1.75 $M$ sucrose enhanced quantum efficiency and slightly shifted the excitation peak of SBFI *in vitro*, effects that were similar to those observed for the dye in the cytoplasm of Jurkat cells (15). Nevertheless, even under extreme physiological conditions cytoplasmic viscosity sensed by a molecule the size of SBFI (molecular mass, 907 Da) is unlikely to be much greater than that of water (25). Therefore, the effects on SBFI fluorescence observed in the cytoplasm of Jurkat cells are likely due to other parameters that mimic the effects of viscosity, perhaps binding of the dye to soluble or fixed cytoplasmic proteins.

## Practical Use of SBFI

### *Loading SBFI into Cells*

As for other polycarboxylate dyes, SBFI free acid does not freely cross cell membranes. Cells can be loaded with the dye by microinjection of the free acid (15, 26). This technique is technically demanding, but nevertheless is the method of choice for loading SBFI into cells that tend to compartmentalize the dye. A more common method to load SBFI into cells is to incubate them with the membrane-permeant acetoxymethyl (AM) ester derivative, SBFI-AM. Cleavage of the ester linkages by cytoplasmic esterases exposes four negative charges, rendering the dye hydrophilic, thereby trapping it within the cell. SBFI is available commercially from several vendors; our experience is with dye purchased from Molecular Probes. A stock solution of the dye is prepared by dissolving it in dry dimethyl sulfoxide (DMSO) at a concentration of 1–10 m*M*. The stock solution should be stored in the dark at −20°C. We have used dye prepared this way after 8 weeks or more in the freezer with no apparent loss of activity. To minimize thawing and refreezing of the stock solution, which may diminish the quality of the dye, the stock solution can be stored in small aliquots sufficient for 1 or 2 days of experiments. Cells can be loaded in suspension or on glass coverslips. Most laboratories, including our own, have found that the nonionic detergent Pluronic F-127 (Molecular Probes) greatly facilitates loading of cells with SBFI-AM (15, 17, 27, 28). Presumably this effect is due to enhancement of the solubility of SBFI-AM in aqueous loading media. An aliquot of the stock solution of SBFI-AM is mixed with equal volumes of Pluronic at a concentration of 10–25% w/v. The SBFI-AM stock/Pluronic mixture is then added to the loading buffer to achieve the final desired concentration of SBFI-AM. To load isolated salivary acinar cells, we mix equal volumes of Pluronic-F127 (Molecular Probes; 25% w/v in DMSO) and 10 m*M* stock solution of SBFI-AM by vigorous vortexing in a microtest tube just prior to addition of dye to the loading solution ($HCO_3$ - or HEPES-buffered balanced salt solution with 1% bovine serum albumin) to achieve a final concentration of SBFI-AM of 7 $\mu M$. The cells are loaded for 60 min at room temperature and then washed in dye-free buffer before the experiment. Cells are used within 1 hr of loading with the dye. In comparison, the loading conditions for fura-2-AM in the same cells are 1 $\mu M$ for 15–20 min at room temperature without addition of Pluronic. Thus, SBFI-AM is generally much more difficult to load into cells than the more widely used fura-2. Furthermore, because SBFI has a relatively low quantum efficiency compared to fura-2, more SBFI must be loaded into cells to provide comparable fluorescence signals. In other cell types, extracellular concentrations up to 22 $\mu M$ and incubation times ranging from 40 min to 3 hr have been used (24, 29–31). As with other anionic dyes such as fura-2 and 2′,7′-bis(carboxyethyl)-5,6-carboxyfluorescein (BCECF), transport of SBFI by cells may contribute to poor cytoplasmic loading or to enhanced intracellular compartmentalization. Organic anion transporters and p-glycoprotein, the multidrug resistance pump, have been demonstrated to transport such dyes (32, 33). Thus, SBFI loading may also be increased in the presence of probenecid and sulfinpyrazone, inhibitors of organic anion transporters, or verapamil, an inhibitor of drug pumping by p-glycoprotein. Incubation with SBFI-AM in the presence of 250 $\mu M$ sulfinpyrazone increased dye loading fourfold in aortic myocytes (34). In contrast, neither probenecid nor sulfinpyrazone prevented SBFI leakage from platelets (35). It is important to note that in most instances the minimum necessary intracellular concentration of SBFI (or any other dye for that matter) should be used. Excessively high intracellular levels of dye enhance the likelihood of incomplete hydrolysis of the acetoxymethyl esters, of compartmentalization, and of toxicity from by-products of hydrolysis (formaldehyde, acetate) and dye illumination (SBFI radicals). Even when precautions are taken, it is advisable to evaluate the physiological competence of cells loaded with the dye. For example, SBFI-loaded smooth muscle cells may become unresponsive to contractile agonists (20), possibly because SBFI is a crown ether and crown ethers are known to exert inotropic effects on muscle. In contrast, SBFI in isolated salivary gland acinar cells is without effect on normal agonist responsiveness, $[Ca^{2+}]_i$ signaling, or $Ca^{2+}$ activation of several ion transport pathways (9, 10). The amount of intracellular dye required depends on the sensitivity of the detection system as well as on intrinsic fluorescence of the particular cell type. Because of the general difficulty of loading SBFI in most cell types, cellular autofluorescence should be evaluated before SBFI loading. SBFI fluorescence should be at least three times above background fluorescence at each wavelength used in experiments, at a $[Na^+]_i$ that produces the lowest SBFI signal at that wavelength. For example, if a rise of $[Na^+]_i$ produces a twofold decrease in the 380-nm signal, then the initial 380-nm signal should be sixfold higher than cell autofluorescence.

## *Compartmentalization*

Intracellular SBFI loaded as the AM derivative has been reported to localize in addition to the cytoplasm to membrane-delimited compartments in cells, including the nucleus (14, 15), acidic organelles (15), and mitochondria (14, 16, 17, 27). In the latter, the compartmentalized dye was $Na^+$ sensitive and could be exploited to report mitochondrial $[Na^+]$ (16, 17, 27). Nevertheless, in most cases compartmentalization is undesirable because the parameter of interest is cytoplasmic $[Na^+]$. Uniformity of fluorescence in salivary acinar cells is improved when loading is performed at room temperature. In contrast, cells loaded at 37°C contained distinct brightly fluorescent punctate spots, indicating compartmentalization in organelles. Increased compartmentalization was also reported for rabbit gastric glands when loading was performed at 37°C (21). A quantitative estimate of the amount of signal due to compartmentalized probe can be made by selective permeabilization of different intracellular compartments (15–17). A widely used protocol to evaluate the fraction of intracellular dye in the cytoplasm is to permeabilize the plasma membrane with digitonin. At an appropriate concentration (17, 36), digitonin selectively permeabilizes the plasma membrane, releasing cytoplasmic dye, while leaving intracellular organellar membranes intact. In adult rat ventricular myocytes, digitonin released only ~50% of the dye (16, 17). The remaining dye was trapped in mitochondria, which make up 35% of the volume of these cells, and was lost on exposure to Triton X-100 (17). However, compartmentalization appears to be less significant in other cell types. The residual fluorescence after digitonin permeabilization was ~10% in vascular smooth muscle cells (36) and 10–15% in guinea pig myocytes (37). In fibroblasts, digitonin released 95% of the indicator from nuclear regions and 73% from cytoplasmic regions (15).

Contribution to the fluorescence signal from compartmentalized cells can be dealt with in two ways. In imaging systems, the ability to resolve spatially intracellular regions of compartmentalized dye permits the exclusion of those areas from analysis. In nonimaging systems, calibration of SBFI fluorescence by means that alter cytoplasmic $[Na^+]_i$ exclusively (discussed below) allows calculation of $[Na^+]_i$ in the face of a background of compartmentalized dye. An assumption is that the compartmentalized signal does not change when $[Na^+]_i$ changes. Harootunian *et al.* (15) reported that most of the compartmentalized dye in fibroblasts was localized in acidic organelles. Because SBFI becomes insensitive to $[Na^+]$ at low pH, the assumption that compartmentalized fluorescence is constant in the face of changing $[Na^+]_i$ in many cases would seem reasonable.

## *Calibration*

As outlined above, SBFI behaves quite differently in solution and within cells (14–16) and there exists considerable variability among cells in the behavior of the dye. Thus, conversion of intracellular SBFI fluorescence ratios to $[Na^+]_i$ must be performed by calibration *in situ* in each cell type. In our experience there may also be day to day variation in dye behavior. Ideally, therefore, calibration should be performed at the end of each experiment, and particularly for each neutral-density filter used in the optical path, for each new stock solution of dye, and whenever any component of the optical system is modified.

*In situ* calibration is performed most conveniently by exposing SBFI-loaded cells to various $[Na^+]$ in the presence of the pore-forming antibiotic gramicidin. Gramicidin is a monovalent nonselective cation ionophore that will collapse $[Na^+]$ as well as $[K^+]$ gradients across the plasma membrane.

It is a simple matter to make up a series of solutions that differ only in their relative concentrations of $Na^+$ and $K^+$ by first making two solutions of equal ionic strength. One contains 90 m*M* $Na^+$ gluconate and 60 m*M* NaCl; the other contains 90 m*M* $K^+$ gluconate and 60 m*M* KCl. Both solutions contain 10 m*M* HEPES, 1.2 m*M* $CaCl_2$, and are titrated to pH 7.4 with KOH. The solutions are mixed to generate concentrations of $Na^+$ ranging from 0 to 150 m*M*. Cells are exposed to these solutions of known $[Na^+]$ in the presence of 10 $\mu M$ gramicidin D at 37°C (same temperature as experiments). Fluorescence ratios are recorded when equilibration is achieved, ~20 min after addition of gramicidin for the first measurement, and then ~10 min after exposure to each different $Na^+$ concentration (Fig. 3A). The concentration of $Cl^-$ is constant in this protocol, and was chosen to be equal to the resting $[Cl^-]_i$ measured in the cells. The rationale is that if the cells possess a $Cl^-$ conductance in the plasma membrane, the presence of the nonselective cation conductance contributed by the gramicidin, by collapsing the membrane potential, will provide a counterion conductance for $Cl^-$ permeation, causing the cell to swell or shrink as extracellular $[Cl^-]$ is increased or decreased, respectively. Excessive cell swelling at high extracellular $[Cl^-]$ may cause

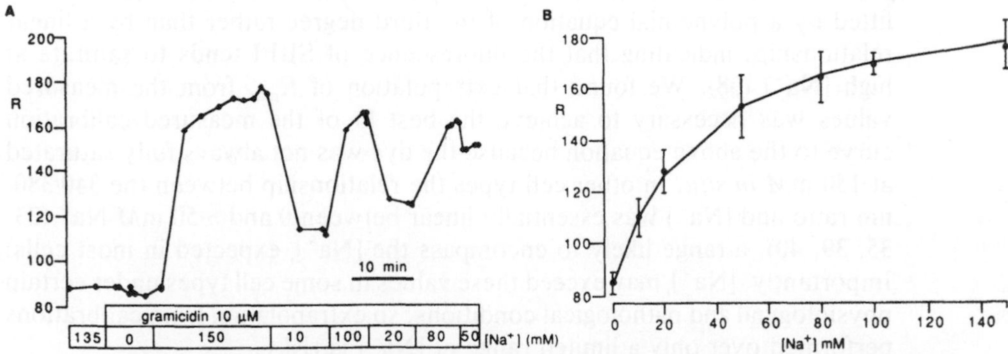


FIG. 3 Intracellular calibration of SBFI fluorescence in a single salivary acinar cell. (A) After control measurements of 340/380-nm fluorescence ratio *(R)* in normal medium, the cell was exposed to a medium containing 10 m*M* HEPES, 90 m*M* $K^+$ gluconate, 60 mM KCl, 1.2 m*M* $CaCl_2$, with 10 $\mu M$ gramicidin (0 m*M* $Na^+$, 37°C, pH 7.4). When a minimum ratio ($R_0$) was obtained, the cell was exposed sequentially to similar solutions containing different $[Na^+]$ replacing $K^+$. (B) Points represent the mean SBFI fluorescence ratio (340/380 = *R*) for each $[Na^+]$ for a group of cells on a single coverslip ($n = 3$). The curve is fitted by the equation $[Na^+] = K_d\beta(R - R_0)/(R_{max} - R)$, where $K_d\beta$ = 35 m*M*, $R_0$ = 85, $R_{max}$ = 200, $\beta$ = 1.8 ($\beta$ is the ratio of the excitation efficiencies of free indicator to $Na^+$-bound indicator at 380 nm). The $K_d$ (dissociation constant for the dye) was ~19 m*M* for these calibration data.

dye leakage, and excessive cell shrinkage at low extracellular [$Cl^-$] may affect the properties of the dye. Because gramicidin does not penetrate into cells, it does not collapse cation gradients across intracellular membranes. Thus, use of gramicidin is desirable if compartmentalization of SBFI is present, as discussed above, because only the cytoplasmic signal will change when extracellular [$Na^+$] is manipulated. Equilibration of $[Na^+]_i$ and $[Na^+]_o$ can also be achieved using the $Na^+/H^+$ ionophore monensin (1–20 $\mu M$) together with the $K^+/H^+$ ionophore nigericin (1–20 $\mu$M) (15, 23). The combination of the two ionophores in essence exchanges $Na_i^+$ for $K_i^+$ at constant $pH_i$. Both ionophores permeate the cell and permeabilize intracellular compartments. Their use may therefore reveal the contribution of SBFI signals in acidic organelles (15).

The values of the measured 340/380-nm ratios obtained during these calibrations are plotted against [$Na^+$] (Fig. 3B). The data obtained from isolated salivary acinar cells fit reasonably well the equation $[Na^+]_i = K_d\beta(R - R_0)/(R_{max} - R)$. $R_0$ and $R_{max}$ are the ratios measured in the absence of $[Na^+]_i$ and in the presence of saturating $[Na^+]_i$, respectively (Fig. 3B). $K_d$ is the $Na^+$ dissociation constant and $\beta$ is the ratio of the excitation efficiencies of free indicator to $Na^+$-bound indicator at 380 nm. In salivary acinar cells the average $R_{max}/R_0$ was ~2.7, which is comparable to that observed for the dye *in vitro* (11). The fluorescence ratios plotted as a function of [$Na^+$] were fitted by a polynomial equation of the third degree rather than by a linear relationship, indicating that the fluorescence of SBFI tends to saturate at high [$Na^+$] (38). We found that extrapolation of $R_{max}$ from the measured values was necessary to achieve the best fit of the measured calibration curve to the above equation because the dye was not always fully saturated at 150 m$M$ *in situ*. In other cell types the relationship between the 340/380-nm ratio and [$Na^+$] was essentially linear between 0 and ~50 m$M$ $Na^+$ (23, 35, 39, 40), a range likely to encompass the $[Na^+]_i$ expected in most cells. Importantly, $[Na^+]_i$ may exceed these values in some cell types under certain physiological and pathological conditions, so extrapolation from calibrations performed over only a limited range of [$Na^+$] is risky.

## Experimental Protocols for Measuring $Na^+$ Transport Using SBFI

SBFI has been used in numerous cell types to estimate the effects of growth factors, hormonal stimulation, and other perturbations, such as hypoxia on $[Na^+]_i$. In fewer studies, the dye has also been exploited to study the regulation and activities of specific $Na^+$ transport mechanisms. Because $[Na^+]_i$ reflects the balance between plasma membrane $Na^+$ influx and efflux activities, the activities of specific $Na^+$ entry pathways can be quantified by measuring $[Na^+]_i$ in the absence of $Na^+$ efflux activity, in most cells contributed by the $Na^+$ pump. The pump can be inhibited by the cardiac glycoside ouabain (1 m*M* in rodent cells, 10 $\mu M$ in other cell types) or the more readily reversible dihydroouabain, or by removal of extracellular $K^+$. The latter is rapidly reversible, but may also substantially and rapidly affect membrane potential. To examine the effects of agonist stimulation on the activities of $Na^+$ influx pathways in the basolateral membrane of salivary acinar cells, we measured $[Na^+]_i$ in single acinar cells in the presence of the $Na^+/K^+$-ATPase inhibitor, ouabain (1 m*M*) (10). Resting $[Na^+]_i$ was ~7 m*M* in these cells. Ouabain was without effect on $[Na^+]_i$ in unstimulated cells for up to 10 min (Fig. 4), evidence that plasma membrane $Na^+$ influx pathways were inactive in resting cells. The muscarinic agonist carbachol (10 $\mu M$) caused a marked, rapid rise of $[Na^+]_i$, indicating that stimulation was associated with activation of $Na^+$ influx pathways (Fig. 4). The unidirectional $Na^+$ flux ($J_{Na^+}$) is a function of both the rate of change of $[Na^+]_i$ and the $Na^+$ buffering capacity (*B*) of the cytoplasm: $J_{Na^+} = d[Na^+]_i/dt\ B$. In gastric gland cells *B* was determined by comparing acid-induced $H^+$ and $Na^+$ fluxes through the $Na^+/H^+$ exchanger using BCECF and SBFI, respectively, and the known

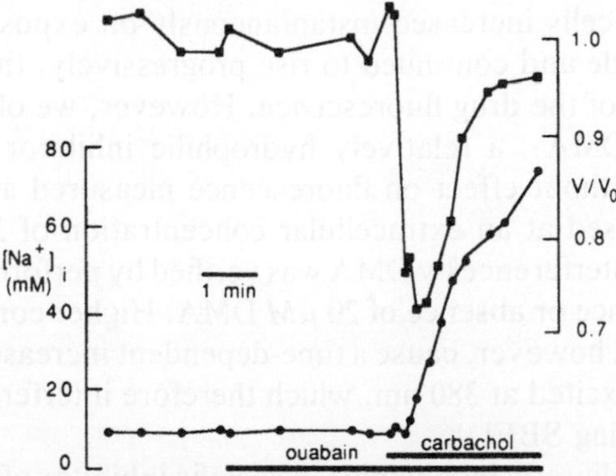


FIG. 4 Intracellular $Na^+$ concentration (●) and cell volume (■) determined simultaneously in a single salivary (parotid) acinar cell during exposure to carbachol (10 $\mu M$) in the presence of ouabain (1 m*M*). Under these conditions, the initial rate of rise of $[Na^+]_i$ provides an estimate of the $Na^+$ permeability, which is activated by the agonist stimulation. The observed carbachol-induced cell shrinkage is due to efflux of $Cl^-$ and $K^+$ through $Ca^{2+}$-activated channels in the plasma membrane, which causes a loss of KCl and osmotically obliged water from the cell. The $Cl^-$ loss results in a pronounced decrease in intracellular chloride concentration, which quantitatively accounts for the shrinkage. Changes in cell volume therefore quantitatively estimate changes in cell $Cl^-$ concentration [J. K. Foskett, *Am. J. Physiol.* **259,** C998 (1990); K. R. Lau and R. M. Case, *Pfluegers Arch.* **411,** 670 (1988)].

$H^+$ buffering capacity of the cytoplasm (21). The $Na^+$ buffering capacity was found to be insignificant. Therefore $d[Na^+]_i/dt$ measured in the presence of ouabain provides a measure of unidirectional $Na^+$ influx. Taken together with estimates of cell surface area, such measurements can provide estimates of plasma membrane $Na^+$ permeability.

To define the pathways involved in carbachol-stimulated $Na^+$ influx in rat salivary cells we performed experiments in the presence of ouabain together with inhibitors of known $Na^+$ transport mechanisms. Amiloride analogs are potent inhibitors of $Na^+/H^+$ exchange. However, amiloride analogs are fluorescent when excited at 380 nm and emission monitored at 500 nm. Therefore, these analogs have the potential to interfere with SBFI determinations of $[Na^+]_i$. At concentrations normally employed to inhibit $Na^+/H^+$ exchange, many of the analogs do not contribute sufficient extracellular fluorescence to cause interference. However, many of the analogs are highly lipophilic, rapidly permeate into cells, and accumulate in the cytoplasm. We found that phenamil (1 $\mu M$), EIPA (5 $\mu M$), and MIBA (5 $\mu M$) rapidly entered isolated salivary acinar cells as evidenced by a rapid, significant, time-dependent enhancement of cell fluorescence excited at 380 nm (M. A. Robertson and J. K. Foskett, 1992, unpublished). These analogs could not be used with SBFI in our system. Haroontunian *et al.* (15) reported that the fluorescence intensity of cells increased instantaneously on exposure of the cells to 0.1 m$M$ amiloride and continued to rise progressively, thus preventing reliable subtraction of the drug fluorescence. However, we observed that dimethylamiloride (DMA), a relatively hydrophilic inhibitor of $Na^+/H^+$ exchange (41), was without effect on fluorescence measured at 380 nm for up to 20 min when used at an extracellular concentration of 20 $\mu M$ (10). A lack of significant interference by DMA was verified by performing SBFI calibrations in the presence or absence of 20 $\mu M$ DMA. Higher concentrations (>50 $\mu M$) of DMA did, however, cause a time-dependent increase in fluorescence when cells were excited at 380 nm, which therefore interfered with the estimation of $[Na^+]_i$ using SBFI.

The loop-diuretic bumetanide, a specific inhibitor of $Na^+/K^+/2Cl^-$ cotransport, is fluorescent when excited at 340 nm (500 nm emission) and may therefore also interfere with $[Na^+]_i$ determinations using SBFI. During exposure of resting SBFI-loaded salivary acinar cells to 100 $\mu M$ bumetanide in the bath, we detected no change in fluorescence ratio, suggesting that background fluorescence of bumetanide was minimal under the conditions of our experiments. However, at higher intensifier-gain settings necessary to observe SBFI fluorescence in cells that had loaded poorly with the dye, bumetanide fluorescence was detected and influenced the SBFI ratio. In contrast to the observations with the amiloride analogs, however, bumetanide did not appear to enter and accumulate in the cells, because the fluorescence contributed by bumetanide was constant and rapidly eliminated by perfusion with a bumetanide-free solution. In situations wherein bumetanide contributes a constant intensity to the fluorescence signal, measurement of the bumetanide fluorescence intensity at 340 nm in the absence of cells should be subtracted from the 340-nm signal from the SBFI-loaded cells before calculation of the 340/380-nm ratio (42).

The activity of the $Na^+/K^+$ pump can be quantified in cells by measuring the total and ouabain-insensitive $Na^+$ effluxes from $Na^+$-loaded cells (21). SBFI-loaded cells are loaded with $Na^+$ by incubation in a $K^+$-deficient medium. Extracellular $K^+$ is then restored and extracellular $Na^+$ is simultaneously removed. The recovery of $[Na^+]_i$ is due to pump-mediated as well as passive pathways. The latter are estimated by repeating the experiment in the presence of ouabain. The difference between the two $[Na^+]_i$ recovery curves reflects the pump-mediated component. The slope at any particular $[Na^+]_i$ provides the pump rate vs. $[Na^+]_i$ dependence.

## References

1. M. Crompton, *in* "Intracellular Calcium Regulation" (F. Bronner, ed.), p. 181. Alan R. Liss, New York, 1990.
2. L. E. Limbird, *Am. J. Physiol.* **247,** E59 (1984).
3. R. L. Watson, K. L. Jacobson, and J. C. Singh, *Biochem. Pharmacol.* **38,** 1069 (1989).
4. R. S. Dunman, R. Z. Terwilliger, E. J. Nesler, and J. F. Tallman, *Biochem. Pharmacol.* **38,** 1909 (1989).
5. A. Marty, *Pfluegers Arch.* **396,** 179 (1983).
6. M. Kameyama, M. Kakei, R. Sato, T. Shibasaki, H. Matsuda, and H. Irisawa, *Nature (London)* **309,** 354 (1984).
7. F. Ibarra, A. Aperira, L. B. Svensson, A. C. Ekloff, and P. Greengard, *Proc. Natl. Acad. Sci. U.S.A.* **90,** 21 (1993).
8. D. B. Jaffe, D. Johnston, N. Lasser-Ross, J. E. Lisman, H. Miyakawa, and W. N. Ross, *Nature (London)* **357,** 244 (1992).
9. M. M. Y. Wong and J. K. Foskett, *Science* **254,** 1014 (1991).
10. M. A. Robertson and J. K. Foskett, *Am. J. Physiol.* **267,** C146 (1994).
11. A. Minta and R. Y. Tsien, *J. Biol. Chem.* **264,** 19449 (1989).
12. R. P. Haughland, "Handbook of Fluorescent Probes and Research Chemicals" (K. D. Larison, ed.), Molecular Probes, Eugene, Oregon, 1992.
13. G. A. Smith, T. R. Hesketh, and J. C. Metcalfe, *Biochem. J.* **250,** 227 (1988).
14. E. D. W. Moore, A. Minta, R. Y. Tsien, and F. S. Fay, *Biophys. J.* **55,** 471a (1989).
15. A. T. Harootunian, J. P. Y. Kao, B. K. Eckert, and R. Y. Tsien, *J. Biol. Chem.* **264,** 19458 (1989).
16. S. Borzak, M. Reers, J. Arruda, V. K. Sharma, S.-S. Sheu, T. W. Smith, and J. D. Marsh, *Am. J. Physiol.* **263,** H866 (1992).

17. P. Donoso, J. G. Mills, S. C. O'Neill, and D. A. Eisner, *J. Physiol.* (*London*) **448,** 493 (1992).
18. S. M. Harrison, E. McCall, and M. R. Boyett, *J. Physiol.* (*London*) **449,** 517 (1992).
19. P. A. Negelscu and T. E. Machen, *in* "Methods in Enzymology" (S. Fleischer and B. Fleischer, eds.), Vol. 192, p. 38. Academic Press, San Diego, 1990.
20. D. E. W. Moore and F. S. Fay, *Proc. Natl. Acad. Sci. U.S.A.* **90,** 8058 (1993).
21. P. A. Negelscu, A. Harootunian, R. Y. Tsien, and T. E. Machen, *Cell Regul.* **1,** 259 (1990).
22. S. M. Harrison, J. E. Frampton, E. McCall, M. R. Boyett, and C. H. Orchard, *Am. J. Physiol.* **262,** C348 (1992).
23. K. Tornquist and E. Ekokoski, *J. Cell. Physiol.* **154,** 608 (1993).
24. M. C. P. Haigney, H. Miyata, E. G. Lakatta, M. D. Stern, and H. S. Silverman, *Circ. Res.* **71,** 547 (1992).
25. K. Jacobson and J. Wajcieszyn, *Proc. Natl. Acad. Sci. U.S.A.* **81,** 6747 (1984).
26. K. Iijima, L. Lin, A. Nasjletti, and M. S. Goligorsky, *Am. J. Physiol.* **260,** C982 (1991).
27. D. W. Jung, L. M. Apel, and G. P. Brierley, *Am. J. Physiol.* **262,** C1047 (1992).
28. Z. Deri and V. Adam-Vizi, *J. Neurochem.* **61,** 818 (1993).
29. S. Ishikawa, G. Fujisawa, K. Okada, and T. Saito, *Biochem. Biophys. Res. Commun.* **194,** 287 (1993).
30. S. E. Ishikawa, K. Okada, and T. Saito, *Endocrinology* (*Baltimore*) **127,** 560 (1990).
31. M. Sorimachi, K. Yamagami, S. Nishimura, and K. Kuramoto, *J. Neurochem.* **59,** 2271 (1992).
32. L. Homolya, Z. Hollo, V. A. Germann, I. Pastan, M. M. Gottesman, and B. Sarkadi, *J. Biol. Chem.* **268,** 21493 (1993).
33. F. DiVirgilio, T. H. Steinberg, and S. C. Silvestein, *Cell Calcium* **11,** 57 (1990).
34. R. Lyu, L. Smith, and J. Bingham Smith, *Am. J. Physiol.* **263,** C628 (1992).
35. M. Borin and W. Siffert, *J. Biol. Chem.* **266,** 13153 (1991).
36. M. L. Borin, W. F. Goldman, and M. P. Blaustein, *Am. J. Physiol.* **264,** C1513 (1993).
37. H. Satoh, H. Hayashi, N. Noda, H. Terada, A. Kobayashi, Y. Yamashita, T. Kawai, M. Hirano, and N. Yamazaki, *Biochem. Biophys. Res. Commun.* **175,** 611 (1991).
38. P. Gilon and J. C. Henquin, *FEBS Lett.* **315,** 353 (1993).
39. L. Ali, E. Grapengiesser, E. Gylfe, B. Helleman, and P. Lund, *Biochem. Biophys. Res. Commun.* **164,** 212 (1989).
40. E. M. Johnson, J. Theler, A. M. Capponi, and M. B. Valloton, *J. Biol. Chem.* **266,** 12618 (1991).
41. T. R. Kleyman and E. J. Cragoe, Jr., *J. Membr. Biol.* **105,** 1 (1988).
42. G. H. Zhang, E. J. Cragoe, Jr., and J. E. Melvin, *Am. J. Physiol.* **264,** C54 (1993).

# [14] Measurement of Cytosolic Sodium Using Ion-Selective Microelectrodes

Joachim W. Deitmer and Thomas Munsch

## Introduction

The determination of intracellular $Na^+$ has been of particular interest since it became evident that the $Na^+$ gradient across cell membranes is an extremely important energy store. Not only does the $Na^+$ gradient provide the electrochemical force for Na ions to move downhill via ion channels into the cell, but it also drives other ions or substrates against their (electro)chemical gradient across the membrane. The first is essential for electrical excitability in many cell types, and the latter is required to maintain ionic homeostasis and/or transport of organic metabolites into and out of cells. Because the energy for such $Na^+$-dependent carrier systems is delivered directly from the $Na^+$ gradient, the kinetics and amplitude of these processes are a function of membrane potential and intracellular $Na^+$ (assuming relatively constant extracellular $Na^+$).

Membrane potential and the chemical gradient of $Na^+$ add up to the electrochemical driving force for $Na^+$ across cell membranes, which simply is the difference between membrane potential and the equilibrium potential for $Na^+$, $E_{Na}$. The electrochemical equilibrium potential for $Na^+$ can be described by the Nernst equation as shown in Eq. (1):

$$E_{Na} = (RT/zF)\ \ln([Na^+]_o/[Na^+]_i), \qquad (1)$$

where $R$ is the gas constant, $T$ is the absolute temperature, $F$ is the Faraday constant, $z$ is the valency of $Na^+$, and $[Na^+]_o$ and $[Na^+]_i$ are the extracellular and intracellular $Na^+$ concentrations, respectively. With a chemical gradient of $Na^+$ across cell membranes of living cells, derived from the ratio of $[Na^+]_o$/$[Na^+]_i$, which ranges between 10 and 20, an $E_{Na}$ of +58 to +76 mV can be calculated according to Eq. (1). Taking into account membrane potential values of between −40 and −90 mV, the electrochemical gradient for Na ions can be as much as 100 to 166 mV across cell membranes. The most variable parameters for the $Na^+$ electrochemical gradient are membrane potential and the intracellular Na concentration, whereas the high external Na concentration is usually relatively constant for a given animal species. As such, marine animals generally have a much higher Na concentration in

*Methods in Neurosciences, Volume 27*

their blood and interstitial fluids (between 350 and 500 m$M$) compared to freshwater or terrestrial animals (between 80 and 160 m$M$). The intracellular Na concentration is accordingly three to five times higher in cells of marine as compared to nonmarine animals, resulting in approximately similar $Na^+$ gradients across cell membranes of all metazoa.

It is essential for the survival of every cell to maintain a low intracellular $Na^+$ concentration and hence a large $Na^+$ gradient. This is achieved by the ubiquitous $Na^+/K^+$ pump, an ATPase, in the cell membrane of all cells, which extrudes $Na^+$ and imports $K^+$. This pump, driven by the energy derived from cleaving ATP at the inside of the membrane, carries both $Na^+$ and $K^+$ against their electrochemical gradients. Because usually more $Na^+$ ions are extruded than $K^+$ ions are imported ($3Na^+/2K^+$ per cycle is the most common stoichiometry), the pump transports net charge across the membrane, and therefore is electrogenic (1, 2).

Hence the $Na^+/K^+$-ATPase is membrane potential sensitive, and its activity is mainly determined by the concentration of intracellular $Na^+$ and extracellular $K^+$. Many of the physiological properties and functions of this $Na^+/K^+$ pump as well as other $Na^+$-dependent transport mechanisms have been studied using $Na^+$-selective microelectrodes in a great variety of different cell types.

## Historical Background

$Na^+$-selective microelectrodes for intracellular measurements have been used for decades, and a variety of different types exist. The original sensor for $Na^+$ was a $Na^+$-sensitive glass (3); this was used to make $Na^+$-selective microelectrodes by fitting a piece of $Na^+$-sensitive glass (NAS 11–18) into the tip of an insulating glass pipette. The $Na^+$-sensitive glass either protruded from the outer pipette tip [''Hinke-type'' electrode (4)] or was recessed into the pipette tip [''Thomas-type'' electrode (5)]. When liquid ion exchanger (LIX) microelectrodes were developed (6), the use of $Na^+$-sensitive LIX microelectrodes became popular for a while (7). This type of electrode was easier to make than the type consisting of the NAS 11–18 $Na^+$-sensitive glass, but had a much poorer selectivity for $Na^+$ over $K^+$.

Other types of $Na^+$ sensors include electrically charged carriers, such as monensin (8), or $Na^+$ sensors based on lipophilic crown ether compounds (9, 10). Synthetic neutral carrier $Na^+$ sensors (ETH 227, ETH 157) have been developed at the Eidgenössische Technische Hochschule in Zürich, Switzerland (11, 12). The manufacture of these microelectrode is similar to that of LIX microelectrodes. At the current state of the art, $Na^+$-selective microelectrodes based on neutral carriers are generally used to measure intracellular $Na^+$ with electrodes.

## Manufacture of $Na^+$-Selective Microelectrodes

Ion-selective microelectrodes suitable for intracellular measurements should have a tip size of less than 2 $\mu$m; often the tip needs to be around 1 $\mu$m or smaller to avoid cell damage. Ion-selective microelectrodes have been used for cells of less than 15 $\mu$m in diameter, for example, mammalian glial cells (13, 14). However, below a tip size of 1 $\mu$m, the selectivity of the electrodes may suffer, and there seems to be a limit as to how small the tip size, i.e., the sensor area exposed to the fluid, may be for a suitable ion-selective microelectrode. As for all liquid ion exchanger and neutral-carrier electrodes, the inside of the glass pipettes must be made hydrophobic to maintain the lipophilic sensor within the tip. This is a crucial step in manufacturing the electrodes, because it determines their sensitivity and the mechanical stability of the sensor inside the tip.

A variety of $Na^+$-selective sensors are available commercially (see above). For neutral-carrier $Na^+$-selective microelectrodes the essential ingredient is a $Na^+$ carrier, which is dissolved in an organic solvent. Such a "Na cocktail" may be purchased, but may also be prepared in the lab, which can be useful (see below) and can save some expense.

Here we describe the steps in manufacturing a $Na^+$-selective microelectrode, including the choice of electrolyte filling solutions. Although we are aware that there are many ways to obtain an operational electrode, we will focus on the methodology used in our laboratory.

### *Preparing Glass Pipettes*

Because intracellular ion-selective microelectrodes not only sense the potential specific to the ion species to be measured, but also all other potential differences, an ion-nonselective electrode must be used in combination with the ion-selective electrode to obtain the ion-specific potential. Hence, at least two electrode barrels, the ion-sensing barrel and the ion-insensitive barrel, or "reference barrel," either as two single-barreled electrodes or as a double-barreled electrode, must be impaled into the same cell (or enter electrically coupled, isopotential, cells).

The assumption is that all ion-nonselective potentials are the same, as measured by the ion-selective barrel and the reference barrel. The subtraction of the potentials measured by the two barrels should give the potential indicative of the ion to be measured (and all potentials of interfering ions).

The preparation of an ion-selective microelectrode hence starts with the decision of whether to use two single-barreled or one double-barreled pipette. In most cases the tips of the two barrels must penetrate the same cell, if there is any doubt that adjacent cells may not have the same membrane potential due to electrical coupling. Hence, the most widely used type of ion-selective electrode is the double-barreled type, wherein the tips of the two barrels are directly side by side.

Having chosen a double-barreled, or even a triple-barreled, micropipette (Fig. 1), one needs to decide whether it is better to use a theta-type glass tubing, wherein a septum divides the single tube into two barrels, or two regular glass capillaries, held together and pulled to a single tip.

Borosilicate glass tubing of 1–2 mm diameter is often used, but some experiments may prefer aluminosilicate or quartz glass tubing, both of which have a higher melting point and are hence harder. The choice of glass and tubing may depend on the sensor used, the preparation, silanization procedure, or individual preference. In any case, silanization of the reference barrel, which would pull the sensor into the tip of the reference barrel, must be avoided. If the sensor is drawn into the reference barrel, the latter becomes ion sensitive and therefore unsuitable as an ion-nonselective reference.

If two capillaries are used to manufacture a double-barreled ion-selective

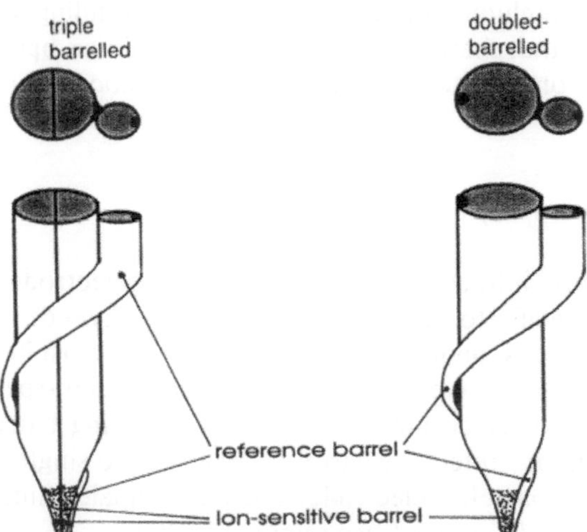


FIG. 1 Basic designs of triple-barreled (left) and double-barreled (right) ion-selective microelectrodes. See text for further details.

microelectrode, 7- to 10-cm-long glass tubes with an inner filament are glued or taped side by side. We use a strip of aluminum paper wrapped around the two capillaries. It may be useful to use glass tubing of two different diameters, e.g., 1.5-mm tubing for the respective ion-sensing barrel and 1.0 mm for the respective reference barrel, which makes it easier to recognize the respective ion-sensitive and reference electrodes.

The side-by-side tubings are then heated and prepulled 2–5 mm. We use a vertical puller (Narishige PE-2) with an adjustable stopper underneath. During this first pull, while the glass is soft, the tubing is twisted by 180°; effectively the thinner tubing (reference electrode) is twisted one-half turn around the thicker, "central" barrel (ion-sensing electrode). After cooling the twisted tubings for 1–2 min, they are recentered and pulled out to a fine tip (0.2–1.5 $\mu$m). Adjustment of heating and pulling force will shape the tip of the pipette.

## *Silanization*

To make the inside of the central ion-sensing barrel hydrophobic, the inside glass wall of the pipette has to be silanized. Only a proper silanization ensures that the lipophilic sensor remains in the tip of the pipette and makes smooth contact with the glass wall, allowing no water/electrolyte microchannels to be formed between sensor and glass wall. This would short-circuit the voltage response of the ion-selective barrel and reduce or abolish the sensitivity of the electrode. A well-silanized $Na^+$-selective microelectrode responds with 50–57 mV for a 10-fold change in the $Na^+$ concentration of physiological saline (above 5 m*M* $Na^+$). Smaller electrode responses may be due to inappropriate sensor, to tip shape or size of the electrode, and/or, most often, to inadequate silanization. Below 5 m*M* $Na^+$ the contribution of interfering ions to the response of the ion-selective microelectrode increases significantly with decreasing $Na^+$ concentration (see below).

Many ways to silanize an electrode have been described by various authors, and surely many more have been tried and abandoned. Here, we describe our method of silanization, shared by many others, who have contributed various steps of the elaborate procedure.

First, a 5% silane solution is produced by adding 0.5 ml tributylchlorosilane to 9.5 ml carbon tetrachloride (99.9% purity). A small droplet of this mixture is sucked up into a glass capillary (diameter ~0.5 mm), from which it is backfilled into the central ion-sensing barrel of the double-barreled micropipette. When the pipette is held vertically, the silane mixture should flow along the inner filament into the very tip of the central barrel (gentle tapping of the barrel shaft can speed this process). The pipette is then baked, either in a hot oven, or, as described by Borelli *et al.* (15), placed on a hot plate, heated for 4–5 min near 500°C. We heat our pipettes for 4.5–4.8 min at 460–475°C on a hot plate. The pipettes are then stored dry and dust free in a closed container, and, after cooling, they are ready to be filled. Once silanized, we use the pipettes within 1 week.

## Choice of Sensor

There are different $Na^+$-selective sensors on the market. At present, neutral carriers for $Na^+$ are mainly used. The relevant properties of a $Na^+$-selective sensor, as is the case with other ion-selective compounds, are its sensitivity to the ion species to be measured and its selectivity for the ion species to be measured over other, interfering, ions. Interfering ions, mainly $K^+$ and $Ca^{2+}$, can contribute considerably to the voltage response of $Na^+$-selective microelectrodes, especially at low $Na^+$ concentrations. Therefore it might be necessary to estimate this interference quantitatively, if large changes in the concentration of the interfering ion(s) during a given experimental protocol cannot be excluded. This can be done by estimating the selectivity factors ($K_{ij}^{pot}$) that are simply a measure of the preference by the sensor for an interfering ion *(j)* relative to the ion *(i)* to be detected, and which are given by the Nicolsky–Eisenman [Nicolsky (16), Eisenman (17)] equation [Eq. (2)]:

$$E = E_0 + s \log\left(a_i \sum_{i \neq j} K_{ij}^{pot} (a_j)^{z_i/z_j}\right), \tag{2}$$

where $E$ is the potential difference measured by the ion-selective electrode; $E_0$ is the concentration-independent potential produced up by boundary and liquid junction potentials; $s$ is the theoretical slope of the electrode potential; $z_i$ and $a_i$ are the valency and activity of the primary ion $i$, respectively; $z_j$ and $a_j$ are the valency and activity of the interfering ion $j$, respectively; and $K_{ij}^{pot}$ is the potentiometric selectivity factor [small selectivity factors indicate a good preference for the ion *(i)* to be detected].

Many of the ionophores used in the past 25 years in ion-selective microelectrodes have been developed by Ammann, Simon, and co-workers at the ETH Zürich. Most often used in microelectrodes are ionophores in the cocktails ETH 157 and ETH 227 (Fig. 2), described by Ammann *et al.* (20), Güggi *et al.* (11), and Steiner *et al.* (18).

For $Na^+$-selective microelectrodes (based on ETH 227), selectivity factors estimated are typically 0.01–0.03 for $K_{NaK}^{pot}$ and 1–4 for $K_{NaCa}^{pot}$ (12, 18). Once determined, these factors can be used for corrections of the actual voltage

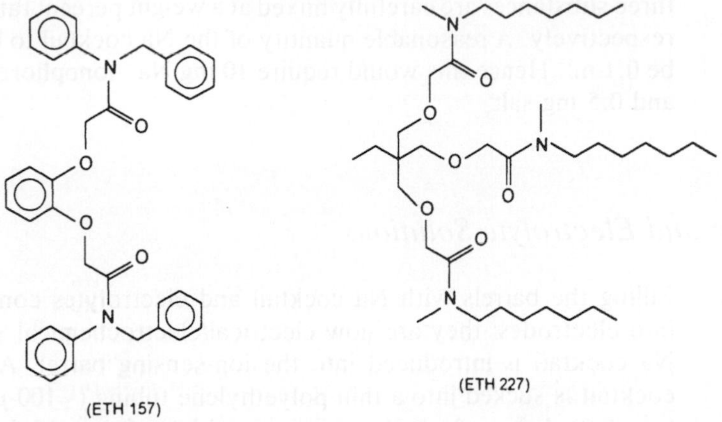

FIG. 2 Chemical structure of neutral carriers for $Na^+$-selective microelectrodes. ETH 157 is *N,N'*-dibenzyl-*N,N'*-diphenyl-1,2-phenylene dioxyacetamide; ETH 227 is 1,1,1-tris[1'-(2'-oxa-4'-oxo-5'-aza-5'-methyl)dodecanyl]propane, (after D. Ammann, "Ion-Selective Microelectrodes," Springer-Verlag, Berlin, 1986).

responses of the microelectrodes, for instance, when intracellular $Ca^{2+}$ changes significantly, and under the assumption that selectivity factors do not change during an experiment (19).

## *Making a Na Cocktail*

Using a $Na^+$ ionophore, an organic solvent, and a lipophilic salt, a Na "cocktail" can easily be prepared in any laboratory. This is particularly useful if larger quantities of the $Na^+$ sensor are needed. Even if $Na^+$-selective electrodes are only occasionally used, there are reasons to produce one's own cocktail. The main reason is that, in our experience, purchased Na cocktails sometimes do not work properly or produce an extremely high resistance of the ion-sensing barrel ($>10^{12}\ \Omega$), resulting in too noisy or unstable recordings; some may have an unacceptably high $K^+$ selectivity. Pitfalls may also occur with a Na cocktail prepared in the laboratory, but trial evaluations are relatively fast and less costly than ordering new Na cocktail each time.

To mix a Na cocktail, three ingredients are needed: the $Na^+$ ionophore [Ionophor VI or ETH 227; Fluka 71739, based on a bis(12-crown-4-methyl)dodecylmethylmalonate], the solvent 2-nitrophenyl octyl ether (*o*-NPOE; Fluka 73732), and a trace of the salt sodium tetraphenyl borate (Fluka 72018). The three substances are carefully mixed at a weight percent ratio of 10 : 89.5 : 0.5, respectively. A reasonable quantity of the Na cocktail to be made up would be 0.1 ml. Hence this would require 10 mg $Na^+$ ionophore, 89.5 mg solvent, and 0.5 mg salt.

## *Filling and Electrolyte Solutions*

Filling the barrels with Na cocktail and electrolytes converts the pipettes into electrodes: they are now electrical/electrochemical sensors. First, the Na cocktail is introduced into the ion-sensing barrel. A small amount of cocktail is sucked into a thin polyethylene tubing (<100-$\mu$m tip size) and is transferred through the large, open end into the tip of the central barrel of the double-barreled pipette. The smallest cocktail quantity possible to handle is that sufficient to fill the tip of the barrel with a column of 0.5–10 mm (depending on tip length). The viscous cocktail takes some minutes to flow into the tip by capillary forces. This, however, can only work, if the tip is not blocked! Sometimes gentle application of moderate heat alongside the pipette tip, e.g., with the help of a soldering iron, helps to remove the air bubble from the tip. However, care must be taken not to overheat the tip, which may result in crystallizing the cocktail.

After the cocktail has reached the tip of the central barrel, the reference barrel should be filled. With a tapered polyethylene tubing, a solution of 3 *M* KCl is introduced into the reference barrel up to 5–10 mm from the top. An electrolyte solution is then placed above the sensor cocktail in the central barrel, making sure that there is no air bubble at the interface. This solution must contain $Na^+$ (the ion to be measured by this barrel); we use a solution of 0.1 *M* NaCl + 10 m*M* 3-*N*-morpholinopropanesulfonic acid (MOPS) buffered to pH 7.0. A chlorided silver wire (diameter ~0.3 mm) is then inserted below the level of electrolyte in each barrel and is sealed inside with a small drop of dental wax.

It may be useful, and sometimes even necessary, to bevel the electrode tip. This also helps to remove dust or grease from the sensor facing the fluid, and/or to unclog the reference barrel (sometimes the very tip of the reference barrel becomes ion sensitive, as described above; this problem is eliminated by gentle beveling). The tip beveling can be accomplished with a jet stream of aluminaoxide powder (particle size 0.05 $\mu$m) in distilled water jetted for about 10–30 sec over the electrode tip at an angle of approximately 30°. A more elaborate technique is to use a rotating disk, onto which the electrode tip is gently pushed for 1–3 sec (21).

## Calibration of Electrodes

Testing ion-selective electrodes and calibrating them at different, defined concentrations of the ion species to be measured should precede and follow each experiment in which the electrodes are used in a cell (Fig. 3). Furthermore, all solutions used during an experiment should be tested in terms of their effect on the response of the ion-selective electrode. Here, we describe two ways of calibrating $Na^+$-selective microelectrodes most commonly used today.

One way to calibrate the microelectrodes is to replace $Na^+$ in the calibration solution with an "inert" membrane-impermeant cation. This can be any organic substitute for Na ions used in physiological experiments, such as trishydroxyethylmethane (TRIS), trimethylamine (TMA), $Li^+$, or choline. The most widely used $Na^+$ substitute today is probably *N*-methyl-D-glucamine (NMDG). NMDG is believed to be nontoxic and membrane impermeant, and no $Na^+$-dependent biological process is known whereby NMDG could

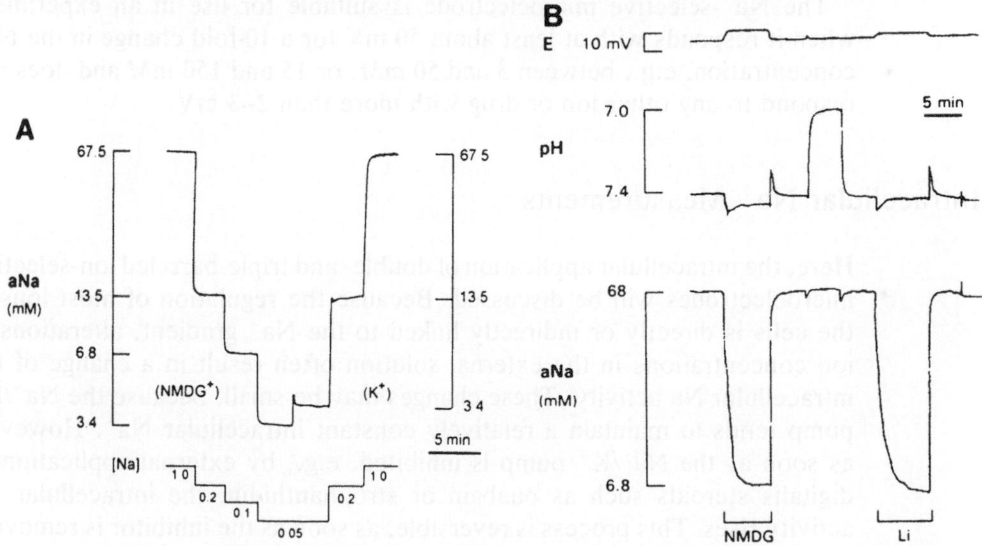


FIG. 3 Calibration of a $Na^+$-selective microelectrode in experimental solutions at varying Na activities (*a*Na). (A) Double-barreled microelectrode: $Na^+$ was substituted with equimolar amounts of $NMDG^+$ (left) and $K^+$ (right). (B) Triple-barreled pH- and $Na^+$-selective microelectrode in solutions of varying pH (7.0 and 7.4) and Na activities (68 and 6.8 m*M*). $Na^+$ was substituted with either $NMDG^+$ (left) or $Li^+$ (right). All solutions contained 2 m*M* $Ca^{2+}$.

replace $Na^+$. $Na^+$-selective electrodes do not respond to NMDG, which is another essential requisite for using NMDG in calibration solutions (Fig. 3).

NaCl of physiological saline is exchanged by equimolar amounts of NMDG–HCl to make calibration solutions containing 20, 10, and 5% of the $Na^+$ concentration in the physiological saline, when testing the electrodes (Fig. 3A). When the calibration procedure is performed routinely before and after experiments, calibration solutions containing 10 and/or 5% are generally used in addition to normal physiological saline (Fig. 3B). Using $Li^+$ as the $Na^+$ substitute usually gives a slightly larger response of the $Na^+$-selective microelectrode, the reason for which is not known. In the "reciprocal" calibration method NaCl is exchanged by KCl. This means that when the $Na^+$ concentration is reduced to, e.g., 10 and 5%, the $K^+$ concentration is increased and reaches values close to those usually found inside cells. Thus, this method provides a condition wherein the $Na^+$ and $K^+$ concentrations of the calibration solutions simulate intracellular conditions, and therefore a realistic $K^+$ interference to the response of the intracellular $Na^+$-selective microelectrode (Fig. 3A). Therefore, the reciprocal calibration method is preferred by many experimentors.

The $Na^+$-selective microelectrode is suitable for use in an experiment when it responds with at least about 50 mV for a 10-fold change in the $Na^+$ concentration, e.g., between 5 and 50 m*M*, or 15 and 150 m*M* and does not respond to any other ion or drug with more than 2–3 mV.

## Intracellular $Na^+$ Measurements

Here, the intracellular application of double- and triple-barreled ion-selective microelectrodes will be discussed. Because the regulation of most ions in the cells is directly or indirectly linked to the $Na^+$ gradient, alterations in ion concentrations in the external solution often result in a change of the intracellular Na activity. These changes may be small, because the $Na^+/K^+$ pump tends to maintain a relatively constant intracellular $Na^+$. However, as soon as the $Na^+/K^+$ pump is inhibited, e.g., by external application of digitalis steroids such as ouabain or strophanthidin, the intracellular Na activity rises. This process is reversible, as soon as the inhibitor is removed. Figure 4 shows a recording with a double-barreled $Na^+$-selective microelectrode in a leech glial cell (a giant, identified glial cell in the neuropile of a segmental ganglion of the annelid worm *Hirudo medicinalis*). The steady-state intracellular Na activity in HEPES-buffered, $CO_2/HCO_3^-$-free saline was near 5 m*M*. After application of ouabain, the intracellular $Na^+$ increased. After removal of the glycoside, the intracellular $Na^+$ rapidly decreased again, presumably due to reactivation of the $Na^+/K^+$ pump.

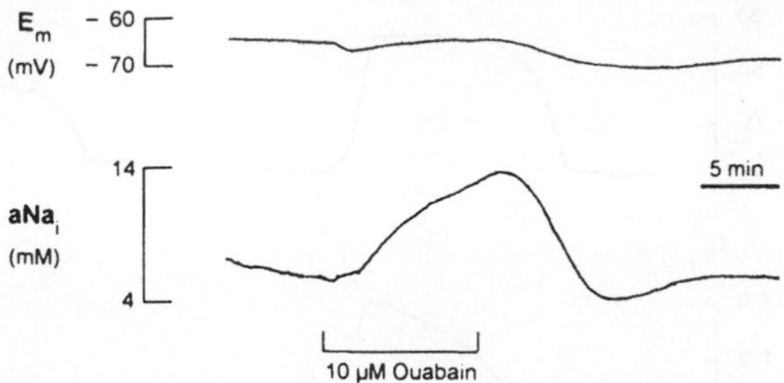


FIG. 4 Membrane potential ($E_m$) and intracellular Na activity ($a$Na$_i$) of a leech giant glial cell measured with a double-barreled $Na^+$-selective microelectode. Addition of the cardiac glycoside ouabain (10 $\mu M$) to the superfusate caused a reversible rise in $a$Na$_i$.

$Na^+$ movements across the cell membrane may be coupled not only to $K^+$, but also to $Ca^{2+}$, $Mg^{2+}$, $H^+$, and/or $HCO_3^-$. Hence, the intracellular regulation of these ions is sensitive to changes in intracellular $Na^+$. In the mammalian heart, for example, the rise of intracellular $Na^+$ reduces Na/Ca exchange, or even reverses this carrier, and subsequently leads to a rise in intracellular $Ca^{2+}$ (22), which results in marked alteration of cardiac muscle contraction. In glial cells, where the intracellular pH is dominated by an electrogenic $Na^+/HCO_3^-$ cotransport (23–25), the intracellular Na activity can change significantly due to the stimulation of this cotransporter. An example of the activity of this electrogenic cotransporter is given in Fig. 5. Changing the pH of the saline in the presence of 5% $CO_2$ and $HCO_3^-$ (i.e., varying the $HCO_3^-$ concentration between 10 and 60 m$M$) induced large shifts of intracellular $Na^+$ and pH and of the membrane potential, due to the activation of inwardly or outwardly directed cotransport of $Na^+$ and $HCO_3^-$ (24, 25). The stoichiometry of this cotransport was determined by the steady-state levels of intra- and extracellular $Na^+$ and $H^+/HCO_3^-$ (23, 25), and by a recent voltage-clamp study (26). It was found that two $HCO_3^-$ are transported with one $Na^+$, and that the equilibrium potential follows the membrane potential of the glial cell (27).

Neurotransmitter agonists act on receptors, which are often coupled to cation channels through which, after binding of the ligand, $Na^+$ moves across the cell membrane along the inwardly directed electrochemical gradient. Leech glial cells possess a variety of neurotransmitter receptors, for example, receptors for excitatory amino acids such as glutamate (28, 29). Figure 6 shows an experiment where the intracellular $Na^+$ measurement was combined with a two-electrode voltage clamp of the glial cell membrane. An additional single-barreled microelectrode was inserted into the cell for current

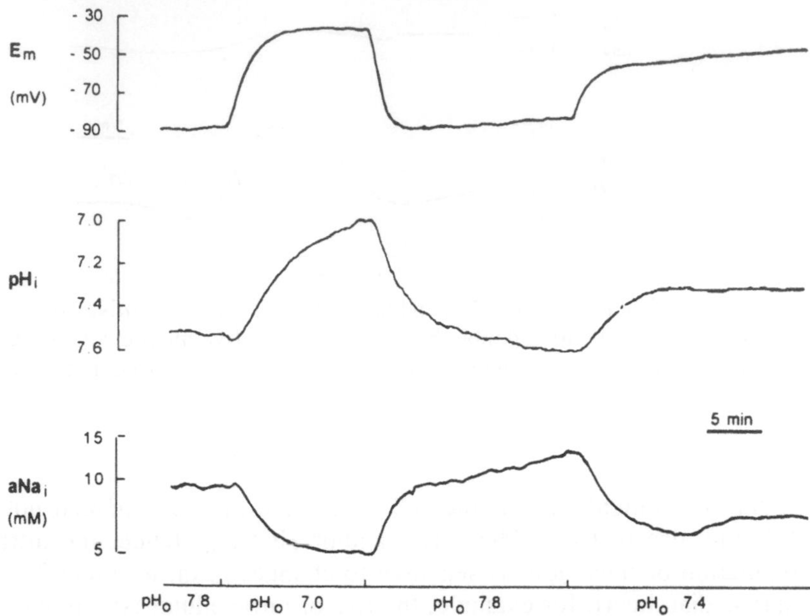


FIG. 5 Membrane potential ($E_m$), intracellular pH ($pH_i$), and intracellular Na activity ($a Na_i$) of a neuropil glial cell measured with a triple-barreled pH- and $Na^+$-selective microelectrode. Reversal of the operation of the $Na^+/HCO_3^-$ cotransport by changing from more alkaline $pH_o$ (7.8) to more acid $pH_o$ (7.0 and 7.4) caused a reversible intracellular acidification and decrease of $a Na_i$ due to the efflux of base equivalents and $Na^+$ [from J. W. Deitmer, *Pfluegers Arch.* **420,** 584 (1992)].

injection and the voltage signal of the reference barrel was used for voltage clamping and for substracting the membrane potential from the signal of the $Na^+$-selective barrel to obtain direct readings of the $Na^+$ activity. At a holding potential of −70 mV the glutamate agonist kainic acid caused a biphasic membrane current that was accompanied by an initial decrease followed by a large increase in the intracellular Na activity. The time course of the recovery of the intracellular Na activity was closely related to the outward current after kainic acid removal. This suggests that the inward current, carried to a large extent by $Na^+$, caused the increase in intracellular $Na^+$, and that the recovery of intracellular Na activity, brought about by electrogenic $Na^+/K^+$ pumping, produced at least part of the outward current (30).

Removal of external $Na^+$ leads to a decrease in intracellular $Na^+$ in most cells, and the rate of $Na^+$ decrease can be considered as a measure of $Na^+$ movement down its electrochemical gradient. Because this movement of

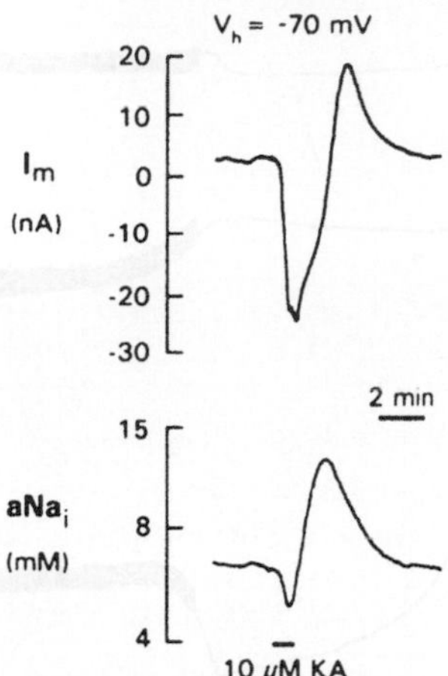


FIG. 6 Membrane current ($I_m$) and intracellular Na activity ($a\mathrm{Na}_i$) of a neuropil glial cell voltage clamped to a holding potential of −70 mV. Addition of 10 $\mu M$ kainic acid (KA) induced a biphasic membrane current accompanied by an initial decrease and a subsequent large increase of $a\mathrm{Na}_i$.

$Na^+$ appears to occur to a large extent via $Na^+$-coupled carriers in the membrane [see, e.g., Deitmer (25)], and not only through "leak" or stochastically open Na channels, other ion species may also change during the decrease of intracellular $Na^+$. Figure 7 shows a recording with a triple-barreled pH- and $Na^+$-selective microelectrode in a Retzius neuron of the leech central nervous system, when external $Na^+$ was replaced by NMDG in the absence and presence of $CO_2/HCO_3^-$. The rate of intracellular $Na^+$ decrease is significantly reduced, whereas the intracellular pH decreases faster, and the membrane does not hyperpolarize in the presence of $CO_2/HCO_3^-$. The mechanisms of these $CO_2/HCO_3^-$-dependent differences are not yet fully understood, but it is clear that the buffer system used has a significant effect on downhill $Na^+$ movements in both neurons and glial cells (25).

## Concluding Remarks

Although a variety of techniques are available today to determine the cytosolic $Na^+$, ion-selective microelectrodes remain a powerful tool for continuous and relatively accurate measurement of the Na activity in cells. They have the disadvantage that they are invasive, are successfully used only for

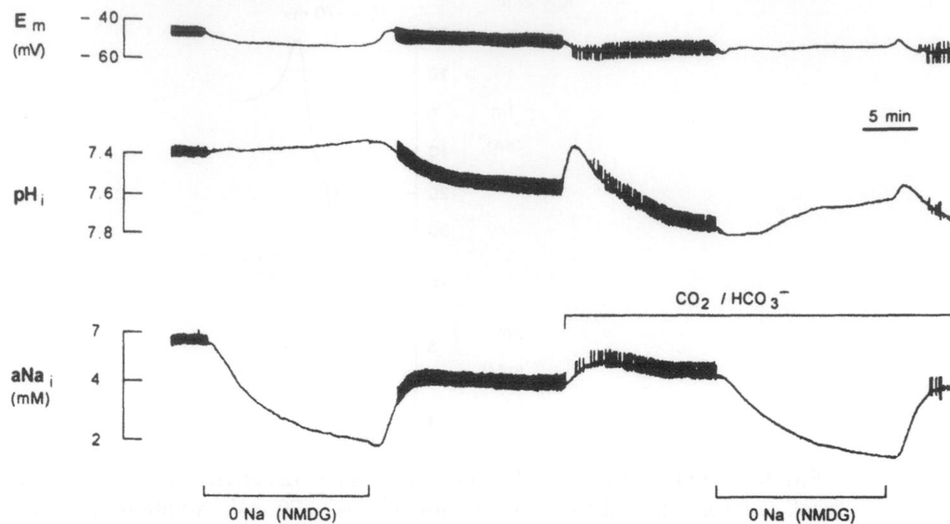


FIG. 7 Membrane potential ($E_m$), intracellular pH ($pH_i$), and intracellular Na activity ($aNa_i$) of a Retzius neuron measured with a triple-barreled pH- and $Na^+$-selective microelectrode during removal of extracellular $Na^+$ (substituted by $NMDG^+$) in the absence (left) and presence (right) of 5% $CO_2$/24 m*M* $HCO_3^-$.

cells ≥10 $\mu$m, and may cause cell damage. Due to their limited selectivity, $Na^+$-selective microelectrodes may be problematic to use when large changes of intracellular $K^+$ and/or $Ca^{2+}$ are to be expected. In the physiological range of these ions, $Na^+$-selective microelectrodes can usually, however, be regarded as reliable probes for determining intracellular $Na^+$. They are easily and rapidly made and are relatively inexpensive; in addition, they can be used to monitor $Na^+$ in single cells, in culture, and *in situ*, for many hours. Whether this technique or another is used to measure intracellular $Na^+$ is finally decided by the preparation and the experimental protocol.

## References

1. R. C. Thomas, *Physiol. Rev.* **62,** 563 (1972).
2. H.-J. Apell, *J. Membr. Biol.* **110,** 103 (1989).
3. G. Eisenman, D. O. Rudin, and J. V. Casby, *Sciences* (*N.Y.*) **126,** 831 (1957).
4. J. A. Hinke, *in* "Experiments in Physiology and Biochemistry" (G. A. Kerkut, ed.), p. 1. Academic Press, London and New York, 1969.
5. R. C. Thomas, *J. Physiol.* (*London*) **210,** 82P (1970).
6. J. L. Walker, *Anal. Chem.* **43,** 89A (1971).

7. L. G. Palmer and M. M. Civan, *J. Membr. Biol.* **33,** 41 (1977).
8. R. P. Kraig and C. Nicholson, *Science* **194,** 725 (1976).
9. T. Shono, M. Okahara, I. Ikeda, K. Kimura, and T. Tamura, *J. Electroanal. Chem.* **132,** 99 (1982).
10. K. Suzuki, K. Hayashi, K. Tohda, K. Watanabe, M. Ouchi, T. Hakushi, and Y. Inoue, *Anal. Lett.* **24,** 1085 (1991).
11. M. Güggi, M. Oehme, E. Pretsch, and W. Simon, *Helv. Chim. Acta* **59,** 2417 (1976).
12. D. Ammann, "Ion-Selective Microelectrodes." Springer-Verlag, Berlin, 1986.
13. K. Ballanyi, P. Grafe, and G. Ten Bruggencate, *J. Physiol.* (*London*) **382,** 159 (1987).
14. K. Ballanyi and H. Kettenmann, *J. Neurosci. Res.* **26,** 455 (1990).
15. M. J. Borelli, W. G. Carlini, W. C. Dewey, and B. R. Ransom, *J. Neurosci. Methods* **15,** 141 (1985).
16. B. P. Nicolsky, *Acta Physicochim. URSS* **7,** 597 (1937) (cited in: G. Eisenman, *Biophys. J.* **2,** 259 (1962).
17. G. Eisenman, *Biophys. J.* **2,** 259 (1962).
18. R. A. Steiner, M. Oehme, D. Ammann, and W. Simon, *Anal. Chem.* **51,** 351 (1979).
19. J. W. Deitmer and W.-R. Schlue, *Pfluegers Arch.* **397,** 195 (1983).
20. D. Ammann, R. Bissig, Z. Cimerman, U. Fiedler, M. Güggi, W. E. Morf, M. Oehme, H. Osswald, E. Pretsch, and W. Simon, *in* "Ion and Enzyme Electrodes in Biology and Medicine" (M. Kessler, L. C. Clark, D. W. Lübbers, I. A. Silver, and W. S. Simon, eds.), p. 22. Urban and Schwarzenberg, Munich, 1976.
21. K. Kaila and J. Voipio, *J. Physiol.* (*London*) **369,** 8P (1985).
22. D. Ellis, J. W. Deitmer, and D. M. Bers, *in* "Progress in Enzyme and Ion-Selective Electrodes" (D. W. Lübbers, H. Acker, R. P. Buck, G. Eisenman, M. Kessler, and W. Simon, eds.), p. 148. Springer-Verlag, Berlin, 1981.
23. J. W. Deitmer and W.-R. Schlue, *J. Physiol.* (*London*), **411,** 179 (1989).
24. J. W. Deitmer, *J. Gen. Physiol.* **98,** 637 (1991).
25. J. W. Deitmer, *Pfluegers Arch.* **420,** 584 (1992).
26. T. Munsch and J. W. Deitmer, *J. Physiol.* (*London*) in press (1994).
27. J. W. Deitmer and H. P. Schneider, *J. Physiol.* (*London*), **485.1,** 157 (1995).
28. K. Ballanyi, R. Dörner, and W.-R. Schlue, *Glia* **2,** 51 (1989).
29. J. W. Deitmer and T. Munsch, *NeuroReport* **3,** 693 (1992).
30. J. W. Deitmer and T. Munsch, *Verh. Dtsch. Zool. Ges.*, **872,** 185 (1994).

# [15] A Practical Guide to the Use of Fluorescent Indicators for the Measurement of Cytosolic Free Magnesium

Elizabeth Murphy and Robert E. London

## Introduction

This chapter covers several practical aspects and limitations in the use of fluorescent magnesium probes. The reader is also directed to several reviews on this topic (1–4). In addition, because the fluorescent magnesium probes are structurally related to the calcium-sensitive fluorescent probes fura-2 and indo-1, the fluorescent magnesium probes are subject to many of the limitations described in the literature for fura-2. The reader is therefore also referred to the extensive literature on pitfalls involved in the use of the fluorescent calcium indicators (5–9), and to Chapters 5 and 6 in this volume.

## Fluorescent Probes for Measuring Magnesium

Table I lists currently available magnesium-sensitive fluorescent probes, their apparent binding constants for calcium and magnesium (10–20). It is perhaps worth emphasizing that most of the cation probes currently in use contain multiple titratable functions. The p$K$ value(s) obtained using fluorescent techniques will generally reflect the titration step(s) that most strongly perturb the fluorophore. In most cases of current interest, this corresponds to the titration of the amino group directly bonded to the fluorophore. The probe is selected based on matching the apparent $K_d$ value, determined under conditions that are typical of the intracellular milieu, to the cytosolic magnesium ion concentration, $Mg_i$. It is also worth noting that in contrast to the case of $Ca_i$, which can undergo changes of orders of magnitude, making the selection of a probe with the appropriate $K_d$ a significant experimental variable, $Mg_i$ levels do not change as dramatically, so that in general, an indicator with $K_d \sim$ basal $Mg_i$ will be suitable for nearly all applications.

Currently available magnesium ion probes include the furan derivatives furaptra and mag-fura-5, indole derivatives such as mag-indo-1, and a fluorescein derivative, magnesium green (Fig. 1). Two quinoline derivatives


*Methods in Neurosciences, Volume 27*

TABLE I $K_d^{Mg}$, $K_d^{Ca}$, and p$K$ Values for Fluorescent Magnesium Indicators

| Indicator | $K_d^{Mg}$ (m$M$) | Ref. | $K_d^{Ca}$ ($\mu M$) | Ref. | p$K$ |
|---|---|---|---|---|---|
| Furaptra | 1.0–5.5 | (10–20) | 17–60 | (3,10,11,13,14,17) | 5.0[a] |
| Mag-fura-5 | 2.6 | (20)[b] | — | —[b] | —[c] |
| Mag-indo-1 | 2.7–3.8 | (11,20) | 8.9 | (11,20) | — |
| Magnesium green | 0.9 | (20) | ~5 | (20) | — |

[a] See Ref. 10.

[b] Illner *et al.* (18) report "experimentally determined dissociation constants" that are the apparent $K_d \times \beta$. The experimentally determined $K_d$ for mag-fura-5 binding to Mg was reported as 14.7 m$M$ and the $K_d$ for mag-fura-5 binding to Ca was reported as 1.8 m$M$. From data in this manuscript, $\beta$ can be estimated to be ~6.7; this would suggest an apparent $K_d^{Mg}$ of 269 $\mu M$.

[c] Illner *et al.* (18) show that the mag-fura-5 $K_d^{Mg}$ is significantly different at pH 6.5 compared to pH 7.2.

proposed to be useful for cytosolic magnesium ion determinations, mag-quin-1 and mag-quin-2, were also sold commercially for a brief period, but do not appear to be currently available. All of the current fluorescent magnesium indicators are built around the basic *o*-aminophenol-*N,N,O*-triacetic acid (APTRA) structure (21), which exhibits adequate $Mg^{2+}$ selectivity for many applications. The quinoline and fluorescein derivatives exhibit changes in magnitude of fluorescence on $Mg^{2+}$ complexation and their use is in most respects analogous to the $Ca_i$ indicator quin-2. Probes with the furan fluorophore undergo an excitation shift on binding $Mg^{2+}$, making these indicators useful for imaging experiments. Probes with the indo-1 fluorophore undergo an emission shift on binding $Mg^{2+}$ and are therefore useful for flow cytometry as well as surface fluorescence of perfused organs. Magnesium green is a longer wavelength indicator that has fluorescence properties similar to fluorescein and calcium green (20). This indicator is useful for experiments in which caged compounds are simultaneously used because the wavelengths required to photolyse the caged compounds do not interfere with measurements using magnesium green. Because magnesium green is a long-wavelength indicator, it is also subject to reduced interference due to autofluorescence. However, magnesium green does not exhibit an excitation or emission shift on binding magnesium, so one cannot ratio two wavelengths to increase dynamic range and correct for variations in probe concentration, as is the case with furaptra, mag-fura-5, and mag-indo-1.

## Measurement of Cytosolic Free $Mg^{2+}$

On binding $Mg^{2+}$, the fluorescent magnesium indicators, for example, furaptra, undergo changes in the excitation fluorescence spectra, forming the

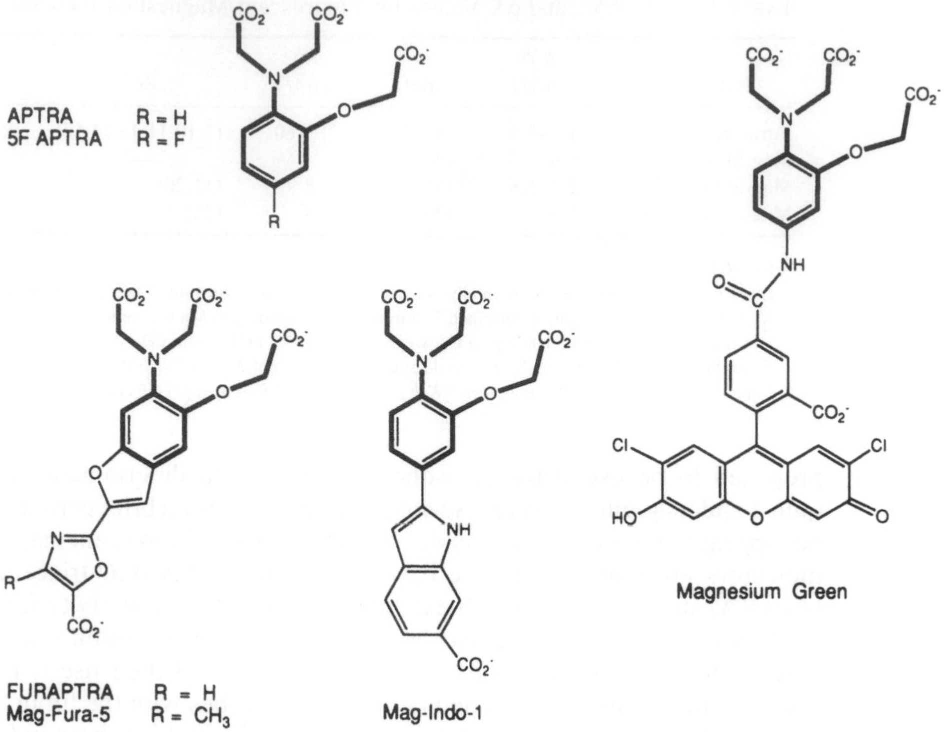

APTRA R = H
5F APTRA R = F

FURAPTRA R = H
Mag-Fura-5 R = $CH_3$

Mag-Indo-1

Magnesium Green

FIG. 1 Structures of APTRA (*o*-aminophenol-*N*,*N*,*O*-triacetic acid) and related fluorescent $Mg^{2+}$ indicators. The common APTRA moiety, which is embedded in all of the indicators, is highlighted in the structures.

basis for measuring $Mg_i$. $Mg^{2+}$ binding to furaptra causes an increase in the excitation at 335 nm and a concurrent decrease at 375 nm, with the emission measured at 510 nm. Thus, $Mg_i$ can be calculated as described originally by Grynkiewicz *et al.* (22) for the determination of $Ca^{2+}$ with fura-2 using Eq. (1):

$$Mg^{2+} = K_d\beta[(R - R_{min})/(R_{max} - R)], \tag{1}$$

where $K_d$ is the apparent dissociation constant; $R$, $R_{min}$, and $R_{max}$ are the fluorescent 340/375-nm ratios corresponding to the sample, to the uncomplexed indicator, and to the fully magnesium complexed indicator, respectively; and $\beta$ is the fluorescence intensity ratio at a 375-nm excitation wavelength of uncomplexed to Mg-complexed indicator. For an indicator such as magnesium green, which does not exhibit an excitation or emission shift on ion complexation, the ion concentration is determined by measuring the fluorescence at a single wavelength according to Eq. (2):

$$Mg^{2+} = K_d[(F - F_{min})/(F_{max} - F)], \tag{2}$$

where $F$, $F_{min}$, and $F_{max}$ correspond to the fluorescence intensities of the sample, the uncomplexed, and the fully complexed indicator, respectively. It should be noted that $Mg^{2+}$ can also be determined from indicators that exhibit an excitation or emission shift by carrying out the measurements at a single wavelength and using Eq. (2).

Measurement of $Mg_i$ requires selectively introducing the fluorescent indicator into the cytoplasm (or the compartment under study). Thus, several points must be considered in the use of fluorescent magnesium indicators: these include strategies for selectively loading the indicator into the cytosol, leakage of the indicator from that compartment, and determination of intracellular $K_d$, $R_{min}$, $R_{max}$, and $\beta$.

## Loading

In order to measure cytosolic free magnesium, the indicator must be localized in the cytosol. Two basic strategies have been employed to load the indicator into the cytosol: introduction of the acetoxymethyl ester and injection.

### *Acetoxymethyl Ester*

A convenient strategy for loading the indicator into the cytosol is based on the introduction of the indicator as a neutral ester that is freely permeable across the plasma membrane. Once inside the cell, naturally occurring esterases cleave the ester, leaving a negatively charged molecule that is trapped inside the cell. Of various ester groups that have been evaluated, the labile acetoxymethyl ester appears to have the most general suitability for this type of loading (23). However, there are also several potential problems with this method of cytosolic loading. First, depending on the location and activity of the esterases, the indicator may be introduced not only into the cytosol but into intracellular organelles as well. Numerous studies with fura-2 and indo-1 have documented this problem with ester loading (5, 9, 24, 25), and similar loading into intracellular organelles occurs with furaptra (26). Several strategies have been proposed to obtain better cytosolic loading of esterified indicators (9); these include varying the concentration of the indicator, the duration and temperature of the loading process, and the addition of solubilizing agents such as Pluronic acid (Molecular Probes, Inc., Eugene, Oregon).

A second potential problem with ester loading relates to incomplete ester hydrolysis. Different cell types contain different levels of esterase activity, and in some cells the ester is cleaved very slowly. As discussed for fura-2 loading (9, 27–29), incomplete hydrolysis of the ester can lead to errors in the calculated ion concentration because the partially deesterified products fluoresce but either do not bind the ion under study or bind only weakly. The time required for complete hydrolysis of the ester varies with cell type. Some cells, such as chick heart cells (30), require as long as 12 hr to achieve complete hydrolysis. Therefore, in setting up studies it is important to determine that the loading protocol results in completely deesterified indicator. This can be accomplished, as described elsewhere (9, 30), by lysing cells in the presence of varying concentrations of $Mg^{2+}$ and determining if the intracellularly generated indicator has the same $K_d$ and fluorescent properties as the free acid.

### *Injection*

Because of the potential problems mentioned above many investigators introduce the indicator into the cytoplasm by direct injection of the acid form of the indicator (17, 31, 32). This avoids problems with incomplete deesterification and minimizes loading into intracellular organelles. However, injection is more technically difficult, and is not feasible when studying large populations of cells such as those in suspension.

## *Leakage*

The carboxylic acid chelators have a propensity to leave the cell. The leakage appears to be due to transport of the indicators on the organic acid carrier, because inhibitors of this transport such as probenecid and sulfapyrazone reduce the rate of leakage (3). In studies with large populations of cells, extracellular indicator can also be produced by lysis of some of the suspended cells. In some cell types the leakage is so rapid that the indicator is transported from the cell with a half-time of minutes. In these cells it is difficult to maintain a sufficient signal-to-noise ratio to perform experiments that last for more than a few minutes. In theory, with indicators such as furaptra and mag-fura-5, calculation of $Mg_i$ using the fluorescence ratio at two wavelengths can correct for differences in indicator concentration. If, however, the indicator concentration in the cell is reduced sufficiently, it may be difficult to make measurements due to a reduced signal-to-noise ratio. Furthermore, as indicator leaks from the cytosol, the fractional fluorescence contribution of any indicator in intracellular organelles, as well as the autofluorescence background will be increased. In addition, if the indicator that leaks from the cell is not removed or is insufficiently diluted, it will contribute to the measurement, resulting in significant errors, because extracellular indicator cannot be distinguished spectrally from intracellular indicator. This problem is generally more serious for studies of cells in suspension. Several strategies have been used to reduce the rate of leakage, including reducing temperature and adding inhibitors of the organic acid transporter such as probenecid (5, 9).

## Intracellular Calibration

As discussed above, $Mg_i$ can be calculated from fluorescent changes of intracellularly loaded indicators such as furaptra using Eq. (1). This calculation, however, requires knowledge of the intracellular $K_d$ and the fluorescence parameters $R_{min}$, $R_{max}$, and $\beta$. There are two general approaches to calculating $Mg_i$. The first approach is to measure the $K_d$ and fluorescent parameters in a solution that mimics the intracellular milieu. Although this approach is probably valid for measuring changes in $Mg_i$, it is very likely that the $K_d$ and fluorescent parameters are not the same in the cell as they are in solution due to differences in viscosity and binding to protein, for example. Thus, it is obviously desirable to measure directly the intracellular $K_d$ and fluorescent parameters; unfortunately, this is a difficult goal to achieve.

### *Equilibration of Intra- and Extracellular Mg*

The measurement of intracellular $K_d$, $R_{min}$, $R_{max}$, and $\beta$ values requires equilibration of intra- and extracellular $Mg^{2+}$. Ionophores are frequently added in an attempt to equilibrate magnesium ions across the plasma membrane. However, cell transport mechanisms work to restore the ion gradients and most ionophores are not very good at equilibrating $Mg^{2+}$. In addition, the determination of $R_{max}$ requires a $Mg^{2+}$ concentration of greater than 20 m*M*, which would be very difficult to obtain *in situ* using ionophores. Approaches have been described for equilibrating $Ca^{2+}$ across the plasma membrane, but it should be noted that only a few micromoles/liter of $Ca^{2+}$ will saturate fura-2, whereas ≥20 m*M* Mg is needed to saturate furaptra or mag-fura-2. Thus, it is significantly more difficult to achieve equilibration of intra- and extracellular $Mg^{2+}$, as is necessary for an *in situ* $K_d$ measurement. A promising alternative approach, described below, is to inject magnesium ions directly into the cell. There are some reports of mitochondrial matrix $K_d$ values (11); however, in this case one can improve equilibration of $Mg^{2+}$ by addition of both uncoupler and ionophore. Values of $K_d$ determined in solutions and in various mitochondrial and *in vivo* preparations are summarized in Table II.

### *Injection of Mg*

Westerblad and Allen (31, 33) pressure injected solutions containing EDTA, saturating $Mg^{2+}$, and intermediate $Mg^{2+}$ concentrations into mouse skeletal muscle in order to obtain *in vivo* values for $R_{min}$, $R_{max}$, $\beta$, and $K_d$. These investigators obtained identical *in vivo* and *in vitro* $R_{min}$ and $R_{max}$ values and the $K_d$ values were 3.8 m*M* *in vitro* and 4.6 m*M* *in vivo*. Westerblad and Allen noted a marked swelling of the fiber during the injection, which they attributed to water crossing the sarcolemmal membrane in order to keep osmotic strength constant. This method has considerable advantages over ionophore equilibration, although it is technically more difficult and membrane transporters will still attempt to return $Mg_i$ to its unperturbed value. In the mouse skeletal muscle studies, Westerblad and Allen (31) measured the change in $Mg_i$ following injection of magnesium into a fiber and found that recovery of $Mg_i$ to basal levels had a half-time of more than 1 hr. Thus, it would appear that injection of Mg and EDTA will allow accurate *in vivo* measurement of $K_d$, $R_{min}$, $R_{max}$, and $\beta$. Using this procedure the *in vivo* $K_d$ was found to be somewhat higher than the *in vitro* $K_d$. This relationship has also been reported for the calcium indicator fura-2 and is reported to result from indicator binding to intracellular protein.

### *Viscosity*

Fluorescence studies in solutions containing agents that modify viscosity have shown significant perturbations of fluorescent behavior for both the calcium indicator fura-2 (8, 9, 22) and the analogous magnesium indicator furaptra (3). It has been observed that the $R_{max}$ of cytoplasmic fura-2 is

TABLE II $Mg^{2+}$ and $Ca^{2+}$ Dissociation Constants for Furaptra Measured *in Vivo* and in Solution

| $K_d^{Mg}$ (m*M*) | | $K_d^{Ca}$ ($\mu$*M*) | | |
|---|---|---|---|---|
| Intracellular | Solution | Intracellular | Solution | Ref. |
| 2.3[a] | 1.3 | 10[a] | 21 | (11) |
| 2.2[b] | 1.4 | 20[b] | 17 | (13) |
| 4.6[b] | 3.8 | — | — | (31) |

[a] Mitochondrial.
[b] Cytosolic.

greater than the $R_{max}$ measured in solution (8, 22), and this discrepancy has been attributed to the increased intracellular viscosity. Grubbs *et al.* (3) have shown that increasing viscosity also enhances the fluorescence of furaptra. Therefore, if one is using *in vitro* calibrations it is prudent to match the viscosity to that *in vivo*. Details of this procedure are described elsewhere (8).

*Binding to Intracellular Proteins*

The intracellular $K_d$ is also altered by binding of the indicator to intracellular proteins. Several groups have shown that in cells ~70% of the total fura-2 or indo-1 concentration is associated with intracellular proteins (25, 34). The intracellular $K_d$ for $Ca^{2+}$ binding to fura-2 is approximately double that measured in solution due to binding of fura-2 to intracellular protein (25, 34, 35). Similarly, Konishi *et al.* (17) have shown that in skeletal muscle approximately 30–40% of furaptra is bound to intracellular proteins. Analogous to the calcium indicators, the $K_d$ for furaptra binding to magnesium is also increased *in vivo* (31); this is illustrated in Table II.

*Summary*

To calculate $Mg_i$ requires knowledge of $K_d$, $R_{min}$, $R_{max}$, and $\beta$. These parameters can be obtained in a solution that mimics the cytosol. This approach suffers from potential problems because the $K_d$ for indicator binding to $Mg^{2+}$ is altered by binding to intracellular protein (17), and the fluorescent properties are altered by viscosity (3). Such effects will therefore result in a systematic underdetermination of $Mg_i$ if not taken into account. In spite of these problems, the use of solution $K_d$, $R_{min}$, $R_{max}$, and $\beta$ frequently offers an acceptable approach to measuring changes in intracellular $Mg_i$. Clearly an *in vivo* $K_d$ determination is desirable; however, using currently available divalent cation ionophores, it is difficult to equilibrate magnesium concentrations, which are necessary to obtain $R_{max}$ ($\geq$20 m*M*) across the plasma membrane. An alternative approach that has been used with success for the calcium indicators is to adjust the *in vitro* conditions (viscosity and added protein) so that they mimic those *in vivo*. Injection of magnesium (and EDTA) is also an attractive alternative that should be pursued.

## Interference from Other Ions

As a general rule, the errors resulting from the interaction of magnesium indicators with other ions can be considered to fall into two categories: (1) interactions that alter the affinity of the indicator for magnesium, but have minimal direct effect on the spectral properties of the indicator, and (2) interactions that alter both the affinity of the indicator for $Mg^{2+}$ and also produce significant spectral perturbations. We have previously termed these type I and type II perturbations (2). Type I perturbations are in principle more readily dealt with by using an "apparent" dissociation constant, which can be varied along with the concentration of the interfering ion, if the latter is independently determined. Type II perturbations, such as the effects of calcium ion fluctuations on $Mg_i$ determination, are more difficult to deal with. The separation of such interference effects into two groups is also suggested by the molecular structure of these indicators (Fig. 1). For example, many of the carboxylate groups are sufficiently removed from the fluorophore such that protonation will have minimal effect on fluorescence properties, but can significantly reduce the affinity of the indicator for magnesium ions. Alternatively, protonation of the amino group that is directly bonded to the polycyclic moiety would be more likely to cause a type II perturbation. Some of the more common interference effects are discussed below.

### *$Ca^{2+}$*

Although furaptra binds magnesium with a $K_d$ of ~1.5 m*M*, it has a greater affinity for calcium, with an apparent $K_d$ for calcium binding in solution of ~50 μ*M* (10). Apparent $K_d$ values for calcium measured in solutions are summarized in Tables I and II and range from 20 to 60 μ*M* (3, 10, 11, 13, 14, 17). Apparent $K_d$ values for calcium measured *in vivo* are 10 and 20 μ*M*. It is important to emphasize here that the utility of an indicator for intracellular determinations is based on the ratio of [cytosolic ion]/$K_d$. Despite the lower $Ca^{2+}$ $K_d$ value of furaptra, the value of the ratio [ion]/$K_d$ is typically ~1 for magnesium, but <0.01 for calcium. Thus, under basal conditions, only approximately 0.2 to 0.4% of the furaptra will be complexed with calcium whereas approximately 33–66% of the furaptra will be magnesium complexed. As discussed by London (2), if calcium and magnesium cause similar spectral changes in furaptra, one can correct for calcium binding to furaptra by using the following equation, where $\beta$, $R$, $R_{max}$, and $R_{min}$, are as described previously:

$$Mg_i = K_d^{Mg}\beta\left(\frac{R - R_{min}}{R_{max} - R}\right) - \frac{K_d^{Mg}}{K_d^{Ca}}[Ca_i], \qquad (3)$$

where $K_d^{Mg}$ and $K_d^{Ca}$ are the apparent $Mg^{2+}$ and $Ca^{2+}$ dissociation constants. Thus, one can calculate that if $Ca_i$ is 100 n*M*, then $Ca^{2+}$ binding to furaptra would add 0.003 m*M* to the calculated $Mg_i$, which would result in a 0.5%

error in $Mg_i$ if basal $Mg_i$ is 0.5 m*M*. Similarly, one can calculate that if $Ca_i$ is 1 $\mu M$, then the correction term is 0.03 m*M*, which would result in a 5% error in $Mg_i$ if basal $Mg_i$ is 0.5 m*M*. However, if $Ca_i$ is 10 $\mu M$ [a value simliar to that measured by Konishi *et al.* (17)], then the correction term is 0.30 m*M*, or a 50% error in the measurement of $Mg^{2+}$ if basal $Mg_i$ is 0.5 m*M*. As mentioned above, this correction assumes that $Ca^{2+}$ and $Mg^{2+}$ cause the same spectral shift. Raju *et al.* (10) found that $Ca^{2+}$ and $Mg^{2+}$ caused similar changes in the fluorescence of furaptra. In contrast, Quamme and Rabkin (14) present data showing that saturating furaptra with calcium results in a greater $R_{max}$ (335/375-nm ratio) than does saturating furaptra with magnesium. The data presented by Quamme and Rabkin, however, suggest that the apparent $K_d$ for furaptra binding to calcium appears quite high; the 340/380-nm ratio is not saturated even at 100 m*M* $Ca^{2+}$. Clearly, further study is needed to establish whether calcium and magnesium have similar effects on the fluorescence of furaptra.

The effect of the intracellular environment on the apparent $K_d^{Ca}$ for furaptra is less clear. The apparent $K_d^{Ca}$ in cells and mitochondria appears to be in the range of 10 to 20 $\mu M$ (11, 13). Thus, in contrast to the apparent $K_d^{Mg}$ for furaptra (3, 10, 11, 13, 14, 17) and the apparent $K_d^{Ca}$ for fura-2 (25, 35) and indo-1 (25, 34), which increase in the intracellular milieu, the apparent $K_d^{Ca}$ for furaptra appears to either stay the same or decrease inside the cell. Further study is needed to confirm this observation.

Because furaptra binds calcium with an apparent $K_d$ inside cells in the range of 10–20 $\mu M$, it is best to measure $Mg_i$ under conditions in which $Ca_i$ is not changing. Unfortunately, this is not always feasible because many agents that alter $Mg_i$ also alter $Ca_i$. It is also important to note that changes in $Ca_i$ may cause changes in $Mg_i$ as a result of competition for common binding sites inside the cell (30). In theory, one can correct for small changes in $Ca_i$ using Eq. (3); however, as noted above, this correction assumes that the spectral changes caused by magnesium and calcium are identical. Although it is relatively straightforward to check the validity of this assumption in solutions, validating this assumption in cells is more difficult. Assuming that one can show that $R_{max}$ and $R_{min}$ are similar for calcium and magnesium binding to furaptra *in vivo*, one can correct the calculated magnesium concentration for changes in calcium binding to furaptra using Eq. (3). Thus, caution should be used in measuring $Mg_i$ with furaptra under conditions wherein $Ca_i$ is elevated or likely to change significantly. Parallel measurements of $Ca_i$, for example, using fura-2, combined with Eq. (3) and the assumption that the spectral changes in furaptra produced by calcium or magnesium ion binding are identical, provide the best means available at present for correcting measured $Mg_i$ values for changes in $Ca_i$.

If apparent $K_d$ values cannot be measured *in vivo*, it is prudent to limit conclusions to directional changes in the absence of changes in $Ca_i$. If measurements are made in the presence of changes in $Ca_i$, one must confirm that the changes observed in fluorescence are not due to changes in $Ca_i$; for example, if the changes go in opposite directions, or if the changes in $Ca_i$ are much too small to account for the observed changes in fluorescence.

### *pH*

Change in pH is another potential interference that must be investigated. There is little change in furaptra fluorescence when pH is lowered to ~6.2. As pH is lowered below pH 6.2, however, an increase in fluorescence at 335 n*M* occurs (and a decrease at 375 n*M*); these data indicate a p*K* of ~5.0 (10), which most probably corresponds to protonation of the amino group. Even though lowering pH to 6.2 has little effect on the fluorescence, lowering pH will in general alter the apparent $K_d$ for magnesium binding to furaptra. To test for this possibility, Freudenrich *et al.* (36) measured the apparent $K_d^{Mg}$ as pH was varied from 6.2 to 7.8; no significant difference in the apparent $K_d^{Mg}$ was observed. Thus, it appears that furaptra is relatively insensitive to changes in pH above 6.2.

### *Other Cations*

As noted above, the effects of various ions on intracellular chelators are related to the ion concentration divided by the dissociation constant for that ion. In general, the low concentrations of other divalent metal ions minimize the potential for interference with $Mg^{2+}$ determination. Earlier studies have shown that furaptra binds $Mn^{2+}$ quite well (10). Indeed, $Mn^{2+}$ binding to furaptra quenches fluorescence and can therefore be added to quench the fluorescence of any furaptra that has leaked from the cell. To use $Mn^{2+}$ to correct for the presence of extracellular indicator, one must be sure that $Mn^{2+}$ does not enter the cell, where it will also quench the fluorescence of intracellular indicator. Effects of other heavy metals, if present, can be minimized as in the case of fura-2 by the use of a heavy metal chelator such as tetrakis(2-pyridylmethyl)ethylenediamine (TPEN), which binds heavy metals without disturbing $Mg^{2+}$ or $Ca^{2+}$ concentrations (37).

Simons recently investigated the sensitivity of furaptra to $Zn^{2+}$ and found an apparent $K_d$ of 20 n*M* (38), a value considerably higher than the dissociation constant for zinc from the calcium indicator indo-1. He also noted that the spectral perturbation produced by $Zn^{2+}$ differs from that produced by $Mg^{2+}$ or $Ca^{2+}$, with a maximum at 323 nm compared with 335 nm for $Ca^{2+}$ or $Mg^{2+}$. It was suggested that simultaneous measurements of $Zn^{2+}$ and either $Ca^{2+}$ or $Mg^{2+}$ could be carried out by using three different excitation wavelengths. However, an indicator with greater selectivity for $Zn^{2+}$ may be more appropriate for such applications (39).

## *Stability and Photobleaching*

The fluorescent magnesium indicators are likely to be susceptible to photobleaching, although they have not been studied in detail, compared to the fluorescent calcium indicators. Photobleaching has been shown to lead to the breakdown of fura-2 to compounds that bind calcium with a much higher $K_d$ than the parent compound fura-2 (27). Clearly the presence of such breakdown products would lead to errors in the calculated intracellular ion concentration. Photobleaching depends on the details of the experimental setup and the cell type, and care should be taken to reduce the intensity of exposure with the use of neutral-density filters. As discussed by Roe *et al.* (9), for fura-2, one can check for photobleaching by monitoring at the pseudoisosbestic wavelength.

As noted in an earlier publication, furaptra can be transformed into a compound that exhibits only low affinity for divalent cations (10). We have observed (1) that solutions freshly made from a newly opened vial of mag-fura-2 exhibited a significantly lower $R_{max}$ than did other samples of furaptra run on the same day in the same solutions on the same instrument. A fresh sample of mag-fura-2 exhibited a shoulder in the excitation spectrum at ~375 nm, even in the presence of 30 m*M* Mg. Normally with saturating levels of $Mg^{2+}$ there is only baseline fluorescence at 375 nm. The presence of significant fluorescence at 375 nm is taken to indicate that there is a significant fraction of furaptra that behaves as uncomplexed furaptra. Given the known propensity of the related compound fura-2 to photobleach (27), the furaptra sample with a shoulder at 375 nm is likely due to a product of photobleaching that has a much higher $K_d^{Mg}$. The important message of these studies is that caution must be exercised in using furaptra, and one must be sure that saturating divalent cations cause a decrease to approximately baseline fluorescence at 375 nm. The presence of transformed product(s) with higher apparent $K_d$ value(s) could account for the apparent differences in apparent $K_d$ measured by some investigators (19, 32).

### *Autofluorescence*

The fluorescent spectrum of furaptra overlaps with that produced by endogenous fluorescent molecules such as NADH. In addition, changes in NADH fluorescence occur under many of the conditions studied with these fluorescent indicators (e.g., metabolic inhibition or anoxia). Thus, it is important to be sure that the fluorescent changes that are measured are due to changes in indicator fluorescence and not due to changes in endogenous cell fluorophores such as NADH. Depending on the level of intracellular indicator, the cell type under study, and the experiment, autofluorescence may or may not be significant. One should perform the experiment on unloaded cells to be sure that changes in autofluorescence are not significantly contributing to the fluorescence of the indicator. In addition, as discussed by Grynkiewicz *et al.* (22), when fluorescence ratios (e.g., 335/370 nm) are used, baseline fluorescence at 335 and 370 nm must be subtracted.

## Summary

Fluorescent derivatives of the chelator APTRA offer sufficient selectivity for $Mg^{2+}$ to allow measurements of intracellular magnesium ion concentration. These techniques are generally best suited to determinations of changes in ionic concentrations, and attempts at absolute quantitation are pushing the limits of the method. Of the various ionic interferences that can occur, the interactions of these chelators with $Ca^{2+}$ are generally the most significant and require the most caution. Although efforts are made to determine apparent $K_d$ values and the other fluorescent parameters in solutions that model the physiological milieu, the difficulties inherent in accurately modeling the complex intracellular environment have led to attempts to determine these parameters directly inside cells and organelles. The difficulties with *in vivo* calibration are even more problematic with $Mg^{2+}$ compared to $Ca^{2+}$ indicators, because in the first case extremely high and nonphysiological ionic levels must be attained in order to carry out such determinations. In all probability, the use of a single $K_d$ value and the implicit assumption of a homogeneous environment for intracellular indicators may be a significant oversimplification. Having said this, it is clearly valuable to continue the efforts to determine how the intracellular environment may perturb the characteristics of the indicators, primarily to develop more valid model solutions for study as well as to determine whether any general trends are apparent.

All of the currently available fluorescent magnesium indicators are based on the APTRA chelator, which binds $Ca^{2+}$ more tightly than it does $Mg^{2+}$. Thus, caution must be used in applying these first-generation magnesium chelators *in vivo* where there are significant changes in $Ca^{2+}$. Changes in $Ca^{2+}$ are adequately accounted for only if they are small or in the opposite direction from magnesium. In spite of this limitation, the fluorescent magnesium indicators have greatly advanced our understanding of the role of magnesium in cell regulation.

## References


1. E. Murphy, C. C. Freudenrich, and M. Lieberman, *Annu. Rev. Physiol.* **53,** 273 (1991).
2. R. E. London, *Annu. Rev. Physiol.* **53,** 241 (1991).
3. R. D. Grubbs, P. A. Beltz, and K. L. Koss, *Magnes. Trace Elem.* **10,** 142 (1991/1992).
4. E. Murphy, *Miner. Electrolyte Metab.* **19,** 250 (1993).
5. F. Di Virgilio, T. H. Steinberg, and S. C. Silverstein, *Cell Calcium* **11,** 57 (1990).
6. R. Y. Tsien and A. T. Harootunian, *Cell Calcium* **11,** 93 (1990).
7. D. A. Williams and F. S. Fay, *Cell Calcium* **11,** 75 (1990).
8. M. Poenie, *Cell Calcium* **11,** 85 (1990).
9. M. W. Roe, J. J. Lemasters, and B. Herman, *Cell Calcium* **11,** 63 (1990).
10. B. Raju, E. Murphy, L. A. Levy, R. D. Hall, and R. E. London, *Am. J. Physiol.* **256,** C540 (1989).
11. G. A. Rutter, N. J. Osbaldeston, J. G. McCormack, and R. M. Denton, *Biochem. J.* **271,** 627 (1990).
12. A. W. Harman, A. L. Nieminen, J. J. Lemasters, and B. Herman, *Biochem. Biophys. Res. Commun.* **170,** 477 (1990).
13. T. W. Hurley, M. P. Ryan, and R. W. Brinck, *Am. J. Physiol.* **263,** C300 (1992).
14. G. A. Quamme and S. W. Rabkin, *Biochem. Biophys. Res. Commun.* **167,** 1406 (1990).
15. L. L. Ng, J. E. Davies, and M. C. Garrido, *Clin. Sci.* **80,** 539 (1991).
16. R. Lennard and J. Singh, *J. Physiol.* (*London*) **435,** 483 (1991).
17. M. Konishi, S. Hollingworth, A. B. Harkins, and S. M. Baylor, *J. Gen. Physiol.* **97,** 271 (1991).
18. H. Illner, J. A. S. McGuigan, and D. Luthi, *Pfluegers Arch.* **422,** 179 (1992).
19. A. Buri, S. Chen, C. H. Fry, H. Illner, E. Kickenweiz, J. A. S. McGuigan, D. Nobel, T. Powell, and V. W. Twist, *Exp. Physiol.* **78,** 221 (1993).
20. R. P. Haugland, "Handbook of Fluorescent Probes and Research Chemicals," 5th ed., p. 148. Molecular Probes catalogue, Eugene, Oregon, 1992.
21. L. A. Levy, E. Murphy, B. Raju, and R. E. London, *Biochemistry* **27,** 4041 (1988).
22. G. Grynkiewicz, M. Poenie, and R. Y. Tsien, *J. Biol. Chem.* **260,** 3440 (1985).
23. R. Y. Tsien, *Nature* (*London*) **290,** 527 (1981).
24. M. H. Davis, R. A. Altschuld, D. W. Jung, and G. P. Brierley, *Biochem. Biophys. Res. Commun.* **149,** 40 (1987).
25. L. A. Blatter and W. G. Wier, *Biophys. J.* **54,** 1089 (1988).
26. M. Hofer and T. E. Machen, *Proc. Natl. Acad. Sci. U.S.A.* **90,** 2598 (1993).

27. M. Scallon, D. A. Williams, and F. Fay, *J. Biol. Chem.* **262,** 6308 (1987).
28. S. G. Oakes, W. J. Martin, C. A. Lisek, and G. Powis, *Anal. Biochem.* **169,** 159 (1988).
29. A. Malgaroli, D. Milani, J. Meldolesi, and T. Pozzan, *J. Cell Biol.* **105,** 2145 (1988).
30. E. Murphy, C. C. Freudenrich, L. A. Levy, R. E. London, and M. Lieberman, *Proc. Natl. Acad. Sci. U.S.A.* **86,** 2981 (1989).
31. H. Westerblad and D. G. Allen, *J. Physiol.* (*London*) **453,** 413 (1992).
32. M. Konishi, N. Suda, and S. Karihara, *Biophys. J.* **64,** 223 (1993).
33. H. Westerblad and D. G. Allen, *Exp. Physiol.* **77,** 733 (1992).
34. L. Hove-Madsen and D. M. Bers, *Biophys. J.* **62,** 89 (1992).
35. M. Konishi, A. Olson, S. Hollingworth, and S. M. Baylor, *Biophys. J.* **54,** 1089 (1988).
36. C. C. Freudenrich, E. Murphy, L. A. Levy, and R. E. London, *Am. J. Physiol.* **262,** C1024 (1991).
37. P. Arslan, F. Di Virgilio, M. Beltrame, R. Y. Tsien, and T. Pozzan, *J. Biol. Chem.* **260,** 2719 (1985).
38. T. J. B. Simons, *J. Biochem. Biophys. Methods* **27,** 25 (1993).
39. C.J. Frederickson, E. J. Kasarskis, D. Ringo, and R. E. Frederickson, *J. Neurosci. Methods* **20,** 91 (1987).

# [16] Measurement of Cytosolic Magnesium by $^{31}$P NMR Spectroscopy

J. A. Helpern and Herbert R. Halvorson

## Introduction

Magnesium is a ubiquitous divalent cation involved in numerous cellular functions. A lack of methods available for its assessment *in vivo* has limited its study. In general, classical chemical assays suffer from strong interference effects with physiological samples, are invasive, or both. Total magnesium content can be determined, for example, after ashing tissue samples, but, because a great many intracellular constituents bind magnesium, the total content becomes a relatively large number of little interest.

$^{31}$P nuclear magnetic resonance (NMR) spectroscopy makes it possible to assess magnesium activity indirectly via the interaction of $Mg^{2+}$ with physiological phosphates, principally adenosine triphosphate (ATP). Although the effect of $Mg^{2+}$ on phosphate chemical shifts has long been known (1) and procedures for estimating $[Mg^{2+}]$ have been described in the literature (2), a complicating factor for clinical studies has been the pH dependence of the binding of $Mg^{2+}$ to ATP and other phosphates. When the pH is under experimental control or is directly measurable, this is not a problem and a suitable titration curve relating ATP chemical shifts to $[Mg^{2+}]$ can be constructed. In the clinical setting, however, it is necessary to estimate both pH and $[Mg^{2+}]$ from the same spectrum, with the knowledge that all the phosphates bind both protons and magnesium ions.

The approach we have developed (3) deals with these multiple interactions explicity and is based on $^{31}$P NMR spectra recorded from solutions of physiological concentrations of phosphates over an array of proton and magnesium concentrations. From the measured stoichiometric concentrations and the known formation constants, the concentration of individual species can be calculated, permitting the estimation of their individual chemical shifts by a modified regression method. Then, given an experimental spectrum, the formation constants, and the chemical shifts, the concentrations of magnesium and protons are determined by treating them as adjustable parameters and maximizing the agreement with the data.

*Methods in Neurosciences, Volume 27*

## Analysis of $^{31}$P NMR Calibration Spectra

The first step involves the analysis of a series of *in vitro* spectra obtained under known conditions. For this purpose, the solutions should be chosen to span the range of compositions that might be encountered *in vivo*. In our approach (3), we used solutions containing 5 m*M* ATP, 5 m*M* phosphocreatine (PCr), 5 m*M* orthophosphate, 100 m*M* KCl, and 0–17 m*M* $MgCl_2$. Each solution was then titrated from a pH of 5.5 to 7.5 in five steps. As a chemical shift reference, a capillary tube containing phenyphosphonic acid was placed inside the NMR tube.

A representative *in vitro* $^{31}$P NMR spectrum, obtained on a Varian VXR-4000 spectrometer (Varian Associates, Palo Alto, California) at 161.9 MHz, is shown in Fig. 1. The high-resolution spectra showed the expected multiplets for the multiple phosphorus nuclei of ATP. The center of each multiplet, corresponding to the peak in the low-resolution spectrum *in vivo*, was taken for the observed chemical shift.

After collecting this series of *in vitro* spectra, the next step is to determine the complete species distribution for each of the calibration solutions. The pertinent interactions of the various chemical species involved in this analysis are presented in Scheme I, and their associated formation constants are presented in Table I. For this assessment, we used the computer program COMICS (4), which takes a set of formation constants and a set of stoichiometric concentrations and then iterates on six equations to produce calculated concentrations for each species:

$$\mathrm{ATP_t} = \sum_i [\mathrm{ATP\ species}]_i, \qquad \mathrm{PCr_t} = \sum_i [\mathrm{PCr\ species}]_i,$$

$$(\mathrm{PO_4})_\mathrm{t} = \sum_i [\mathrm{PO_4\ species}]_i, \qquad \mathrm{Mg_t} = \sum_i [\mathrm{Mg\ species}]_i,$$

$$\mathrm{K_t} = \sum_i [\mathrm{K\ species}]_i, \qquad \mathrm{Cl_t} = [\mathrm{Cl^-}] + [\mathrm{MgCl^+}].$$

The primary information gained here is the value of $[Mg^{2+}]$ to be used in the subsequent spectral analysis.

The next step of the analysis is to determine the chemical shifts of the individual species presented in Scheme I. For each component (e.g., $PO_4$, PCr, and ATP) one has

$$\delta_{\mathrm{component}} = \sum_i \delta_i f_i,$$

where $f_i$ is the mole fraction of the component that is a particular species

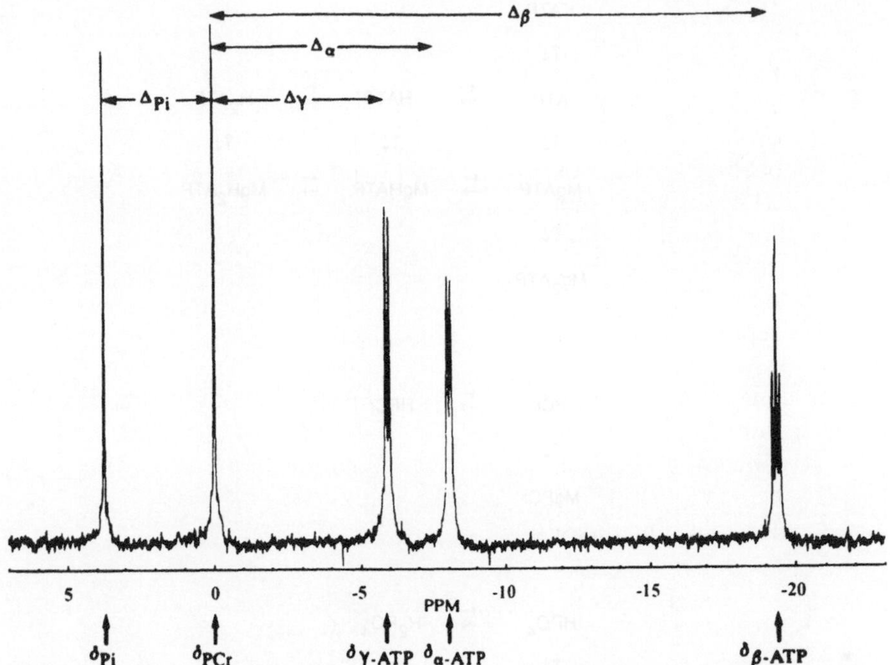


FIG. 1 High-resolution spectrum of a mixture of orthophosphate, phosphocreatine, and ATP. The symbol $\Delta_x$ denotes the chemical shift of a particular phosphorus nucleus relative to the position of the phosphocreatine resonance. These measures were used in the analysis of the *in vivo* specra. Reproduced with the kind permission of John Wiley & Sons, Ltd. from H. R. Halvorson, A. M. Q. Vande Linde, J. A. Helpern, and K. M. A. Welch, Assessment of Magnesium Concentrations by $^{31}$P NMR *In Vivo*, *NMR in Biomedicine* **5**, 53–58 (1992), copyright 1992, John Wiley & Sons, Inc.

(there are three such expressions for ATP, one for each phosphorus). Given the set of $\delta_{obs}$ and $f_i$ for the experimental spectra, the unknown species chemical shifts $\delta_i$ are, in principle, the coefficients of a multiple linear regression.

Some species, however, are only weakly populated under all experimental conditions. Other species undergo minimal changes in their populations over the span of conditions. Additionally, some species have similar chemical shifts. This means that a straightforward multiple linear regression will founder on singularities. To avoid this problem, we used the technique of singular value decomposition (5), which assigns a weight of zero to noninfluential variables. The analysis is then repeated for the resolvable species only. The results of this analysis are given in Table II.

KATP

↑↓

ATP ⇆ HATP ⇆ $H_2ATP$

↑↓ ↑↓ ↑↓

MgATP ⇆ MgHATP ⇆ $MgH_2ATP$

↑↓

$Mg_2ATP$

PCr ⇆ HPCr

↑↓

MgPCr

$HPO_4$ ⇆ $H_2PO_4$

↑↓

$MgHPO_4$

SCHEME I Pertinent equilibria involving PCr, ATP, Pi, $Mg^{2+}$, and $H^+$. Reproduced with the kind permission of John Wiley & Sons, Ltd. from H. R. Halvorson, A. M. Q. Vande Linde, J. A. Helpern, and K. M. A. Welch, Assessment of Magnesium Concentrations by $^{31}P$ NMR *In Vivo*, *NMR in Biomedicine* **5**, 53–58 (1992), copyright 1992, John Wiley & Sons, Inc.

## Analysis of *in Vivo* $^{31}P$ NMR Spectra

The analysis of spectra taken *in vivo* represents an inversion of the problem of analyzing the *in vitro* spectra. That is, given an observed chemical shift and the sets of species chemical shifts calculted from the *in vitro* spectra along with their formation constants, what values of the ionic concentrations are consistent with the observation? The actual concentrations of $H^+$ and $Mg^{2+}$, however, differ by a factor of $10^3$. Hence, to impart numerical stability to the least-squares analysis, the problem was cast in terms of the negative common logarithms of the concentrations (i.e., pH and pMg). Further, ob-

TABLE I Selected Formation Constants[a]

| Reaction | $K_f$ (25°C) | $K_f$ (37°C)[b] |
|---|---|---|
| $H^+ + ATP^{4-} \rightleftharpoons HATP^{3-}$ | $1.09 \times 10^7$ | $1.16 \times 10^7$ |
| $H^+ + HATP^{3-} \rightleftharpoons H_2ATP^{2-}$ | $8.51 \times 10^3$ | $(8.51 \times 10^3)$ |
| $K^+ + ATP^{4-} \rightleftharpoons KATP^{3-}$ | 14 | (14) |
| $Mg^{2+} + ATP^{4-} \rightleftharpoons MgATP^{2-}$ | $3.84 \times 10^4$ | $4.86 \times 10^4$ |
| $Mg^{2+} + HATP^{3-} \rightleftharpoons MgHATP^-$ | $5.42 \times 10^2$ | $5.98 \times 10^2$ |
| $Mg^{2+} + H_2ATP^{2-} \rightleftharpoons MgH_2ATP$ | 20 | —[c] |
| $Mg^{2+} + MgATP^{2-} \rightleftharpoons Mg_2ATP$ | 40 | (40) |
| $Mg^{2+} + Cl^- \rightleftharpoons MgCl^+$ | 3.4 | NA |
| $H^+ + PO_4^{3-} \rightleftharpoons HPO_4^{2-}$ | $8.00 \times 10^{11}$ | NA |
| $H^+ + HPO_4^{2-} \rightleftharpoons H_2PO_4^-$ | $8.20 \times 10^6$ | $7.83 \times 10^6$ |
| $H^+ + H_2PO_4^- \rightleftharpoons H_3PO_4$ | $1.13 \times 10^2$ | —[c] |
| $Mg^{2+} + PO_4^{3-} \rightleftharpoons MgPO_4^-$ | $3.16 \times 10^2$ | —[c] |
| $H^+ + PCr \rightleftharpoons HPCr$ | $1.64 \times 10^9$ | $1.64 \times 10^4$ |
| $H^+ + HPCr \rightleftharpoons H_2PCr$ | $4.0 \times 10^2$ | —[c] |
| $K^+ + PCr^{2-} \rightleftharpoons KPCr$ | 20 | (20) |
| $Mg^{2+} + PCr^{2-} \rightleftharpoons MgPCr$ | 40 | (40) |

[a] From H. R. Halvorson, A. M. Q. Vande Linde, J. A. Helpern, and K. M. A. Welch, Assessment of Magnesium Concentrations by $^{31}$P NMR *In Vivo*, *NMR in Biomedicine* 5, 53–58 (1992), with the kind permission of John Wiley & Sons, Ltd.

[b] Value for $K_f$ at 25°C, used in the absence of known temperature dependence.

[c] Not used because the species was never appreciably populated.

served changes (differences) for these variables correspond to relative changes in concentrations, and greater importance was attached to this information than to the actual value of the concentration (that is, precision was more important than accuracy).

TABLE II Chemical Shifts (ppm) of Various Species Relative to Phosphocreatine[a]

| **R** | — | H | Mg | MgH | $Mg_2$ |
|---|---|---|---|---|---|
| PCr**R** | 0.0 | — | −0.69 | — | — |
| $HPO_4$**R** | 5.69 | 3.20 | 5.82 | — | — |
| γ-ATP**R** | −2.32 | −7.80 | −2.01 | −3.52 | −4.15 |
| α-APT**R** | −7.80 | −8.09 | −7.35 | −7.65 | −8.35 |
| β-ATP**R** | −18.58 | −20.01 | −15.57 | −16.17 | −17.51 |

[a] Reproduced with the kind permission of John Wiley & Sons, Ltd. from H. R. Halvorson, A. M. Q. Vande Linde, J. A. Helpern, and K. M. A. Welch, Assessment of Magnesium Concentrations by $^{31}$P NMR *In Vivo*, *NMR in Biomedicine*, 5, 53–58 (1992).

To avoid the additional uncertainty associated with determining absolute chemical shifts, the observations were expressed as four chemical shifts relative to the PCr peak (see Fig. 1):

$$\Delta_P = \delta(PO_4) - \delta(PCr), \qquad \Delta_\alpha = \delta(ATP_\alpha) - \delta(PCr),$$
$$\Delta_\beta = \delta(ATP_\beta) - \delta(PCr), \qquad \Delta_\gamma = \delta(ATP_\gamma) - \delta(PCr),$$

where δ refers to the function described in the analysis of the *in vitro* spectra.

The expressions, cast in terms of two unknown parameters, pH and pMg, are then adjusted using a nonlinear least-squares algorithm (a modified Marquardt–Levenberg algorithm), so as to minimize the sum of the squares of the residuals:

$$SSR = [\Sigma\ \Delta(obs)_j - \Delta(calc)_j]^2,$$

where the sum is over the four observable relative chemical shifts.

## Examples of *in Vivo* $^{31}$P NMR Spectral Analysis

Some examples of the application of this technique are presented in Fig. 2. Although an extensive consideration of the physiological significance of magnesium concentrations is not appropriate here, it is appropriate to note some specific observations.

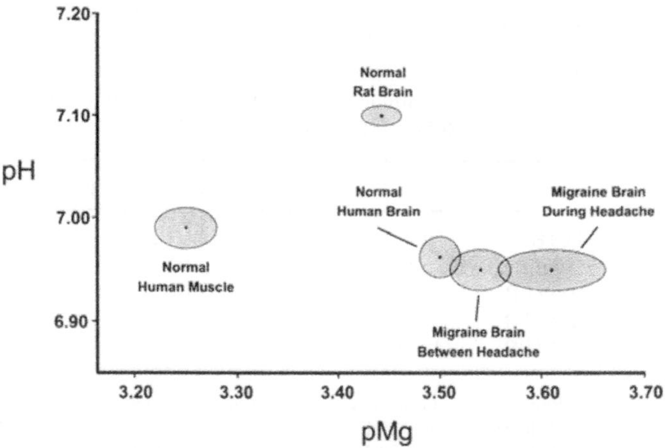


FIG. 2 The pH and pMg values for normal human muscle, normal rat brain, normal human brain, and brains of patients with migraine both during and between headache.

First, the finding of distinctly different values for tissues of different origin (e.g., rat brain, human brain, and human muscle) is reassuring and adds some credence the technique. Second, when spectra from migraineurs were subdivided into spectra obtained during a migraine attack and spectra obtained between migraine attacks, the difference between group means was statistically significant (6). Of particular note here is the fact that the pH did not differ between these two groups or between migraineurs and normal controls.

We have also reported changes in pMg values obtained from stroke patients beginning within hours following the stroke to several weeks after the stroke (7). Under conditions of metabolic acidosis, as in the early period after a stroke, the effective affinity of ATP for $Mg^{2+}$ is reduced, leading to an elevated [$Mg^{2+}$] (low pH, low pMg) and a reduction in the concentration of Mg–ATP complexes. This reduction may be significant to the enzymes of energy metabolism. When the reformation of ATP is impaired as a result of ischemia, the resulting decrease in total ATP also releases magnesium ions (because ADP binds $Mg^{2+}$ less tightly). This release will promote proton release because of the linkage involving the remaining magnesium binding sites. Thus there exists at least the potential for a vicious cycle of metabolic perturbation.

The extent to which the values for the pH and pMg obtained using this method are accurate cannot be determined without a comparison to some other technique. Because the alternative techniques currently available are profoundly invasive, it is unlikely that the question will be answered soon. However, the finding of a free magnesium concentration in human brain that is about 0.3 m*M* is in line with recent work that has steadily revised downward the estimate of intracellular free magnesium (8).

### *Selectivity, Sensitivity, and Precision*

Because of the chemical linkage between $H^+$ and $Mg^{2+}$ there is a natural tendency for pH and pMg to vary colinearly. The NMR signal responds differently to these two variables, permitting some separate resolution. The extent to which this is possible can be explored through first-order nonlinear sensitivity analysis (9). Because of the nonlinearity, the sensitivity matrix (partial derivatives of the signals with respect to pH and pMg) is a function of pH and pMg and must be evaluated at a particular set of experimental conditions.

For convenience, let us consider pH 7.0 and pMg 3.6. At these conditions, and under the linear approximation, one can deal with three separate issues concerning the analytical method: selectivity, sensitivity, and precision. The selectivity, $\Xi$, expresses the ability to estimate the value of one component in the presence of the other. As a numerical parameter it ranges from 0 (impossible to resolve) to 1 (no interference). For the situation at hand, $\Xi$ = 0.78. (An equivalent perspective is that the estimates of pH and pMg are correlated with $r = 0.63$ at these concentrations. For two components, $\Xi^2 = 1 - r^2$.) The sensitivity $H$ is the root-mean-square change in signal (chemical shift) associated with unit change in a component ($H_{pH} = 1.83$; $H_{Mg} = 1.87$ at these conditions). The precision of the estimates of pH and pMg is then a function of sensitivity, selectivity, and the precision of the data (standard deviation of chemical shift measurements). For these conditions (pH = 7.0, pMg = 3.6), $\sigma$(pH) = $\sigma$(pMg) = 0.55 $\sigma(\delta)$. It can be seen that changes of about 0.05 in pH or pMg are about the limit of resolution for a single *in vivo* spectrum [$\sigma(\delta) = 0.1$]. The situation can be improved by simultaneously considering groups of spectra. These may be repeated spectra from the same individuals within the same clinical category.

### *Assumptions and Limitations*

There are some assumptions involved in extrapolating from *in vitro* spectra to conditions *in vivo*. For instance, it was assumed that differences in bulk magnetic susceptibility between solutions and brain would not prevent the use of a common set of chemical shifts for spectra obtained *in vitro* and *in vivo*. Additionally, although the temperature dependence of the equilibrium constants is readily accounted for, it was also assumed that the temperature dependence of the intrinsic (species) chemical shifts was negligible.

It was also assumed that orthophosphate, phosphocreatine, and ATP would be found in a common physiological compartment, so that they would be reporting a common milieu. The analysis is independent of the binding of ATP to various enzymes if the bound ATP is sufficiently immobilized that its resonances become too broad to observe. The presence of a significant amount of ATP bound with sufficient mobility to be observable, however, would introduce an unknown error in the concentrations because enzymes interact preferentially with specific ATP species and alter the effective ion binding equilibria. Relative concentration changes (differences in pMg) are less sensitive to this problem.

Finally, the technique requires chemical shift information from the five resonance signals of $P_i$, PCr, and the three phosphate groups of ATP. Conditions under which there is a reduction in the amount of ATP and concomitant increase in ADP (which is not included in the algorithm) degrade both the quality and the meaning of chemical shifts extracted from these spectra. In addition, the concurrent decrease in both pH and pMg makes the selectivity go to zero (correlation approaches unity), a nonlinear manifestation of the linkage. The interaction of these two phenomena then seriously reduces the information content.

## References

1. M. Cohn and T. R. Hughes, Jr., *J. Biol. Chem.* **237,** 176 (1962).
2. R. K. Gupta and R. D. Moore, *J. Biol. Chem.* **255,** 3987 (1980).
3. H. R. Halvorson, A. M. Q. Vande Linde, J. A. Helpern, and K. M. A. Welch, *NMR Biomed.* **5,** 53 (1992).
4. D. D. Perrin and I. J. Sayce, *Talanta* **14,** 833 (1967).
5. G. E. Forsythe, M. A. Malcolm, and C. B. Moler, "Computer Methods for Mathematical Computations." Prentice-Hall, Englewood Cliffs, New Jersey, 1977.
6. N. M. Ramadan, H. R. Halvorson, A. M. Q. Vande-Linde, S. R. Levine, J. A. Helpern, and K. M. A. Welch, *Headache* **29,** 590 (1989).
7. J. A. Helpern, A. M. Q. Vande Linde, K. M. A. Welch, S. R. Levine, L. R. Schultz, R. J. Ordidge, J. W. Hugg, and H. R. Halvorson, *Neurology* **43,** 1577 (1993).
8. L. Garfinkel, R. A. Altschuld, and D. Garfinkel, *J. Mol. Cell. Cardiol.* **18,** 1003 (1986).
9. G. Bergmann, B. von Oepen, and P. Zinn, *Anal. Chem.* **59,** 2522 (1987).

# [17] Chloride-Sensitive Fluorescent Indicators

A. S. Verkman and Joachim Biwersi

## Introduction

The chloride ion is the major anionic species in biological systems. Chloride transport and its regulation play an important role in cell volume and pH regulation and in transepithelial salt and volume movement in fluid-transporting epithelia. Chloride-sensitive fluorescent indicators provide a complementary and sometimes exclusive approach to study chloride activity and transport processes in extracellular, cytoplasmic, and subcellular compartments, as well as in cell-free vesicle systems and proteoliposomes reconstituted with chloride transporters. The purpose of this chapter is to review briefly the chemistry of chloride-sensitive fluorophores, and, by example, to demonstrate the utility of these indicators in biological systems.

Information in the chemistry literature has been available for many years that certain heterocyclic compounds have anion-sensitive fluorescence properties (1). To determine whether any of the existing compounds fulfilled the criteria necessary for biological applications, including high chloride sensitivity and selectivity, rapid response, low toxicity, bright fluorescence, and low cell membrane permeability, a series of compounds was screened (2). The compound 6-methoxy-*N*-3′-sulfopropylquinolinium (SPQ) has many of the required properties and has been used widely over the past few years for measurements of chloride activity and transporting mechanisms in living cells and cell-free vesicle preparations. Based on a series of synthesis studies to define the structure–activity relationships of chloride-sensitive fluorescent indicators, new compounds have been synthesized with high chloride sensitivity, improved optical properties, membrane impermeability (for labeling extracellular and endocytic compartments), and membrane permeability (for loading cell cytoplasm). Applications of chloride-sensitive fluorescent indicators and a critical assessment of experimental limitations of available indicators are provided below.

## Chemistry of Chloride-Sensitive Fluorescent Indicators

All chloride indicators are based on the sensitivity of the fluorescence of heterocyclic compounds with a quaternary nitrogen (e.g., quinolinium) to quenching by halides (2, 3). Fluorescence is quenched by a collisional mecha-


*Methods in Neurosciences, Volume 27*

nism with a linear Stern–Volmer relation, $F_0/F = 1 + K_{hal}[hal]$, where $F_0$ is the fluorescence in the absence of halide (hal), $F$ is the fluorescence in the presence of halide, and $K_{hal}$ is the Stern–Volmer quenching constant (in inverse molar). Collisional quenching decreases fluorescence intensity and lifetime without a change in spectral shape; therefore, multiwavelength ratiometric measurement of halide concentration is not possible. Because the kinetics of collisional quenching is diffusion limited, the indicators provide submillisecond time resolution. The fluorescence of the chloride indicators is not altered by cations, phosphate, nitrate, and sulfate, and is quenched weakly by other monovalent anions, including citrate, acetate, gluconate, and bicarbonate. The first and still most widely used fluorescent chloride indicator is SPQ (Fig. 1). It is easily synthesized in high yield by reaction of equimolar amounts of 6-methoxyquinoline and propanesultone at 90°C for 45 min. SPQ is excited at ultraviolet wavelengths with maxima at 318 and 350 nm and molar absorbances of 5430 and 3470 $M^{-1}$, respectively. It has a single broad emission peak centered at 450 nm with a quantum yield of 0.55 in the absence of halides. The Stern–Volmer quenching constant for chloride in aqueous buffers is 118 $M^{-1}$, giving a 50% decrease in fluorescence at 8 m$M$ chloride.

## *Structure–Activity Relationships*

To understand the structure–activity relationships of the chloride indicators, the effects of heterocyclic backbone and nature and position of substituents on chloride sensitivity and spectral properties were examined for a series of SPQ analogs (3–5). A positively charged quaternized nitrogen in the heterocyclic ring is necessary for chloride sensitivity. Indicators with the bicyclic quinoline backbone (Fig. 1, structure I) showed the highest chloride sensitivity compared to isoquinoline (compound II), 5,6-benzoquinoline (III), acridine (IV), and phenantridine (V). The tricyclic fluorophores had red-shifted excitation and/or emission spectra but low chloride sensitivity. The position and the nature of the substituents on the quinoline ring altered the chloride sensitivity remarkably. Substitution of *N*-sulfopropylquinolinium with the electron-donating substituents methyl and methoxy at positions 2–6 (see compound I) increased chloride sensitivity, whereas substitution with electron-withdrawing groups (e.g., Cl and Br) decreased sensitivity. Substitution at positions 7 and 8 greatly reduced chloride sensitivity. $K_{Cl}$ was also changed dramatically by the charge of the N substituent, probably because the charge density alters the chloride density near the "quenching center." Replacement of the 3′-sulfopropyl group of 6-methoxyquinolinium by an uncharged ethyl group increased chloride sensitivity 1.3-fold; replacement by 3′-aminopropyl

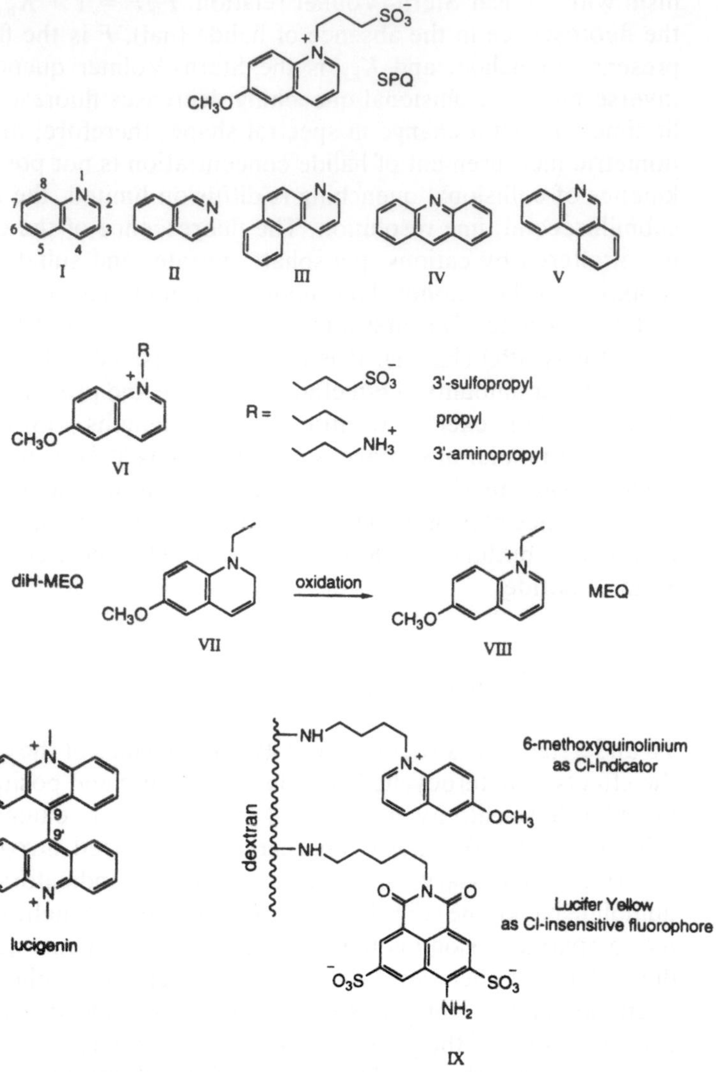

SPQ

I II III IV V

VI

R = 3'-sulfopropyl propyl 3'-aminopropyl

diH-MEQ oxidation MEQ

VII VIII

lucigenin

dextran

6-methoxyquinolinium as Cl-Indicator

Lucifer Yellow as Cl-insensitive fluorophore

IX

FIG. 1 Structure of chloride-sensitive fluorescent indicators. Structures I–V represent different heterocycles containing a single nitrogen. Compound VI is a quaternary N-substituted 6-methoxyquinoline. Compounds VII and VIII are the reduced and oxidized forms of the cell-permeable chloride indicator diH-MEQ, and compound IX is a "dual-wavelength" dextran conjugated with chloride-sensitive and -insensitive chromophores (see text for details).

increased chloride sensitivity by 3-fold when the amino group was positively charged (pH <8) (see compound VI). Figure 2 shows the effect of the charge of the N substituent on the Stern–Volmer quenching constant. The charge density hypothesis was supported by the observation that $K_{Cl}$ decreased as the alkyl chain separating the two positively charged nitrogens was lengthened.

## *Cell-Permeable Chloride Indicator*

A limitation of quinolinium-based indicators is the necessity of invasive or slow diffusive cell loading because the fixed, positively charged nitrogen that is necessary for chloride sensitivity confers high polarity and membrane impermeability. For rapid noninvasive loading into living cells, the positive charge can be masked by reduction of the quinolinium to the uncharged 1,2-dihydroquinoline (6). Reduction of 6-methoxy-*N*-ethylquinolinium (MEQ, compound VIII) with $NaBH_4$ gave the chloride-insensitive compound 6-methoxy-*N*-ethyl-1,2-dihydroxyquinoline (diH-MEQ, compound VII). diH-MEQ is nonpolar and enters the cell rapidly. diH-MEQ is readily oxidized in the cytosolic compartment to the membrane-impermeable and chloride-sensitive MEQ that is trapped in the cell.

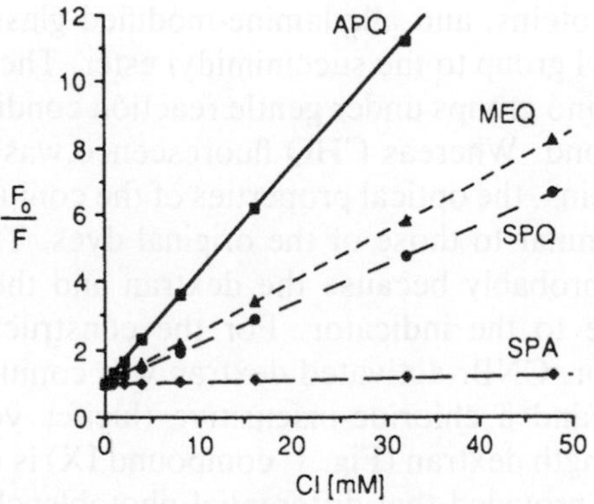


FIG. 2 Stern–Volmer plots for quenching of chloride-sensitive fluorescent indicators. $F_0$ and $F$ are fluorescence measured in the absence and presence of chloride, respectively. Stern–Volmer constants are 320 $M^{-1}$ (APQ, 6-methoxy-*N*-aminopropylquinolinium), 150 $M^{-1}$ (MEQ, 6-methoxy-*N*-ethylquinolinium), 119 $M^{-1}$ (SPQ, 6-methoxy-*N*-sulfopropylquinolinium), and 9 $M^{-1}$ (SPA, *N*-sulfopropylacridinium).

### Long Wavelength Chloride Indicators

Legg and Hercules (6a) reported that the fluorescence of N,N′-dimethyl-9,9′-bisacridinium dinitrate (lucigenin, Fig. 1) is very Cl-sensitive. Lucigenin has excitation and emission maxima of 455 and 505 nm, respectively, and a $K_{Cl}$ of 390 $M^{-1}$. Based on the lucigenin stucture, a series of 9-substituted acridinium indicators were synthesized (6b). All synthesized compounds had red shifted fluorescence spectra compared to SPQ. The 9-carbonyl substituted acridinium indicators showed green fluorescence with high Cl-sensitivity ($K_{Cl}$: 90 -225 $M^{-1}$). Unfortunately these synthesized acridinium indicators are converted in the cell cytoplasm to a Cl-insensitive product by addition of a hydroxyl group at the reactive 9-position. However, because of their superior optical properties over SPQ, the long-wavelength acridinium indicators are useful for Cl-transport measurements in liposomes and membrane vesicles.

### Chloride-Sensitive Conjugates

For conjugation of chloride-sensitive indicators to substrates such as dextrans, proteins, or glass beads (for the construction of a chloride-sensitive fiber-optic sensor), 6-methoxyquinoline was reacted with substituents bearing reactive amino or carboxyl groups. CNBr-activated neutral dextran was reacted with *N*-(4-aminobutyl)-6-methoxyquinolinium (ABQ) to give chloride-sensitive dextrans with a stable imino carbonic acid linkage (5). *N*-(7-Carboxyheptyl)-6-methoxyquinolinium (CHQ) was conjugated to aminodextran, proteins, and alkylamine-modified glass beads after activation of the carboxyl group to the succinimidyl ester. The succinimidyl ester reacts with free amino groups under gentle reaction conditions to form a stable carboxamide bond. Whereas CHQ fluorescence was quenched 50–70% on binding to proteins, the optical properties of the conjugated dextrans and glass beads were similar to those of the original dyes. The chloride sensitivity was reduced probably because the dextran and the glass bead restrict access of chloride to the indicator. For the construction of a ratiometric chloride indicator, CNBr-activated dextran was conjugated with a chloride-sensitive (ABQ) and a chloride-insensitive (lucifer yellow) fluorophore. The dual-wavelength dextran (Fig. 1, compound IX) is suitable for ratio imaging applications provided that differential photobleaching effects are avoided.

## Applications in Cell-Free Vesicle and Liposome Systems

Cell-free studies of chloride transport in fractionated subcellular vesicles and reconstituted proteoliposomes are often used to define chloride transport mechanisms. Whereas multiple factors contribute to chloride fluxes in intact cells, the ionic and hormonal conditions can be specified in cell-free experiments to focus attention on selected transporting proteins. Chloride transport studies in reconstituted proteoliposomes are also useful to screen the function of putative chloride-transporting proteins during isolation and purification procedures. Compared to conventional isotopic uptake methods, the use of entrapped chloride-sensitive fluorescent indicators provides a rapid and continuous measure of chloride activity in very small samples.

SPQ and related compounds have been used to define chloride transport mechanisms in isolated membrane vesicles (7–12). An example of a measurement protocol is vesicle isolation, diffusive loading with indicator (e.g., 5 m*M* SPQ, overnight at 4°C or 30 min at 37°C), removal of extravesicular indicator (e.g., centrifugation or exclusion chromatography), and fluorimetric measurement of transport. Because of the hyperbolic relationship between indicator fluorescence and chloride activity, signal-to-noise ratio is best for influx experiments in which external chloride is added to vesicles containing a 0 chloride content buffer (e.g., chloride replaced by nitrate). The time course of decreasing indicator fluorescence provides a quantitative measurement of chloride influx. The addition of external chloride can be accomplished by standard cuvette fluorimetry, or for faster processes (1 msec–1 sec), by stopped-flow mixing methods. Transport mechanisms are defined from the effects on chloride fluxes of imposed ion gradients, diffusion potentials, and selected inhibitors. This general approach has been used to study chloride/anion exchange in erythrocytes (2), chloride conductance in apical plasma membrane vesicles from epithelial cells (8–10), and NaCl solvent drag (7). In liposomes, the membrane-impermeable indicator is added at the time of liposome formation (e.g., detergent dialysis, extrusion) and extraliposomal indicator is removed by exclusion chromatography (13). SPQ has been used to characterize chloride transport in liposomes reconstituted with several types of purified chloride channels (14–17).

An interesting application of SPQ was the demonstration of a protein kinase A-regulated chloride conductance in endocytic vesicles from kidney proximal tubule (18, 19). Endocytic vesicles were labeled *in vivo* by intravenous injection of rabbits with SPQ. The indicator was filtered by the kidney glomerulus and taken up into proximal tubule epithelial cells by fluid-phase endocytosis. The animal was killed after 10–20 min and the kidney cortex was dissected, homogenized, and differentially centrifuged. A crude microsomal pellet contained SPQ-labeled endocytic vesicles that had internalized the indicator between the time of intravenous injection and animal death. Chloride transport was then measured selectively in the fluorescent endocytic vesicles by stopped-flow fluorimetry. Figure 3 (curve a) shows that fluorescence remains constant in the absence of chloride. Fluorescence decreases rapidly (curve b) in response to a 50 m*M* inwardly directed chloride gradient

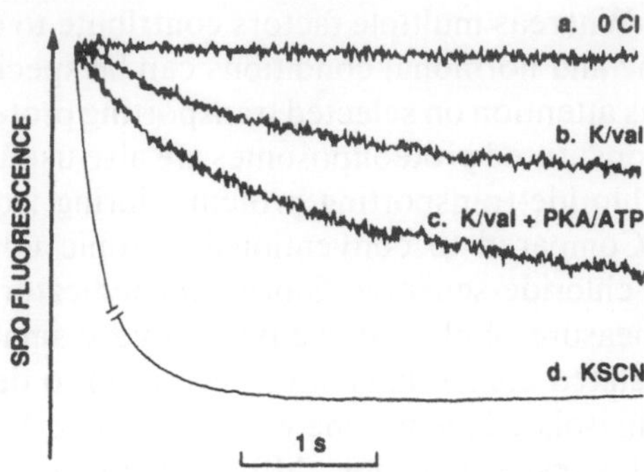


FIG. 3 Time course of chloride influx in SPQ-labeled endocytic vesicles from rat kidney proximal tubule. Vesicles were labeled with SPQ *in vivo* as described in the text and measurements were carried out using a microsomal preparation by stopped-flow fluorimetry. (a) Stopped-flow experiment in the absence of a chloride gradient. (b and c) Microsomes were subjected to a 50 m*M* inwardly directed gradient of chloride in the presence of potassium and valinomycin without and with transporter phosphorylation by ATP and the purified catalytic subunit of protein kinase A. (d) SPQ quenching by a KSCN gradient.

in the presence of potassium and valinomycin (to clamp membrane potential). Transporter phosphorylation by Mg–ATP and the catalytic subunit of protein kinase A increased chloride conductance (curve c), and dephosphorylation by alkaline phosphatase inhibited chloride conductance (not shown). Figure 3 (curve d) shows the rapid, complete quenching of SPQ fluorescence by KSCN. In preliminary imaging studies of fluorescently labeled vesicles immobilized on a polylysine-coated coverslip (20), it was possible to measure chloride transport in individual fluorescent vesicles.

## Measurement of Cytosolic Chloride in Intact Cells

In addition to the requirements for use of fluorescent indicators in cell-free preparations, indicators for use in intact cells should be nontoxic, distributed uniformly in the cytosolic compartment, and cell trappable. SPQ and related indicators work well (21). SPQ does not affect cell viability and growth, and, where tested, SPQ did not affect various transport, signaling, and regulatory processes (21–23). SPQ and related indicators with low membrane permeability can be loaded effectively into cells by slow passive diffusion (e.g., 5 m*M* SPQ overnight in culture medium), hypotonic shock (e.g., 50% hypotonic medium containing SPQ for 3 min), or direct microinjection. Recently, the cell-permeable compound diH-MEQ was developed for rapid loading of living cells (6). diH-MEQ is nonpolar and chloride insensitive; after entering cells diffusively, diH-MEQ becomes oxidized to the polar chloride-sensitive fluorophore MEQ. Although SPQ is probably the indicator of choice for most measurements of cytosolic chloride activity, diH-MEQ is useful for studies in intact tissues in which passive or invasive loading procedures cannot be used.

Figure 4 shows representative experiments in which cAMP-stimulated chloride conductance is measured in T84 colonic epithelial cells. Cells were loaded by an overnight incubation with 5 m*M* SPQ and mounted in a perfusion chamber on the stage of an inverted epifluorescence microscope. SPQ fluorescence was excited at 365 nm and detected at >410 nm using a barrier filter. The cells were perfused initially with a physiological buffer containing 110 m*M* chloride. Chloride was then replaced by nitrate, an anion that is

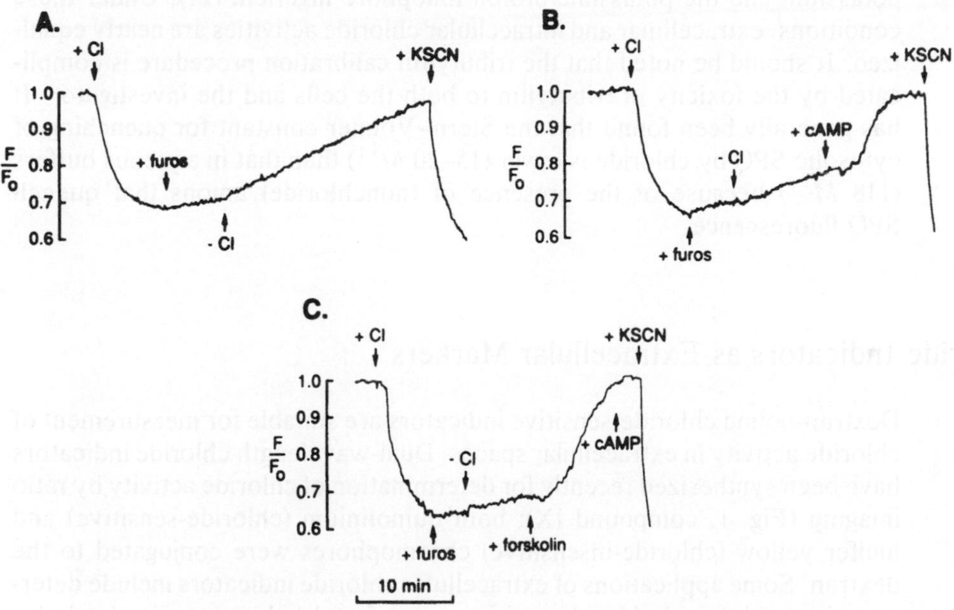


FIG. 4 Cyclic-AMP-stimulated chloride conductance in T84 epithelial cells. T84 cells were loaded with SPQ and fluorescence from 5–10 cells was monitored continuously as described in the text. Perfusion solutions consisted of physiological buffer (phosphate-buffered saline) containing 110 m*M* Cl (+Cl) or 110 m*M* $NO_3$ instead of Cl (−Cl). Where indicated, 0.1 m*M* furosemide (+furos), 0.5 m*M* CPT–cAMP (+cAMP), 0.02 m*M* forskolin (+forskolin), and/or 150 m*M* KSCN (+KSCN) were added.

transported by the cystic fibrosis transmembrane conductance regulator (CFTR) chloride channel but does not quench SPQ fluorescence. Furosemide was added to inhibit transport of chloride by the $Na^+/K^+/2Cl^-$ symporter. In curve A (Fig. 4), there was a slow increase in SPQ fluorescence in response to chloride replacement; in curve B, addition of the cell-permeable cAMP analog chlorophenylthio-cAMP caused a prompt increase in fluorescence corresponding to chloride efflux. Similar results were obtained by stimulation of cAMP production by forskolin (curve C). SPQ fluorescence was quenched by KSCN at the end of the experiment to provide a baseline value. SPQ and related compounds have been used in a variety of measurements of chloride transport in cultured cells (e.g., Refs. 21–29) and intact epithelia (e.g., Refs. 30–32).

If absolute cytosolic chloride activities and chloride transport rates are required, a calibration of indicator fluorescence vs. chloride activity must be established. Intracellular calibration has been accomplished using the chloride/hydroxyl ionophore tributyltin together with buffers containing high potassium and the potassium/proton ionophore nigericin (21). Under these conditions, extracellular and intracellular chloride activities are nearly equalized. It should be noted that the tributyltin calibration procedure is complicated by the toxicity of tributyltin to both the cells and the investigator. It has generally been found that the Stern–Volmer constant for quenching of cytosolic SPQ by chloride is lower (15–20 $M^{-1}$) than that in aqueous buffers (118 $M^{-1}$) because of the presence of (nonchloride) anions that quench SPQ fluorescence.

## Chloride Indicators as Extracellular Markers

Dextran-bound chloride-sensitive indicators are suitable for measurement of chloride activity in extracellular spaces. Dual-wavelength chloride indicators have been synthesized recently for determination of chloride activity by ratio imaging (Fig. 1, compound IX); both quinolinium (chloride-sensitive) and lucifer yellow (chloride-insensitive) chromophores were conjugated to the dextran. Some applications of extracellular chloride indicators include determination of luminal chloride activity in perfused kidney proximal tubules (32) and measurement of chloride activity in the lateral intercellular space in epithelial cells (38). Membrane-impermeable dextran indicators are also suitable for labeling of the endocytic compartment of living cells for determination of intraendosomal chloride activity (5, 33).

Another application of an extracellular chloride-sensitive fluorescent indicator is use as a remote sensor—for example, the monitoring of chloride activity in blood *in vivo*, and the measurement of chloride content in chemical reaction vessels and sewage treatment plants. An iodide/bromide sensor was reported based on the quenching of acridinium and quinolinium indicators, which were immobilized onto a glass surface (34). Recently, the compound CHQ (see above) has been bound to a surface alkylamine-modified 150-$\mu$m diameter porous glass (35). The glass was cemented to the end of a single fiber-optic cable for use as a remote sensor. Sensor fluorescence was sensitive to solution chloride activity with 50% quenching at 250 m*M* chloride. It may be possible to construct smaller chloride-sensitive fiber-optic sensors for use in interstitial spaces and cytosol in living tissues.

## Critique and Future Directions

SPQ and the related quinolinium chloride-sensitive fluorescent indicators have well-established applications in the measurement of chloride activity and transport in cells, cell-free vesicles, and reconstituted proteoliposomes (36). The chloride sensitivity and response kinetics of available indicators are excellent. The chloride selectivity is good but not perfect; because SPQ is quenched strongly by other halides and weakly by organic anions, including proteins and some buffers (37), changes in indicator fluorescence must be interpreted cautiously. Chloride-sensitive indicators have adequate molar absorbance and quantum yields to give acceptable fluorescence signal intensities for most applications; however, the requirement of ultraviolet excitation makes possible photodynamic cell injury and signal contributions from cell autofluorescence. Low illumination intensities are therefore recommended in microscopy applications. Finally, except for the dual-wavelength dextran indicators, the available chloride-sensitive indicators have single fluorescence excitation and emission peaks that do not undergo chloride-dependent wavelength shift. It will be important to develop long-wavelength chloride indicators suitable for ratio imaging.

Although chloride-sensitive fluorescent indicators have many applications in cell-free and intact cell studies described in this chapter, there is important information that cannot be obtained by these methods. Measurements of the single-channel properties of chloride channels require electrophysiological methods. The determination of instantaneous transmembrane and transepithelial chloride conductances is accomplished better by electrical than by fluorescence methods because fluorescence methods provide only the time-integrated flux of chloride through both conductive and nonconductive pathways. The parallel use of fluorescent indicators and electrophysiological (or isotopic) methods may be advantageous in some experiments.

A number of novel applications of chloride-sensitive fluorescent indicators have yet to be explored. The measurement of chloride activities in subcellular compartments, such as endocytic vesicles, golgi, and the nucleus, has just begun. The existence of localized chloride concentration gradients in cell cytosol has been proposed but not tested. The chloride activity in specific extracellular spaces, such as that in the renal medullary interstitium, is of great physiological importance but has not been measured directly. Finally, *in vivo* clinical applications in the continuous monitoring of body fluid compositions should be possible.

## References


1. O. S. Wolfbeis and E. Urbano, *J. Heterocycl. Chem.* **19,** 841 (1982).
2. N. P. Illsley and A. S. Verkman, *Biochemistry* **26,** 1215 (1987).
3. R. Krapf, N. P. Illsley, H. C. Tseng, and A. S. Verkman, *Anal. Biochem.* **169,** 142 (1988).
4. A. S. Verkman, M. Sellers, A. C. Chao, T. Leung, and R. Ketcham, *Anal. Biochem.* **178,** 355 (1989).
5. J. Biwersi, N. Farah, Y.-X. Wang, R. Ketcham, and A. S. Verkman, *Am. J. Physiol.* **262,** C243 (1992).
6. J. Biwersi and A. S. Verkman, *Biochemistry* **30,** 7879 (1991).

6a. K. D. Legg and D. M. Hercules, *J. Phys. Chem.* **74,** 2114–2121 (1970).

6b. J. Biwersi, B. Tulk, and A. S. Verkman, *Anal. Biochem.* **219,** 139–143 (1994).

7. D. Pearce and A. S. Verkman, *Biophys. J.* **55,** 1251 (1989).
8. P.-Y. Chen and A. S. Verkman, *Biochemistry* **27,** 655 (1988).
9. P. Fong, N. P. Illsley, J. H. Widdicombe, and A. S. Verkman, *J. Membr. Biol.* **104,** 233 (1988).
10. P. Placchi, R. Lombardo, A. Tamanini, P. Brusa, G. Berton, and G. Cabrini, *J. Membr. Biol.* **119,** 25 (1991).
11. J. Cuppoletti, D. Aures-Fischer, and G. Sachs, *Biochim. Biophys. Acta* **899,** 276 (1987).
12. M. C. Dechecchi and G. Cabrini, *Biochim. Biophys. Acta* **945,** 113 (1988).
13. A. S. Verkman, R. Takla, B. Sefton, C. Basbaum, and J. H. Widdicombe, *Biochemistry* **28,** 4240 (1989).
14. A. E. Mulberg, B. M. Tulk, and M. Forgac, *J. Biol. Chem.* **266,** 20590 (1991).
15. A. E. Engblom, I. Holopainen, and K. E. Akerman, *Neurosci. Lett.* **104,** 326 (1989).
16. S. M. Dunn, C. Martin, M. W. Agey, and R. Miyazaki, *Biochemistry* **28,** 2545 (1989).
17. M. Garcia-Calvo, A. Ruiz-Gomez, J. Vasquez, E. Morato, F. Valdivieso, and F. Mayor, *Biochemistry* **28,** 6405 (1981).
18. H.-R. Bae and A. S. Verkman, *Nature (London)* **348,** 635 (1990).
19. W. Reenstra, I. Sabolic, H.-R. Bae, and A. S. Verkman, *Biochemistry* **31,** 175 (1992).

20. L.-B. Shi, K. Fushimi, H.-R. Bae, and A. S. Verkman, *Biophys. J.* **59,** 1208 (1991).
21. A. C. Chao, J. A. Dix, M. Sellers, and A. S. Verkman, *Biophys. J.* **56,** 1071 (1989).
22. A. C. Chao, J. H. Widdicombe, and A. S. Verkman, *J. Membr. Biol.* **113,** 193 (1990).
23. A. S. Verkman, A. C. Chao, and T. Hartman, *Am. J. Physiol.* **262,** C23 (1992).
24. T. M. Dwyer and J. M. Farley, *Life Sci.* **48,** 2119 (1991).
25. J. K. Foskett, *Am. J. Physiol.* **259,** F998 (1990).
26. M. C. Dechecchi, A. Tamanini, G. Berton, and G. Cabrini, *J. Biol. Chem.* **268,** 11321 (1993).
27. L. J. MacVinish, T. Reancharoen, and A. W. Cuthbert, *Br. J. Pharmacol.* **108,** 469 (1993).
28. S. H. Cheng, D. P. Rich, J. Marshall, R. J. Gregory, M. J. Welsh, and A. E. Smith, *Cell (Cambridge, Mass.)* **66,** 1027 (1991).
29. B. C. Tilly, M. Kansen, P. G. van Gageldonk, N. van den Berghe, H. Galjaard, J. Bijman, and H. R. de Jonge, *J. Biol. Chem.* **266,** 2036 (1991).
30. R. Krapf, C. A. Berry, and A. S. Verkman, *Biophys. J.* **53,** 955 (1988).
31. S. J. Ram and K. L. Kirk, *Proc. Natl. Acad. Sci. U.S.A.* **24,** 10166 (1989).
32. L.-B. Shi, K. Fushimi, and A. S. Verkman, *J. Gen. Physiol.* **98,** 379 (1991).
33. J. Biwersi, K. Zen, L.-B. Shi, and A. S. Verkman, *Biophys. J.* **64,** A194 (1993).
34. E. Urbano, H. Offenbacher, and O. Wolfbeis, *Anal. Chem.* **56,** 427 (1984).
35. H. P. Kao, J. Biwersi, and A. S. Verkman, *Proc. SPIE Int. Soc. Opt. Eng.* **1648,** 194 (1992).
36. A. S. Verkman, *Am. J. Physiol.* **257,** C837 (1989).
37. M. Vasseur, R. Frangne, and F. Alvarado, *Am. J. Physiol.* **264,** C27 (1993).
38. P. Xia, B. E. Persson, and K. R. Spring, *J. Membr. Biol.* **144,** 21 (1995).

# [18] Measurement of Cytosolic Chloride Activity by Ion-Selective Microelectrodes

Jonathan A. Coles

## Introduction

Two techniques are available for making continuous measurements of intracellular $Cl^-$ activity ($aCl_i$) in single cells. Readers who have never used an intracellular microelectrode to measure the membrane potential ($V_m$) of a cell under study are advised first to consult Verkman and Biwersi, Chapter 17, this volume, to see if fluorescent $Cl^-$ indicators would be appropriate for the problem they wish to analyze. For readers already making intracellular voltage recordings, use of $Cl^-$-selective microelectrodes will involve only modest changes to their experimental apparatus and techniques (essentially, an amplifier head stage with a high-input impedance, and more difficulty in impaling cells). Compared to fluorescent indicators, an advantage of ion-selective microelectrodes is that $V_m$ is measured simultaneously with the ion activity (see Chapter 2, this volume). At the very least, this gives an indication of the health of the cell, and, because ion fluxes depend on $V_m$, this nearly always contributes valuable information. If the cell of interest is surrounded by other, different cells, as in a tissue slice, then a fluorescent indicator would have to be injected from a microelectrode into the cell. In this case it may be simpler to use a $Cl^-$-selective microelectrode.

$Cl^-$-selective microelectrodes give absolute values of $aCl_i$ (as distinct from qualitative changes) much more directly than do currently available fluorescent indicators. On the other hand, the signal recorded by currently available $Cl^-$ microelectrodes can, under some circumstances, be affected not only by $Cl^-$, but also by certain other biological anions and by drugs that would be useful for certain experiments. The situations in which such interference might arise have been known for many years and can usually be avoided. The problem has been reduced by the development of a new sensor for $Cl^-$ microelectrodes (1), and there is hope that further improvements will be made in the near future. At the same time, progress is being made in developing fluorescent indicators that will give a direct measurement of absolute values of $aCl_i$ (Chapter 7, this volume).

With both optical and electrode techniques it is straightforward to switch from measuring $Cl^-$ to measuring other ions ($Na^+$, $K^+$, $H^+$, and $Ca^{2+}$, for example). For a more encyclopedic review of $Cl^-$ microelectrodes, with

*Methods in Neurosciences, Volume 27*

extensive references, the reader is advised to consult Alvarez-Leefmans *et al.* (2).

## Causes of Change in Intracellular Chloride Concentration

In many cells, chloride is not far from being in electrochemical equilibrium across the cell membrane, so its concentration is on the order of a tenth of that in the extracellular fluid. Because this concentration is low, modest transmembrane fluxes produce large fractional changes in $a\mathrm{Cl}_i$. Changes occur physiologically or pathologically in three main ways:

1. As a passive consequence of a change in membrane potential. If the potential change is caused by a change in extracellular $K^+$ concentration, both $K^+$ and $Cl^-$ will cross the membrane in the same direction (3, 4). Water will also move in the same direction and the cell volume will change. This was described in detail by Boyle and Conway (5). Indeed, in all mechanisms of volume regulation that involve charged ions, $Cl^-$ is likely to be involved because (together with $HCO_3^-$) it is the major extracellular anion, and anions as well as cations must cross the cell membrane.
2. As a consequence of a change in $Cl^-$ conductance. In most cells, $Cl^-$ is not passively distributed across the membrane, but, according to the cell type, $a\mathrm{Cl}_i$ is either greater or less than the equilibrium value. Hence, activation of a $Cl^-$ conductance may either depolarize or hyperpolarize cells. There appear to be no safe rules for predicting *a priori* the direction of the electrochemical gradient of $Cl^-$across the membrane of a given type of cell; determination of the direction has often been difficult to establish unequivocally without direct measurements of $a\mathrm{Cl}_i$ (6). $Cl^-$ conductances may be activated by neurotransmitters, directly as in the case of $\gamma$-aminobutyric acid (GABA), glycine (6), and histamine (7), or indirectly via increases in intracellular free $[Ca^{2+}]$ (8) or cAMP (9). $Cl^-$ conductances activated by cell swelling have been reported (10).
3. As a consequence of $Cl^-$ transport. Transport of $Cl^-$ coupled to transport of other ions is involved in the regulation of intracellular pH (see Chapter 12, this volume) (11) and cell volume (12) and in ion transport through epithelia (13).

In understanding all these processes, direct measurement of $a\mathrm{Cl}_i$ is very useful. As an example, a recent paper by Kaila *et al.* (14) shows how a hyperpolarizing response to GABA in crayfish stretch receptor neurons can be masked by a depolarizing current associated with GABA uptake: intracellular $Cl^-$ electrodes proved very useful in analyzing this system.

During cell volume changes and coupled transmembrane transport involving $Cl^-$, the changes in $aCl_i$ are slow enough to be recorded accurately by ion-selective microelectrodes. $Cl^-$ fluxes at synapses can be brief, so an electrode may not follow the immediate changes in $aCl_i$, although it would follow the recovery to baseline.

## Types of $Cl^-$ Microelectrodes

Two basic types of $Cl^-$ electrodes have been used, solid state and liquid membrane. For solid-state microelectrodes, either a bare Ag wire or a chlorided Ag wire is introduced into the cytoplasm. These electrodes have poor selectivity and it is difficult to derive absolute values of $aCl_i$ (2, 15).

For liquid membrane electrodes, a column of a hydrophobic ion-selective liquid sensor is lodged in the open tip of a micropipette and electrical contact is made with the back surface of the column, usually through a backing electrolyte solution (see Chapter 2, this volume). A convenient feature of liquid membrane ion-selective microelectrodes is that electrodes selective for different ions can be made by the same technique, simply with the use of different sensors and backing solutions. I will treat first the construction of liquid membrane ion-selective microelectrodes, then discuss sensor solutions for $Cl^-$, and finally the use of $Cl^-$ microelectrodes.

## Construction of Liquid Membrane $Cl^-$ Microelectrodes: General Principles

If the glass surface inside the tip of the pipette is left in its usual hydrophilic state, it will attract aqueous solutions; as soon as the electrode tip is immersed in a physiological saline solution the sensor column will be displaced upward away from the tip. To avoid this problem (and also to reduce electrical leaks in parallel with the sensor column), the inside of the tip is made hydrophobic, usually by treatment with a silane (a class of silicon compounds; see Ref. 16). Some of the principles involved in making liquid membrane ion-selective electrodes will be discussed, and then a precise procedure for their construction will be given.

### *Capillaries*

If a cell is large enough for more than one electrode to be inserted, then a single-barreled $Cl^-$ electrode used with a separate intracellular reference electrode is usually the simplest solution. A single-barreled electrode alone will, if it has a perfectly Nernstian slope (see below), give the electrochemical potential of intracellular $Cl^-$ and hence indicate whether opening a $Cl^-$ conductance will hyperpolarize or depolarize the cell. Standard electrode capillaries have an internal filament that aids filling by allowing liquid to run to the tip in the crevices between the filament and the capillary wall. One might expect that the same phenomenon would make it easier for an aqueous solution to enter the tip of an ion-selective microelectrode and displace the sensor column. In practice, this seems not to be the case, probably because at the tip the filament fuses smoothly into the wall.

If the reference barrel is to be incorporated into a double-barreled electrode, then theta section tubing can be used (Fig. 1C; cf. Ref. 1). The partition must be well sealed into the wall so that there are no cracks. In the past, this has not been true of all commercially available theta tubing. It is easy to check tubing by breaking off a piece several millimeters long and examining

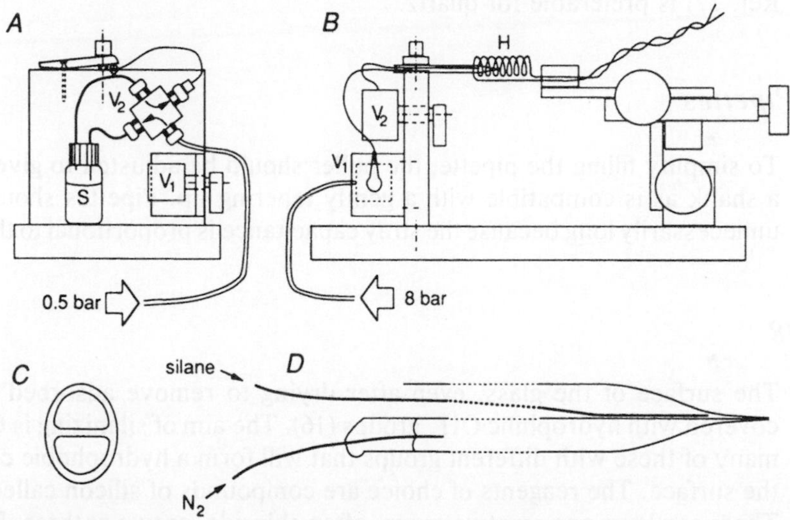


FIG. 1 An apparatus for silanizing double-barreled micropipettes. (A) The end view; (B) the side view. The micropipette with a theta section (C) is clamped horizontally. The end of a syringe needle is sealed into one barrel (the lower one in D) with shellac. Nitrogen at 8 bar is applied through on/off valve $V_1$. The heater coil H, wound from 0.9-mm nichrome and powered at 1.4 V AC, is advanced by means of a manipulator to cover the tip of the pipette. Two-way valve $V_2$ (Hamilton HVP with plug 86779) is set to pass nitrogen at 0.5 bar through a second needle that is inserted into the other barrel as far as a few millimeters from the tip (upper barrel in D). Valve $V_2$ is then turned so that nitrogen flows over the silane S and carries silane vapor into the pipette.

it end-on under a microscope: the joint between the glass of the partition and that of the wall should be visible as no more than a fine, curved line.

Quartz has several advantages over the usual borosilicate (Pyrex) glass for ion-selective microelectrodes (17). Quartz electrodes appear to remain usable for a longer time, either in storage or in use. They will penetrate tough tissue (18) and are less likely to be broken accidently; if a quartz electrode is dropped it is worth examining it to see if the point is unbroken. The disadvantages are that a special high-temperature puller is necessary and, as far as I know, there is no commercial source of quartz theta capillaries, which require special equipment for their fabrication. Presumably, if quartz theta tubing becomes available, the manufacturers of pullers for quartz will know about it. These manufacturers include Sutter Instrument Company [40 Leveroni Court, Novato California 94949, fax (+1) 415-883-0572] and De Marco Engineering [5 rue du Cardinal-Journet, CH-1217 Meyrin, Switzerland, fax. (+41) 22 782 81 27]. It has been claimed (J.-L. Munoz, personal communication, 1990) that a different method of silanization (described in Ref. 17) is preferable for quartz.

### *Pulling Pipettes*

To simplify filling the pipette, the puller should be adjusted to give as short a shank as is compatible with a gently tapering tip. Pipettes should not be unnecessarily long because the stray capacitance is proportional to the length.

### *Silanizing*

The surface of the glass, even after drying to remove adsorbed water, is covered with hydrophilic $OH^-$ groups (16). The aim of silanizing is to replace many of these with different groups that will form a hydrophobic coating on the surface. The reagents of choice are compounds of silicon called silanes. These can have one reactive group, often chloride, or two or three. Examples are trimethylchlorosilane, dimethyldichlorosilane, and methyltrichlorosilane, respectively. If there are two or three reactive groups, then, in the presence of water vapor, the silane molecules can link up with each other to form a silicone rubber. This may produce a highly hydrophobic coating, but tends to block the tip of an intracellular electrode. It is easier, and apparently at least as effective, to use a silane with one reactive group ("monofunctional") and only these are considered here. I know of no paper that reports a systematic comparison of different methods of silanization on electrode performance. However, measurements have been made of three properties of a silanized surface that are presumably important: hydrophobicity, electrical resistivity (along the plane of the surface), and the fraction of the surface hydroxyls that are replaced. The results suggest at least a good starting point for developing a method of silanization.

*The Reactive Group*

Two reactive groups have been widely used: chloride (as in trimethylchlorosilane) and dimethylamine. The latter gives better results in tests of silanization quality (16). Chloride has the additional inconvenience that the reaction product, HCl, is highly corrosive.

*The Hydrophobic Group*

The hydrophobic group should have a diameter in the plane of the glass surface large enough to make an umbrella that covers the space between the hydroxyls. If the diameter is greater, then the silanes will not be able to react with all the hydroxyls. In fact, even the smallest usefully hydrophobic umbrella, three methyl groups, is somewhat too big and allows only about 60% of the hydroxyls to react (19, 20). Put together with a reactive amino group, this gives trimethyl(dimethylamino)silane, known as TMDMAS or TMSDMA and sold as Fluka product 41716. The greatest hydrophobicity has been reported for a silane that should give two superimposed layers of methyl groups: dimethyl-(3,3-dimethylbutyl)dimethylaminosilane (Fluka 39853; Refs. 16 and 21).

*Reaction Conditions*

Tests suggest that 5 min at 300°C gives near-optimum coverage (16). There is no evidence that increasing the reaction time or baking the silanized pipette helps. To make double-barreled electrodes, the trick is to silanize the active barrel but not the reference barrel.

## *Making and Using Double-Barreled $Cl^-$ Microelectrodes*

Excellent experimental results have been obtained by workers who used liquid membrane $Cl^-$ microelectrodes made according to a number of different recipes (see Chapter 2, this volume). Without working in the various laboratories, it is difficult to know which methods give the highest yields. The method to be described is based on that of Coles and Tsacopoulos (22). Despite a constant search for improvements, their method has changed little except in the direction of simplification. The silanization apparatus requires the drilling of 12 holes for its construction; readers not prepared to make such an investment might consider the methods of Kondo *et al.* (1) or Thomas (21).

*Materials*

Style IA theta tubing 1.8–2.2 mm OD [R & D Scientific Glass Company, 15931 Batson Road, Spencerville, Maryland 20868, tel. (+1) 301 421 9719] or TCG200 (Clark Electromedical Instruments, P.O. Box 8, Pangbourne, Reading RG8 THU, England).
Diamond triangular file (much nicer to use than an ordinary file).
Old, valueless, watchmaker's forceps.
Touch shellac (Elephant brand, Thew, Arnott & Co, 270 London Road, Wallington, Surrey SM6 7DJ, England).
Cauterizing iron with a foot switch, to melt the shellac
A silanizing apparatus; my present one is shown in Fig. 1.
Thermocouple reading to 350°C to set the temperature in the heating coil.
Silane: Fluka 41716 (cheaper) or 39853 (probably better).
Fume hood (desirable because the silane and its reaction by-product smell unpleasant).
Stop clock.
Two syringes fitted with tubing fine enough to enter the micropipette barrel; I use Hamilton syringes (25 $\mu$l, Teflon piston) and drawn-out polyethylene tubing.
Chlorided Ag wires.

*Method*

1. Cut 4-cm lengths of capillary, not specially cleaned. Break back one side at each end for about 2 mm with watchmaker's forceps, and tap out any glass particles. The barrel that is broken back will become the active barrel. Pull micropipettes. If you have trouble with capillaries cracking in the puller clamp, shrink 3-mm lengths of heat-shrink tubing on each end.

2. Clamp a micropipette in the silanizer (Fig. 1). Insert the 0.3-mm needle from tap $V_1$ a short way into the reference barrel and seal with shellac. Turn on the high pressure (8 bar) nitrogen gas at tap $V_1$ and check that no leak can be heard. The heating coil, which is left permanently on, is advanced over the micropipette tip. Insert the other needle from tap $V_2$ as far as conveniently possible into the active barrel. Nitrogen at low pressure (0.5 bar) is flowing permanently through this. Turn tap $V_2$ to divert the gas flow to pass over the silane for 7 min. Turn off the silane at tap $V_2$, pull the silane needle out of the micropipette, withdraw the heating coil, soften the shellac, and pull the gas needle out of the reference barrel. Inject enough sensor into

the active barrel to fill most of the taper. Fill the reference barrel with a suitable electrolyte. As discussed below, 3 *M* KCl will do for a start. The reference barrel should fill to the tip spontaneously. The sensor may need some help, depending on the exact shape of the tip. (If a long tip fills very easily, it may mean that the silanization is poor.) A rapid method is to use a vacuum. First protect the reference barrel by completing it, i.e., sealing in a chlorided Ag wire with shellac. Then place a batch of electrodes, tip downward, in a chamber that is evacuated for about 2 min with a rotary vacuum pump (16). Any remaining bubbles should dissolve within tens of minutes. Alternatively, it is usually sufficient simply to leave the electrodes overnight. Or the sensor column can be poked with a freshly pulled glass filament, or a rat's whisker, to get sensor to the extreme tip.

3. Inspect the tip for clean filling, preferably under a microscope objective corrected for no coverslip (marked "0" and not the usual "0.17"). Fill the back of the active barrel with a solution containing $Cl^-$ (10 to 200 m*M*). Seal in chlorided Ag wires with shellac. Bevel the tip, e.g., on a miniature grindstone [De Marco Engineering, 5 rue Cardinal Journet, CH-1217 Meyrin, Switzerland, fax: (+41) 22 782 81 27]. Alternatively, it may be possible to break the tip just before use on a tough part of the biological preparation, such as connective tissue. For tip diameters below 1 $\mu$m there is no simple relationship between diameter and the ability to impale a cell.

4. Check that you have two or three working electrodes; there is little point in making full calibrations at this time. Set up a preparation. Position an electrode in the superfusion solution, change to a test solution that is not harmful to the preparation, change back, and wait for steady baseline. Impale your cell (tapping may help more than buzzing). Immediately after a good recording is obtained, withdraw to establish the baseline and then retest and calibrate the electrode.

*Expected Success Rate*

Of electrodes that appear cleanly filled under the microscope at least four out of five should respond after beveling. There is a trade-off between low noise and small tip diameter. If reference barrels respond to $Cl^-$, then reduce the silanization time. If the sensor column retracts from the tip so that the response becomes very slow, increase the silanization time. If the electrode potential appears noisy, check that the bias current of the head stage amplifier is quiet (fluctuations <10 fA) by observing the voltage with a $10^{11}$-$\Omega$ resistor inserted between the input and ground. If fine-tipped, double-barreled electrodes do not work, start by making a single-barreled electrode with a 2-$\mu$m tip.

## *Choice of a Sensor Solution for Chloride*

A major factor in evaluating the merits of a sensor solution is its selectivity for the ion to be measured. This selectivity is expressed by a series of selectivity coefficients.

### *Selectivity Coefficients*

The ion voltage signal of a $Cl^-$ electrode (see Chapter 2, this volume) in a solution of chloride salts depends only on *a*Cl, but if the solution contains other anions these will contribute to the signal and interfere with the measurement. The degree to which an anion A tends to interfere can be specified by a selectivity coefficient $k_{Cl,A}$ (23). For example, if a particular electrode has a selectivity coefficient for $HCO_3^-$ of 0.1, then the addition of 10 m*M* $HCO_3^-$ to a solution will produce the same voltage change as the addition of 1 m*M* $Cl^-$. Accurately determined selectivity coefficients are essential for chemists attempting to develop better macroelectrodes. In microelectrodes, selectivity coefficients vary from one electrode to another, and in intracellular measurements the identities and concentrations of interfering ions are usually unknown. Physiologists therefore try to find experimental ruses to avoid using selectivity coefficients to calculate results; what is necessary are approximate values of selectivity coefficients that indicate which interfering ions are likely to give trouble. If there is any reasonable probability that interference might occur in an experiment, then this should be tested for in conditions as close as possible to those of the experiment. Some published values of selectivity coefficients for $Cl^-$ electrodes made with three kinds of sensors are given in Table I. The ion-exchanger sensors have been used for longest, so selectivity coefficients have been measured by many different authors (see Ref. 2). Different authors have obtained values differing by factors of up to more than 2 and the numerical values in Table I are only indicative. They merely suggest ions that might cause problems, and are quite inadequate for correcting an electrode reading. The interfering ions are grouped approximately as "biological anions," which might be present in cells; "replacement anions," which might be useful for replacing $Cl^-$ in solutions; and certain "drugs," which are often useful in studying $Cl^-$ movements. Among the biological anions, $HCO_3^-$ is the one most likely to interfere with measurement of changes in $aCl_i$. $Br^-$, $I^-$, and isethionate may be present in significant concentrations in marine organisms. Among the drugs in Table I, it is clear, for example, that 4-acetamido-4′-isothiocyanostilbene-2,2′-disulfonic acid (SITS), an inhibitor of $Cl^-/HCO_3^-$ exchange, must be used with great care with an ion-exchanger electrode. It is not yet known whether it would be usable with the other two sensors. Gluconate and sulfate might be worth trying as replacements for $Cl^-$.

TABLE I Indicative Values of Selectivity Coefficients for $Cl^-$ Microelectrodes and Macroelectrodes[a]

| Interfering ion | Selectivity coefficient $k_{Cl,A}$ MnTPP[b] | Ion exchangers[c] | ETH 9009[d] |
|---|---|---|---|
| **Biological anions** | | | |
| $SCN^-$ | 3000 | 10–120 | 0.6 |
| Salicylate | 1000 | 100 | 0.4 |
| $I^-$ | 250 | 46–94 | 10 |
| $NO_3^-$ | 14 | 7.6 | <0.001 |
| $Br^-$ | 8 | 2.7–4.2 | 1 |
| $NO_2^-$ | 5.5 | 2.9 | — |
| Lactate | 0.1 | — | — |
| Acetate | 0.06 | 0.3 | — |
| Isethionate | 0.05 | 0.2 | |
| $HCO_3^-$ | 0.03 | 0.04–0.2 | <0.00001 |
| **Replacement anions** | | | |
| Gluconate | 0.004 | <0.03 | — |
| Sulfate | 0.002 | 0.1 | <0.000001 |
| **Drugs** | | | |
| Furosemide | — | 200 | — |
| Bumetamide | | 151 | |
| SITS | — | 1000 | — |

[a] The microelectrodes contain sensors based on MnTPP or ion exchangers; the macroelectrodes are based on ETH 9009.

[b] MnTPP is 5,10,15,20-tetraphenyl-21*H*,23*H*-porphine manganese(III) chloride (Aldrich product 25,475-4). The composition of the sensor and the selectivity coefficients are given in Y. Kondo, T. Bührer, K. Seiler, E. Frömter, and W. Simon, *Pfluegers Arch.* **414**, 663 (1989).

[c] See Ref. 1 and references in F. J. Alvarez-Leefmans, F. Giraldez, and J. M. Russell, *in* "Chloride Channels and Carriers in Nerve, Muscle and Glial Cells" (F. J. Alvarez-Leefmans and J. M. Russell, eds.), p. 3. Plenum, New York, 1990.

[d] M. Rothmeier and W. Simon, *Anal. Chim. Acta* **271**, 135 (1993).

### *Kinds of Sensor Solutions*

There are two major classes of sensor solutions for ion-selective microelectrodes, those based on an ion exchanger and those based on carrier molecules (24). In general, ion-exchanger sensors have poor selectivity but have the advantage of having a lower electrical resistivity.

### *Ion Exchangers for Chloride*

Wegmann *et al.* (25) tested 120 different membrane compositions containing various ion exchangers for anions. They all showed the same sequence of selectivities, known as the "Hofmeister series":

Lipophilic anions > $ClO_4^-$ > $SCN^-$ > $I^-$ > $NO_3^-$
> $Cl^-$ > $HCO_3^-$ > $SO_4^{2-}$ > $HPO_4^{2-}$

Although anion-exchanger electrodes are more sensitive to some other ions than to $Cl^-$ (to $ClO_4^-$, $SCN^-$, $I^-$, etc.), $a$Cl$_i$ usually dominates the signal in cytoplasm and nearly all measurements of $a$Cl$_i$ have been made using ion-exchanger sensors. Anion exchangers are available commercially from W.P.I. (IE 170), Orion, or Corning [447913, which is the same as their earlier exchanger 447315 but with five times more solute (26)]. These manufacturers are reluctant to disclose the composition of their ion exchangers, but it is thought they they all consist of Aliquat 336 in 2-nitrophenyl octyl ether. In the course of work for Coles *et al.* (4) some attempts were made to use components described by Wegmann *et al.* (25), but the results were not strikingly better than with commercially prepared ion-exchanger sensors.

*Sensors Based on Carrier Molecules*

The laboratory of the late W. Simon synthesized several anion-selective carrier molecules. In view of clinical applications these were tested first in polyvinylchloride (PVC) membranes, but at least one, a tetraphenylporphyrin manganese derivative, has been used in a liquid sensor for microelectrodes (1). This sensor is available from Fluka (product 24902), who follow an admirable policy of publishing the composition of all their products for ion-selective electrodes. Unfortunately, the sensor based on porphyrin manganese is better than the ion exchangers in only some respects. As seen in Table I, it has better discrimination against $HCO_3^-$ but is more sensitive to $SCN^-$. Rothmeier and Simon (27) have recently tested some mercury organic compounds. Two of these, ETH 9009 and ETH 9011, when incorporated in PVC membranes and used in macroelectrodes, had selectivities that were a vast improvement over earlier sensors (Table I). If a way could be found of using one of these compounds in microelectrodes, as has been done for carrier molecules for cations, then a highly selective $Cl^-$ microelectrode might become a reality. Unfortunately, it is not obvious how microelectrodes with a usably low electrical resistance could be made: ETH 9009 and 9011 both have a low solubility in the polar solvent usually used in sensor solutions (2-nitrophenyl octyl ether). Therefore, less polar solvents would have to be used, and these would give a higher electrical resistance. To make things worse, lipophilic salts such as tridodecylmethylammonium chloride, which might be added to lower the resistivity, also have a low solubility in these less polar solvents (E. Pretsch, personal communication, 1994).

## *Reference Electrodes*

The reference barrel of a double-barreled $Cl^-$ electrode has three distinguishable functions. (1) It should measure the membrane potential ($V_m$). (2) It should measure a potential that is subtracted from the potential of the active barrel to give an ion signal that can be translated into $aCl_i$. (3) It should be stable, so that changes in $aCl_i$ can be compared accurately. In addition, it should not disturb the cell by damaging it or by leaking $Cl^-$ into it. In principle, all these conditions can be met quite well by a reference barrel with a fine tip filled with 3 *M* KCl. An ideal 3 *M* KCl electrode experiences a change in junction potential of about +2.6 mV when it passes from Ringer solution into the myoplasm of frog muscle (28). Compared to other likely errors, this is modest. However, to avoid significant leakage of $Cl^-$ into the cell, the tip must be fine (2, 29). Fine-tipped KCl micropipettes can have an additional tip potential of up to about 10 mV (28, 30).* If this tip potential changes on entering a cell it will introduce errors: Hironaka and Morimoto (28) found, in muscle cells, that it did not; Adrian (30) found that it could change by up to about 15 mV in the direction indicating a smaller $V_m$. Because the error is in the direction of an apparently reduced membrane potential, selecting recordings with larger membrane potentials will also tend to select for those made with reference barrels with small tip potentials.

Most useful physiological results with $Cl^-$ electrodes have been obtained by concentrating on only one or two of the three functions listed above. If the changes in junction or other tip potentials on the reference barrel were the same on passing from Ringer solution either into a calibration solution or into cytoplasm, then $aCl_i$ would be measured correctly even if the apparent $V_m$ were in error. If this is allowed, the filling solution for the reference barrel can be selected for other criteria, such as low noise and absence of blocking (e.g., a lower concentration of KCl) or replacement of most of the $Cl^-$ by another anion such as $SO_4^{2-}$ or gluconate so as to be sure that $Cl^-$ leaks affect neither the cell nor the active barrel (4, 30). Reference barrels (silanized) have been filled with a cation exchanger [2% potassium tetrakis (*p*-chlorophenyl) borate in *n*-octanol] to produce what is known as an RLIE electrode (29). RLIE electrodes avoid the possibility of a $Cl^-$ leak. If they were equally selective for $Na^+$ and $K^+$ and if the sum ($a$Na + $a$K) were the

*The tip potential in Ringer solution can be measured by connecting an agar/3 *M* KCl bridge to the amplifier input, taking the reading with this agar bridge in the Ringer as zero, and then applying the agar bridge to the back of the KCl column in the micropipette (28). Hence, electrodes could be selected for negligible tip potential, but this would lead to discarding many otherwise good ion-selective microelectrodes.

same in the Ringer solution and in the cytoplasm, then RLIE electrodes would also measure $V_m$ correctly. Direct comparisons of values of $V_m$ measured with KCl and RLIE electrodes have shown that in at least some cells the latter give longer lasting, more stable recordings and with a mean value not significantly different from that measured by KCl electrodes (29, 31). A disadvantage of RLIE electrodes is their high electrical resistance, which makes them unsuitable for recording rapid electrical activity, such as action potentials.

The question, often an important one, of the direction of the transmembrane electrochemical gradient of $Cl^-$, might best be answered by observing only the potential of the active barrel (4, 32). Provided $a\mathrm{Cl}_i$ is sufficiently large for the electrode to be in a nearly Nernstian range, the error by this method is likely to be smaller than the error in the measurement of $V_m$. For example, direct measurement of the electrochemical potential in this way may be more accurate than calculating it from electrochemical potential $= E_{Cl} - V_m$, where $E_{Cl}$ is the equilibrium potential $= RT/F \ln(a\mathrm{Cl}_i/a\mathrm{Cl}_o)$. This is illustrated by the middle trace in Fig. 3. When the cell was depolarized (by a light stimulus), the electrochemical gradient for $Cl^-$ changed so that passive movement of $Cl^-$ would be into the cell.

## Calibration

### *Activity and Free Concentration*

Ions in cells can be sequestered into organelles, tightly bound to large molecules (chelated) or in free solution in the cytosol. Only in the last condition can they directly affect an ion-selective microelectrode. The property of ions in solution that affects the potential of an ion-selective microelectrode is defined as activity. Determination of the activity in a cell requires knowing the activity in calibration solutions. Unless some other definition is given, it is assumed that the activity scale is that used by physical chemists: in a very dilute solution, activity (in, e.g., millimoles/liter) is defined as equal to the concentration. As the concentration of the solution is increased into the physiological range, activity is found to increase progressively more slowly than the concentration; the ions interfere with each other so that they have less effect on an electrode. Hence, by extrapolating a calibration curve made with very low concentrations, a perfect $Cl^-$ electrode with a perfect reference electrode could be used to measure activity in solutions of higher concentrations. Measured values of $a_{Cl}$ in pure solutions of KCl or NaCl at 25°C are given in tables published by the United States National Bureau of Standards (33, 34). The ratio (activity/concentration) is called the activity coefficient,

$y_{Cl}$.[†] Values of $y_{Cl}$ in NaCl solutions calculated from these tables are plotted in Fig. 2.

Approximate values for $y_{Cl}$ at other temperatures can be calculated (e.g., Ref. 36). The activity coefficient of an ion varies little from one solution to another if the solutions have the same ionic strength:

$$I = \frac{1}{2} \sum_j c_j z_j^2. \tag{1}$$

where $c_j$ is the concentration of ion $j$ and $z_j$ is its charge. For example, a solution consisting of 10 m$M$ $MgCl_2$ and 100 m$M$ Na gluconate has an ionic strength of $\frac{1}{2}$ (10 × 4 + 20 + 100 + 100) = 130 m$M$. Hence, to measure $y_{Cl}$ in a physiological solution (e.g., a Ringer solution), use a large diameter $Cl^-$ microelectrode and a good, separate, reference electrode to compare its activity with that in an NaCl solution of about the same ionic strength (e.g., Ref. 31). Alternatively, with less precision, $y_{Cl}$ in the physiological solution can be taken to be equal to that in NaCl of the same ionic strength.

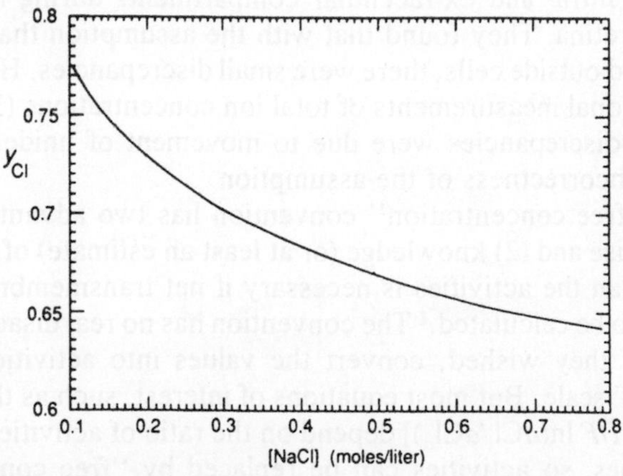


FIG. 2 A graph of the activity coefficient of chloride ($y_{Cl}$) in pure solutions of NaCl at 25°C drawn through values calculated from the table of measured values in B.R. Staples, Certificate for Standard Reference Material 2201, United States National Bureau of Standards, Washington (1971).

[†]The recommended symbol is $y$ if concentration is given, as physiologists usually do, in millimoles/(liter of solution) (35); $\gamma$ is used for the molal scale.

## An Alternative Convention

Many physiological publications use an experimentally more convenient scale in which the electrode is calibrated in terms of the concentration of the ion in the calibration solutions, rather than its activity. As Tsien (37) pointed out, this is equivalent to defining the activity coefficient in the Ringer solution as equal to unity ($y_{Cl} = 1$). This definition is heretical for physical chemists, so it needs to be made clear in the text: one convention is to call the values obtained in this way "free concentrations." This would be strictly true for intracellular $Cl^-$ only if $y_{Cl}$ in the cytoplasm were equal to that in the Ringer solution. In a careful study on smooth muscle, Aickin and Brading (31) found that with this assumption the value of $aCl_i$ measured by an ion-selective electrode agreed closely with those calculated from ion analysis and $^{36}Cl$ efflux, not only for normal extracellular $Cl^-$ activity ($aCl_o$) but for a range of values of $aCl_o$. It therefore seems likely that $Cl^-$ is uniformly distributed in the cell water, where it has an activity coefficient equal to that in Ringer solution. Other possibilities cannot be totally excluded. For example, that $Cl^-$ is distributed in only part of the cell water, where it has a low activity coefficient. In a different approach, Coles *et al.* (38) tried to check the conservation of mass and charge as $Na^+$, $K^+$, and $Cl^-$ moved between intra- and extracellular compartments during light stimulation in the bee retina. They found that with the assumption that $y_{Cl}$ was the same inside and outside cells, there were small discrepancies. However, supported by additional measurements of total ion concentrations (39), they suggested that the discrepancies were due to movement of unidentified ions, rather than to incorrectness of the assumption.

The "free concentration" convention has two advantages: (1) it is easy and precise and (2) knowledge (or at least an estimate) of the concentrations rather than the activities is necessary if net transmembrane movements of $Cl^-$ are to be calculated.[‡] The convention has no real disadvantages. Readers could, if they wished, convert the values into activities on the physical chemists' scale. But most equations of interest, such as the Nernst equation [$E_{Cl} = RT/F \ln(aCl_i/aCl_o)$] depend on the ratio of activities rather than absolute values, so activities can be replaced by "free concentrations" without error.

[‡]It should be borne in mind that because $Cl^-$ movements are usually associated with changes in cell volume, calculation of a change in the quantity of intracellular $Cl^-$ requires knowing the change in cell volume (2, 38).

### *Calibration Solutions*

The zero of the potential scale for the reference barrel is taken as that in the defined physiological solution (Ringer, Krebs, etc.) that bathes the biological preparation. It is convenient to define the ion voltage signal as zero in this solution too. The other calibration solutions are designed to mimic the cytosol. In the absence of better information about the cytosol, these solutions could consist of mixtures of Ringer solution and a solution of the same ionic strength of the $K^+$ salt of a noninterfering anion. The values of $[Cl^-]$ should be chosen to allow a curve to be drawn that includes the expected range of $[Cl^-]_i$. Some noninterfering anions are listed in Table I and are discussed below in the section "Replacement Anions"; gluconate and glucuronate are commonly used. Determination of a complete calibration curve extending to 0 $Cl^-$ is usually only necessary when experiments with 0 $Cl^-$ solutions are undertaken.

## Use of $Cl^-$ Microelectrodes

### *Response Time*

As far as is known, the response time of a $Cl^-$ microelectrode, like that of a $K^+$ electrode (40), is determined almost entirely by the electrical time constant of the electrode and amplifier. This is approximately given by $RC$, where $R$ is the resistance of the electrode and $C$ is the sum of the stray capacitances of the electrode, the wire to the amplifier, and the input capacitance of the amplifier. $C$ is usually dominated by the amplifier capacitance and the capacitance of the length of electrode that is immersed in fluid or tissue. If a short response time is important (<1 sec), then $C$ should be reduced by keeping the bath shallow, and partially compensated by electronic feedback. $R$ could be reduced by reducing the length of the sensor column, but little is gained until the length is below about 100 $\mu$m (41). The effective length of the column can be reduced by making electrical contact with the sensor through a sharpened Ag wire advanced close to the electrode tip (42).

### *Coping with Interfering Ions*

It is always necessary to consider interference with respect to a particular sensor, a particular cell, and a particular experiment, and often with respect to a particular microelectrode. Intracellular interfering ions are much more of a nuisance in cells, such as many types of neurons, that maintain $a\mathrm{Cl_i}$ at values lower than the equilibrium value, and conversely, they can often be shown to have a negligible effect in cells, such as many types of muscle cells, that maintain a high $a\mathrm{Cl_i}$. The indicative values of selectivity coefficients given in Table I show that, for the sensors that have been used in microelectrodes, $HCO_3^-$ is likely to be the main biological interfering ion. Some experiments can be done in the absence of exogenous $HCO_3^-$ and $CO_2$, but because $Cl^-$ transport is often linked to $HCO_3^-$ transport (11) and $Cl^-$ conductances are also permeable to $HCO_3^-$ (14), the results must then be interpreted with caution. The situation may change dramatically for the better if the new carrier molecules ETH 9009 and ETH 9011 (27) can be used in microelectrodes.

### *Measurement of Baseline* $a\mathrm{Cl_i}$

When a cell is bathed in 0 $Cl^-$ solution, the reading given by an intracellular $Cl^-$ electrode decreases over tens of minutes to a steady value. For ion-exchanger electrodes, the steady value is usually equivalent to a few millimoles/liter of $Cl^-$. Because $a\mathrm{Cl_i}$ is presumably zero under these conditions, the residual reading is attributed to interfering ions (4, 31, 43). Some possible interfering ions are listed in Table I and others are in the extensive table given by Alvarez-Leefmans *et al.* (2). If it is assumed that the concentration of interfering ions does not change when the external $Cl^-$ is changed, then a correction can be made.

## Replacement Anions

In the calibration solutions, and in many experimental solutions, it is necessary to replace some or all of the $Cl^-$. Anions that produce little interference with an anion-exchanger electrode, and which are readily available as salts, include gluconate, glucuronate, $SO_4^{2-}$, and methyl sulfate (see Table I). When a normal $[Cl^-]$ solution perfusing a chamber is replaced by a solution in which a substantial proportion of the $Cl^-$ is replaced by one of these ions, a junction potential on the order of 10 mV is set up at the interface between the two solutions. This potential will appear on the voltage recorded by the reference barrel for the time that the interface is moving between the reference barrel and the bath electrode. Normally this time is brief. However, if the recording is made in tissue at some depth from the superfusate, the potential can be present for some minutes, until the new solution has diffused to the recording site (4). In addition, a salt bridge containing 3 *M* KCl used as a bath electrode is likely to suffer from changes in junction potential on the order of 10 mV, unless it is in good condition. Changes in junction potential can readily be measured against an RLIE electrode.

Another point to keep in mind in using these anions is that to a greater or lesser extent they all chelate $Ca^{2+}$. If a reduction in free [$Ca^{2+}$] has significant effects on the physiology of the preparation, then free [$Ca^{2+}$] in the solutions must be restored by adding an appropropriate extra quantity of $Ca^{2+}$ (44). If precision is necessary, then free [$Ca^{2+}$] can be measured with a $Ca^{2+}$ electrode and total [$Ca^{2+}$] adjusted as required. In such measurements, an RLIE electrode makes a good reference electrode, with no change in junction potential, because the concentrations of $Na^+$ and $K^+$ remain constant.

## Analysis of a Typical Recording

Figure 3 shows three traces obtained from a double-barreled ion-exchanger $Cl^-$ microelectrode in a bee photoreceptor: apparent membrane potential, not corrected for junction potentials and possibly suffering from additional tip potentials; the potential of the active barrel, $V_{act}$; and the potential difference between the two barrels with the sign chosen so that more $Cl^-$ is upward [$V$(reference) − $V_{act}$]. This last trace is calibrated as [$Cl^-$] because the activity coefficient in the Ringer solution was defined as unity. Its accuracy depends on three assumptions: (1) that $y_{Cl}$ in the cytoplasm was equal to that in the calibrating solutions (this appears not to be wildly wrong in bee photoreceptors) (38, 39); (2) that the junction and tip potentials of the reference barrel were the same in the cytoplasm as in the calibration solution with the same [$Cl^-$]; and (3) that interfering ions did not contribute significantly. When bee photoreceptor cells were bathed in 0 $Cl^-$ solution, the residual intracellular signal indicated a contribution from interfering ions equivalent to 2.6 m$M$ $Cl^-$ (4); the ordinate scale has not been corrected for this. The middle trace shows $V_{act}$, the potential between the active barrel and the bath reference electrode. The electrode calibration curve was within 2 mV of Nernstian at concentrations down to the intracellular one, so this record should indicate the electrochemical potential of intracellular $Cl^-$ with respect to the bath to within 2 mV. It is seen that initially $V_{act}$ was negative, indicating that $Cl^-$ was transported into the cell against an electrochemical gradient. When the retina was stimulated with light, the cell depolarized and $V_{act}$ changed in the positive direction so that passive fluxes of $Cl^-$ would be into the cell, as shown in the third trace. It was a reasonable working hypothesis that the increase in [$Cl^-$]$_i$ at the onset of illumination was mainly due to a passive movement of $Cl^-$. The cell had a membrane potential less than the 50 mV

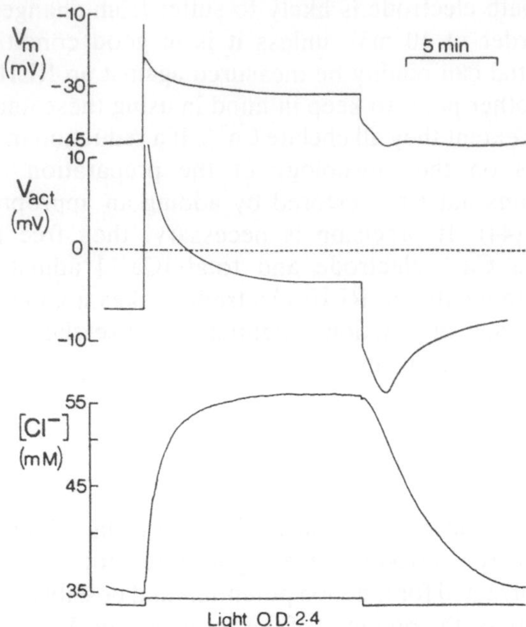


FIG. 3 Change in $aCl_i$ induced by light stimulation in a bee photoreceptor cell recorded with a double-barreled ion-selective microelectrode. The top trace is the membrane potential ($V_m$) recorded by the reference barrel. The second trace is the potential recorded by $Cl^-$-selective barrel ($V_{act}$) referred to the potential in the Ringer solution. The third trace is the difference between the first and the second ($V_m - V_{act}$). The ordinate scale is labeled $[Cl^-]$ because the activity coefficient in the Ringer solution was defined as unity and it is assumed that it was the same in the cytoplasm (see text, Calibration). The deflection on the bottom line indicates when the photoreceptors were stimulated with continuous light. Filters of optical density (O.D.) 2.4 were placed in the light beam to give an intensity in the physiological range. Recording by J. A. Coles and R. K. Orkand following methods described in Ref. 4.

that is usual in these cells, so it may have been leaky. During illumination, there was current flow between the bath and the cells, which created extracellular field potentials; $V_{act}$ has not been corrected for this.

The $[Cl^-]$ in the bee Ringer solution was 230.7 m$M$ so that at the beginning of the record $E_{Cl}$ was 25.3 $\ln_e(35/230.7)$ mV = −47.7 mV. Subtracting the apparent $V_m$ from this gives −2.7 mV. If everything were perfect, this would equal $V_{act}$, which it does not; correcting for errors mentioned above would reduce this difference.

## Conclusion

A $Cl^-$ microelectrode is a powerful and elegant tool that allows a direct correlation of electrical activity and changes in $aCl_i$. In large cells, such as snail neurons, the power can be increased by simultaneous use of an additional electrode to measure a second ion (e.g., Ref. 11). The reader should not be put off by excessive concern with junction and tip potentials: all the achievements of intracellular electrophysiology have been made despite them. Poor selectivity has been an occasional inconvenience of $Cl^-$ microelectrodes, but improvements have been made, and more are to be expected. Obtaining a clean, low-noise recording of $aCl_i$ in a small cell requires perserverance, but the experimenter has the satisfaction that success is at least partly related to skill.

## References

1. Y. Kondo, T. Bührer, K. Seiler, E. Frömter, and W. Simon, *Pfluegers Arch.* **414,** 663 (1989).
2. F. J. Alvarez-Leefmans, F. Giraldez, and J. M. Russell, *in* "Chloride Channels and Carriers in Nerve, Muscle and Glial Cells" (F. J. Alvarez-Leefmans and J. M. Russell, eds.), p. 3. Plenum, New York, 1990.
3. K. Ballanyi, P. Grafe, and G. ten Bruggencate, *J. Physiol.* (*London*) **382,** 159 (1987).
4. J. A. Coles, R. K. Orkand, and C. L. Yamate, *Glia* **2,** 287 (1989).
5. P. J. Boyle and E. J. Conway, *J. Physiol.* (*London*) **100,** 1 (1941).
6. F. J. Alvarez-Leefmans, *in* "Chloride Channels and Carriers in Nerve, Muscle and Glial Cells" (F. J. Alvarez-Leefmans and J. M. Russell, eds.), p. 109. Plenum, New York, 1990.
7. R. C. Hardie, *Nature* (*London*) **339,** 704 (1988).
8. M. G. Evans and A. Marty, *J. Physiol.* (*London*) **387,** 437 (1986).
9. A. Bahinski, A. C. Nairn, P. Greengard, and D. C. Gadsby, *Nature* (*London*) **340,** 718 (1989).
10. J. Zhang, R. L. Rasmussen, S. K. Hall, and M. Lieberman, *J. Physiol.* (*London*) **472,** 801 (1993).
11. R. C. Thomas, *J. Physiol.* (*London*) **354,** 3P (1984).
12. A. Soler, R. Rota, P. Hannaert, E. J. Cragoe, Jr., and R. P. Garay, *J. Physiol.* (*London*) **465,** 387 (1993).
13. B. J. Harvey and B. Lahlou, *J. Physiol.* (*London*) **370,** 467 (1986).
14. K. Kaila, B. Rydqvist, M. Asternack, and J. Voipio, *J. Physiol.* (*London*) **453,** 627 (1992).
15. T. O. Nield and R. C. Thomas, *J. Physiol.* (*London*) **242,** 453 (1974).
16. J. L. Munoz, F. Deyhimi, and J. A. Coles, *J. Neurosci. Methods* **8,** 231 (1983).
17. J. L. Munoz and J. A. Coles, *J. Neurosci. Methods* **22,** 57 (1987).

18. H. Widmer, S. Poitry, and M. Tsacopoulos, *J. Gen. Physiol.* **96,** 83 (1990).
19. L. Boksanyi, O. Liardon, and E. sz. Kováts, *Adv. Colloid Interface Sci.* **6,** 95 (1976).
20. K. Szabó, N. L. Ha, P. Schneider, P. Zeltner, and E. sz. Kováts, *Helv. Chim. Acta* **67,** 2128 (1984).
21. R. C. Thomas, *J. Physiol. (London)* **476,** 9P (1994).
22. J. A. Coles and M. Tsacopoulos, *J. Physiol. (London)* **270,** 13P (1977).
23. International Union of Pure and Applied Chemistry, *Pure Appl. Chem.* **48,** 127 (1976).
24. D. Ammann, "Ion-Selective Microelectrodes." Springer-Verlag, Berlin, 1986.
25. D. Wegmann, H. Weiss, D. Ammann, W. E. Morf, E. Pretsch, K. Sugahara, and W. Simon, *Mikrochim. Acta* **3,** 1 (1984).
26. C. M. Baumgarten, *Am. J. Physiol.* **241,** C258 (1981).
27. M. Rothmeier and W. Simon, *Anal. Chem. Acta* **271,** 135 (1993).
28. T. Hironaka and S. Morimoto, *J. Physiol. (London)* **297,** 1 (1979).
29. R. C. Thomas and C. J. Cohen, *Pfluegers Arch.* **390,** 96 (1981).
30. R. H. Adrian, *J. Physiol. (London)* **133,** 631 (1956).
31. C. C. Aickin and A. F. Brading, *J. Physiol. (London)* **326,** 129 (1982).
32. A. Mauro, *Biol. Bull. Mar. Biol. Lab. Woods Hole* **105,** 378 (1953).
33. B. R. Staples, Certificate for Standard Reference Material 2201, U.S. National Bureau of Standards, Washington, D.C., 1971.
34. B. R. Staples, Certificate for Standard Reference Material 2202, U.S. National Bureau of Standards, Washington, D.C., 1971.
35. Symbols Committee of the Royal Society, "Quantities, Units, and Symbols." The Royal Society, London, 1975.
36. R. G. Bates, B. R. Staples, and R. A. Robinson, *Anal. Chem.* **42,** 867 (1970).
37. R. Y. Tsien, *Annu. Rev. Biophys. Bioeng.* **12,** 91 (1983).
38. J. A. Coles, R. K. Orkand, C. L. Yamate, and M. Tsacopoulos, *Ann. N.Y. Acad. Sci.* **481,** 303 (1986).
39. J. A. Coles and R. Rick, *J. Comp. Physiol.* **156,** 213 (1985).
40. J. A. Coles and M. Tsacopoulos, *J. Physiol. (London)* **290,** 525 (1979).
41. E. Ujec, E. E. O. Keller, N. Kříž, V. Pavlík, and J. Machek. *Bioelectrochem. Bioenerg.* **7,** 363 (1980).
42. J. Janus and A. Lehmenkühler, *Pfluegers Arch.* **389,** R32 (1981).
43. R D. Vaughan-Jones, *J. Physiol. (London)* **295,** 83 (1979).
44. A. L. Hodgkin and P. Horowicz, *J. Physiol. (London)* **148,** 127 (1959).

# [19] Use of Ion-Selective Microelectrodes and Fluorescent Probes to Measure Cell Volume

Francisco J. Alvarez-Leefmans, Julio Altamirano, and William E. Crowe

## Introduction

### *Importance of Measuring Cell Volume Changes in Neurons and Other Cells*

A vital function of animal cells is their ability to maintain their volume constant in isosmotic media (such as the interstitial fluid) in the face of the Donnan effect. In addition, most cells regulate their volume when exposed to anisosmotic media (1, 2). Irreversible disruption of the membrane transport mechanisms underlying osmotic and ionic balance results in lethal cell injury (3), hence the importance of studying cell volume regulation and maintenance.

Cell volume regulation refers to the mechanisms whereby a cell is able to restore its volume to control levels during exposure to anisosmotic solutions. Two types of regulatory responses have been described: regulatory volume decrease (RVD) and regulatory volume increase (RVI). Cell volume regulation is usually studied *in vitro* by exposing cells to rapid changes in external osmolality. Most cells, including neurons, are seldom exposed to such osmotic challenges. However, this paradigm has proved useful to characterize the mechanisms of water transport and volume control in many cells (1, 2). Cell volume maintenance (CVM) refers to the mechanisms whereby cells keep their volume constant in isosmotic media. The study of CVM aims to elucidate the mechanisms of extrusion of fluid and osmolytes gained by the cells in isosmotic media, due to Donnan forces.

Neurons are particularly vulnerable to changes in intracellular and extracellular solute composition under both physiological and pathological conditions. This is a consequence of their small volume/surface ratio in conjunction with their incessant synaptic and action potential activity and the paucity of the extracellular space of the central nervous system (4). In spite of their physiological, clinical, and therapeutic implications, little is known about the cellular and molecular mechanisms with which nerve cells maintain their water volume within narrow limits in the face of the Donnan effect and substantial net solute and water fluxes across their plasma membranes (5–7).

*Methods in Neurosciences, Volume 27*

Changes in brain volume resulting from failure of cellular osmoregulatory mechanisms can lead to permanent neurological injury and death (8). Neuronal and glial cell swelling (cytotoxic edema), a dreaded complication of ischemia, trauma, seizures, or metabolic disorders, is believed to be the result of a loss of control of cell volume (9, 10). Hence it is important to develop appropriate methods to study cell volume.

## *Overview on Methods for Measuring Cell Volume Changes*

A variety of methods exist for measuring cell volume changes in cell populations and in single cells (11). In the present chapter, only some techniques developed or adapted in our laboratory for continuous measurement of rapid cell volume changes in single neuronal cells are considered. Obviously, these techniques may be used in cells other than neurons.

Current techniques for measuring cell volume changes in single cells fall within two categories: (a) those involving electrophysiological (12) or optical (13) measurement of changes in the concentration of impermeant substances introduced into the cells and (b) those based on morphometric methods, involving three-dimensional reconstruction of serial sections (14) or measurement of changes in one or two dimensions. In the present account we will concentrate on the former types of techniques. The reader interested in morphometric methods should consult the excellent work of Foskett (15) or Strange and Spring (16). These morphometric techniques yield measurements of total cell volume, which is the sum of cell water volume (CWV) and the cell solids volume (CSV). These techniques have poorer time resolution than the ones described below, they are better suited to large cells having simple geometry, and require more complicated and expensive equipment.

## *General Principles of the Techniques for Measuring Changes in Water Volume in Single Cells*

The techniques considered here are based on the general principle that relative CWV can be quantitatively assessed by introducing an impermeant probe, a volume marker, into cells and measuring its changes in concentration. If the intracellular content ($Q_m$) of the probe is constant, changes in its concentration ($C_m$) reflect changes in CWV [Eq. (1) and Fig. 1],

$$\mathrm{CWV} = Q_m/C_m. \tag{1}$$

For Eq. (1) to be valid, the probe should fulfill the following criteria: (a) it

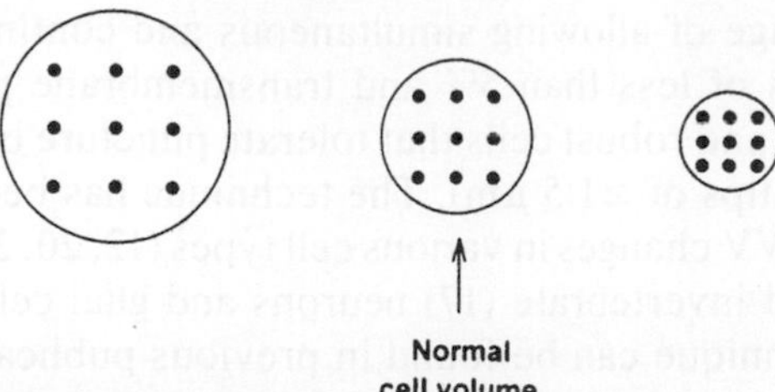


FIG. 1 Basic principle of the technique for measuring cell water volume by introducing an impermeant probe into cells. Each large circle represents a cell loaded with a volume marker (•). Left: the cell is in a hypo-osmotic solution. Middle: normal cell volume (the cell is in an isosmotic solution). Right: the cell is in a hyperosmotic solution. The intracellular content of the marker $Q_m$, represented by nine solid dots within the cell, remains constant. The cell water volume changes in each condition, and the concentration of the probe $C_m$ changes following Eq. (1). By measuring $C_m$, it is possible to determine the cell water volume.

must be confined to the cytosolic water space and it must be distributed homogeneously within this compartment; (b) it must not be transported by the membrane limiting the compartment, either by active or passive mechanisms, between this and other intracellular compartments or to the outside of the cell; (c) in addition, the probe must be chemically stable in the compartment, i.e., it must not be produced (synthesized) or consumed (metabolized) by the cell, and ideally it should remain unbound, i.e., it should be freely diluted in the cell water compartment; (d) the probe should be nontoxic and should not interfere with the mechanisms involved in cell volume regulation and maintenance. Two types of intracellular probes have been successfully used to measure changes in CWV: impermeant ions, such as tetramethylammonium ($TMA^+$) or choline, and fluorescent dyes, such as calcein and fura-2.

## Use of Tetramethylammonium to Assess Changes in Cell Water Volume

This technique was pioneered by Reuss (12) to study CWV in epithelial cells and later adapted to measure CWV changes in neurons (17). Tetramethylammonium is used as the volume indicator, and $TMA^+$-selective microelectrodes are used to monitor its intracellular activity during experimental perturbations (12, 17, 18). Others (19) have used choline or decamethonium instead of $TMA^+$, but the principle is the same. This technique has the unique advantage of allowing simultaneous and continuous measurement of CWV changes of less than 5% and transmembrane potential ($E_m$), but is limited to large and robust cells that tolerate puncture by double-barreled microelectrodes (tips of ≥1.5 μm). The technique has been successfully used to measure CWV changes in various cell types (12, 20, 21), including large vertebrate (18) and invertebrate (17) neurons and glial cells (22). Detailed accounts of the technique can be found in previous publications (17, 20).

## Preparation of $TMA^+$-Selective Double-Barreled Microelectrodes

Double-barreled microelectrodes are used to measure intracellular $TMA^+$ activity and transmembrane potentials. $TMA^+$-selective microelectrodes are based on the standard liquid cation exchanger commonly used to measure $K^+$ activity. This liquid cation exchanger is made up of 5 mg potassium tetrakis(*p*-chlorophenyl) borate (Fluka Chemie; Buchs Switzerland, product number 60591) dissolved in 0.1 ml 3-nitro-*O*-xylene (Aldrich; Milwaukee, Wisconsin, product number 13,030-3). Microelectrodes made with this liquid sensor are far more sensitive to $TMA^+$ than to $K^+$ or $Na^+$. Their selectivity ratios for $TMA^+/K^+$ are $10^2$ to $10^3$, and for $TMA^+/Na^+$, $10^3$ to $10^4$, respectively (17, 20). To prepare them (Fig. 2), segments of a 9.5-cm length of double-barreled laterally fused borosilicate glass with an inner filament (each barrel 1.0 mm OD, 0.43 mm ID; Hilgenberg, Malsfeld, Germany) are cleaned by partial immersion in nitric acid and addition of small aliquots of ethanol at intervals to maintain a vigorous but controlled evolution of $NO_2$ (caution: the reaction beaker should be less than one-tenth full and be placed in an efficient hood). The glass is rinsed and boiled in five changes of deionized water. After drying in an oven, each fused doublet is held on a horizontal puller (PD-5, Narishige, Tokyo), heated and twisted 360°, allowed to cool down for 20–30 sec, and then pulled (Fig. 2a–c). In order to adjust the puller settings, it is recommended to fill some micropipettes with 3 *M* KCl and check their tip resistances, which should be 50–100 MΩ in 0.1 *M* KCl. The back of one of the barrels is then broken by inserting a sharp blade between both barrels. The barrel with the long end of each microelectrode is inserted into one of several holes drilled across the top surface of a silanization chamber made from the bottom part of a 50-mm-diameter plastic Petri dish (Fig. 2d). Each of the holes is fitted with Teflon gaskets to establish a tight seal with the body of the micropipettes, preventing the vapors of the silanizing agent from escaping from the chamber. The hollowed cap containing the electrodes to be silanized is screwed on the top of a small bottle containing 3 ml DDS (dimethyldichlorosilane; Serva, Feinbiochemica, Heidelberg, Germany, product #20255). The DDS vapors are allowed to act for 3 to 5 min

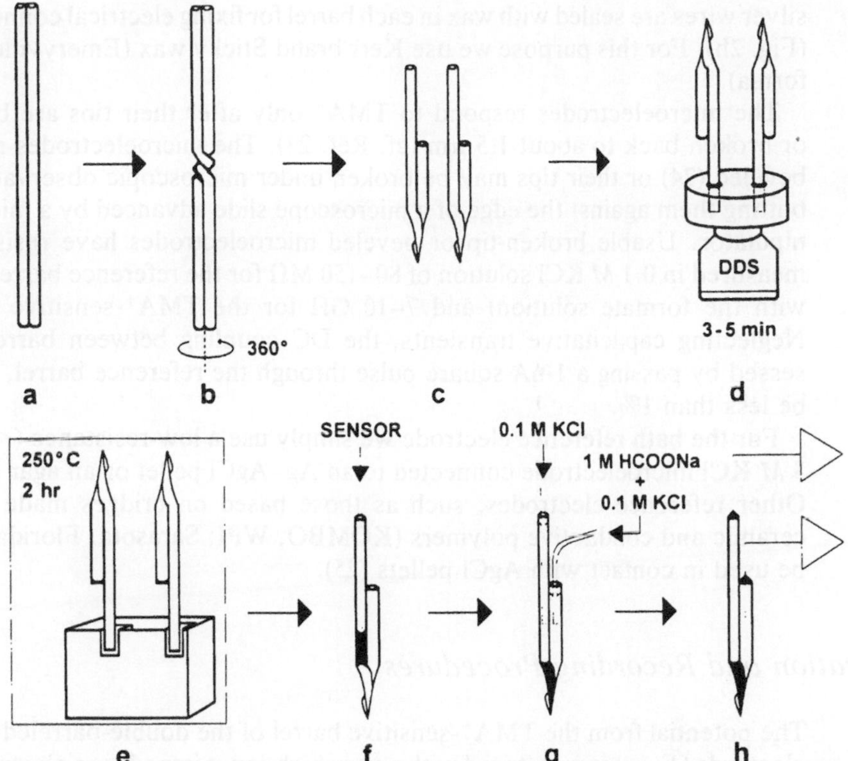


FIG. 2 Procedure for preparing double-barreled $TMA^+$-selective microelectrodes to measure cell water volume changes, as explained in the text. DDS, Dimethyldichlorosilane.

in a hood. The microelectrodes are then placed, tip upward, on the holes of an aluminum cube and transferred to a preheated (250°C) oven for 2 hr to complete the silanization of the long barrel (Fig. 2e). The cube with the pipettes is then removed from the oven and allowed to cool. The silanized barrel of each micropipette is injected with a drop of liquid cation exchanger, by means of a capillary tube pulled to approximately a 5-cm-long shank, with a diameter of about 75 to 100 $\mu$m, attached to a syringe (Fig. 2f). The remaining part of the barrel is similarly backfilled with 0.1 *M* KCl. The reference barrel, which is used to record the transmembrane potential ($E_m$), is then filled in a similar manner with a 1 *M* sodium formate/0.01 *M* KCl solution (Fig. 2g). It is important to make sure that all the electrolyte filling solutions are passed through Millipore filters (e.g., Millex-GS 0.22-$\mu$m filter unit; Millipore Corporation, Bedford, Massachusetts). When necessary, bubbles may be removed by gentle local heating with a microforge. Chloridized silver wires are sealed with wax in each barrel for fixing electrical connections (Fig. 2h). For this purpose we use Kerr brand Sticky wax (Emeryville, California).

The microelectrodes respond to $TMA^+$ only after their tips are beveled or broken back to about 1.5 $\mu$m (cf. Ref. 23). The microelectrodes may be beveled (24) or their tips may be broken under microscopic observation by butting them against the edge of a microscope slide advanced by a micromanipulator. Usable broken-tip or beveled microelectrodes have resistances measured in 0.1 *M* KCl solution of 80–150 MΩ for the reference barrel (filled with the formate solution) and 7–10 GΩ for the $TMA^+$-sensitive barrel. Neglecting capacitative transients, the DC coupling between barrels, assessed by passing a 1-nA square pulse through the reference barrel, should be less than 1%.

For the bath reference electrode we simply use a low-resistance (<1 mΩ) 3 *M* KCl microelectrode connected to an Ag–AgCl pellet or an agar bridge. Other reference electrodes, such as those based on bridges made out of ceramic and conductive polymers (KOMBO, WPI; Sarasota, Florida), may be used in contact with AgCl pellets (25).

## *Calibration and Recording Procedures*

The potential from the $TMA^+$-sensitive barrel of the double-barreled microelectrode ($E_{TMA}$) is monitored with a very high-input impedance electrometer (e.g., FD-223 WPI; Sarasota, Florida). The potential of the reference barrel ($E_{ref}$) is recorded using an ordinary electrometer for microelectrode work (e.g., Cyto 721 or Intra 767; WPI, Sarasota, Florida) and subtracted electronically from $E_{TMA}$ to give the differential signal ($E_{TMA} - E_{ref}$), which is proportional to [$TMA^+$]. The output of the electrometers is low-pass filtered (0–5.3 Hz), displayed on a chart recorder, digitized with an analog-to-digital converter, and displayed and stored on a PC computer for subsequent analysis. Each channel is usually sampled at 10 Hz. Data acquisition can be made with commercial software (e.g., Asyst; Macmillan Software Co., New York, New York). Recorded signals may be analyzed with customized programs or commercial software such as Sigma Plot (Jandel Scientific; San Rafael, California) to calculate changes in cell water volume (see below) and other parameters.

The $TMA^+$-sensitive microelectrodes must be calibrated before and after each impalement to assess drift and general performance. Calibration solutions contain 100 or 120 m*M* KCl, depending on cell type (e.g., for terrestrial invertebrates use 100 m*M* KCl and for vertebrate cells use 120 m*M* KCl) plus 0.1–20 m*M* TMACl. Slopes usually range between 53 and

65 mV/log[$TMA^+$], for [$TMA^+$] between 0.5 and 20 m*M*. Between 0.1 and 0.5 m*M* [$TMA^+$], slopes range from 25 to 35 mV/log[$TMA^+$]. In a previous study (17) it was found that the mean slope for the precalibration curves in the relevant range of intracellular $TMA^+$ concentrations (0.5–20 m*M*) was 59.6 ± 0.6 mV/log[$TMA^+$] ($n$ = 21).

Figure 3 shows a typical calibration of one such electrode. Given the selectivity ratios of these microelectrodes for $TMA^+/K^+$ ($10^2$ to $10^3$) and for $TMA^+/Na^+$ ($10^3$ to $10^4$) and the fact that resting $[Na^+]_i \approx$ 5–10 m*M* and $[K^+]_i \approx$ 100–120 m*M* for vertebrate cells or terrestrial invertebrates, for $[TMA^+]_i \geq 0.5$ m*M*, the ion-sensitive barrel is expected to respond only to $TMA^+$. However, during certain experimental treatments (e.g., pharmacological inhibition of the sodium pump) cells may gain $Na^+$ and $Ca^{2+}$ and lose $K^+$. Hence appropriate controls must be made under each experimental situation to avoid changes in the potential of the ion-sensitive barrel that may be erroneously interpreted as changes in [$TMA^+$], and hence in CWV. As expected for selectivity coefficient measurements, reducing [$K^+$] from 100 to 75 m*M* and increasing $Na^+$ from 5 to 30 m*M* against a background [$TMA^+$] of 2 m*M* give no measurable response of the ion-selective microelectrode (Fig. 3A). Possible interference from $Ca^{2+}$ on $TMA^+$-sensitive microelectrode response has also been assessed. Varying free [$Ca^{2+}$] between $10^{-7}$ and $10^{-4}$ *M* in EGTA-buffered solutions having a background of 100 m*M* $K^+$ plus 2 m*M* TMACl gave no measurable response of the $TMA^+$-sensitive microelectrode (17).

## *Loading of Cells with $TMA^+$ and Recording Procedures*

Depending on plasma membrane permeability to $TMA^+$, different techniques can be applied for loading cells with $TMA^+$. In some cells, the native membrane permeability for $TMA^+$ is undetectably small, e.g., *Necturus* gallbladder epithelial cells (12, 20). Hence, in order to introduce $TMA^+$ into the intracellular compartment, it is necessary to increase transiently the plasma membrane permeability to $TMA^+$. This is done by exposing the cells to a solution containing high [$TMA^+$] and the antibiotic nystatin (12). Details of this procedure can be found in the work of Reuss and colleagues (12, 20). Frog motoneurons have been loaded via $TMA^+$ diffusion from the reference barrel of a microelectrode (18). Muscle cells may be loaded by iontophoresis (19). In contrast, *Helix* neurons, leech glial cells, and retinal pigment epithelial cells are permeable to $TMA^+$ and hence they can be simply loaded by bath exposure (17, 19, 21). In all these procedures the final intracellular $TMA^+$ concentration, $[TMA^+]_i$, ranges from 0.5 to 20 m*M*. Within this range of $[TMA^+]_i$ the electrodes do not respond to changes in intracellular ions other

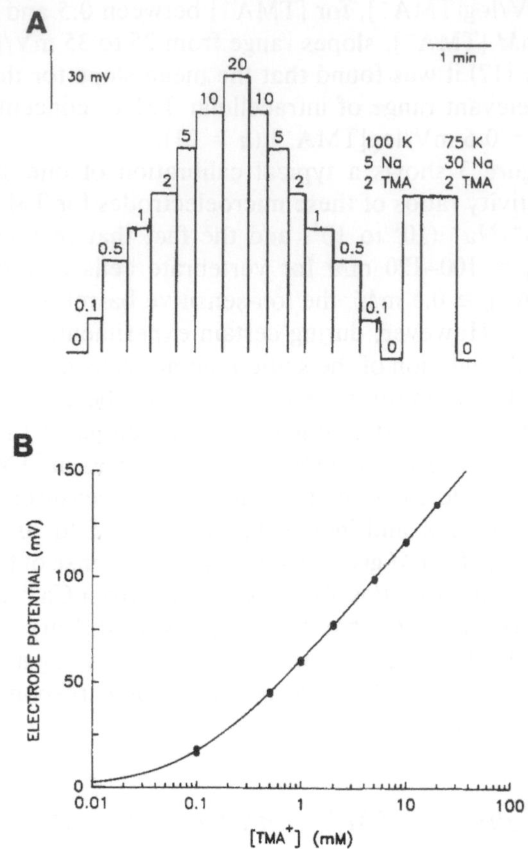


FIG. 3 Calibration and selectivity test of a double-barreled $TMA^+$-selective microelectrode, tip external diameter approximately 1 $\mu$m. (A) Differential signal $E_{TMA} - E_{ref}$ recorded in various calibration solutions. The numbers indicate the $TMA^+$ concentration (m*M*) of the solutions, which contained a background [KCl] of 100 m*M*. The differential signal recorded in the $TMA^+$-free solution containing 100 m*M* KCl is denoted 0. Just after the calibration test, the selectivity of the microelectrode for $K^+$ and $Na^+$ was assessed with solutions whose composition (m*M*) is indicated. (B) Calibration plot of electrode potential against $[TMA^+]$. The continuous line is a fit of the Nikolsky–Eisenman equation by nonlinear least-squares regression: $E = E_0 + S \log([TMA^+] + k_{TMA,K}[K^+])$, where $E$ is the electrode potential, $E_0$ is the constant reference potential, $S$ is the electrode potential change for a 10-fold increase in $[TMA^+]$ in the absence of interfering ions, and $k_{TMA,K}$ is the selectivity coefficient. Because ionic strengths were kept within 0.1–0.12 *M*, concentrations rather than activities were used. The linear part of the plot had a slope of 56.4 mV/log$[TMA^+]$ (modified from Ref. 17).

than $TMA^+$. A detailed account of $TMA^+$ loading of *Helix* neurons, which may be used for other neuronal types, has been described (17). Briefly, once calibrated, the double-barreled ion-sensitive microelectrode is positioned in the control bathing solution and the potentials with respect to the bath reference electrode are taken as zero. Then the cell is impaled. The transmembrane potential difference ($E_m$), measured with the reference barrel, is subtracted from the voltage recorded with the $TMA^+$-sensitive barrel ($E_{TMA}$), giving the differential signal ($E_{TMA} - E_m$), which indicates the $[TMA^+]_i$. Once the records are stable, cells are loaded with $TMA^+$ by exposure to the $TMA^+$-loading solution. The latter is made by isosmolar replacement of NaCl of the control solution (e.g., Ringer) with TMACl. We load the neurons to a final concentration of about 2 to 4 m*M*. Once the cells are loaded to the desired $[TMA^+]_i$ the bathing fluid is changed to a control solution to obtain readings of stable baseline. In *Helix* neurons $TMA^+$ is virtually trapped, as required for a good intracellular water marker. $TMA^+$ leaks out at a rate that results in a concentration change of about 0.5 m$M$ $hr^{-1}$. This gives a baseline on which volume changes, assessed from changes in $[TMA^+]_i$, can be continuously measured over periods of several minutes. The noise levels of the reference and the ion-sensitive barrels allow for measurements with a sensitivity of 1 mV or less in the differential voltage trace, even in spiking neurons. In the latter case measurements can be made in the interspike intervals. Cell volume changes of less than 5% can be detected. An example is shown in Fig. 4.

## *Calculation of Cell Water Volume*

The $TMA^+$ technique allows measurement of changes in the fraction of the total cell volume corresponding to solvent or osmotically active water. This fraction is referred to as the cell water volume, which of course does not include the so-called nonsolvent or structured water (26). For practical purposes the latter is considered to be part of the nonaqueous volume of the cell. In other words, the measured parameter with this technique is the difference between total cell volume and nonaqueous volume, the latter being the part of the cellular volume that does not participate in the experimentally induced osmotic swelling or shrinkage.

Normalized cell water volume changes ($V_t/V_0$) are computed from changes in $[TMA^+]_i$ elicited by exposure to anisosmotic solutions or agents such as neurotransmitters or inhibitors of the sodium pump, according to Eq. (2):

$$V_t/V_0 = [TMA^+]_0/[TMA^+]_t, \qquad (2)$$

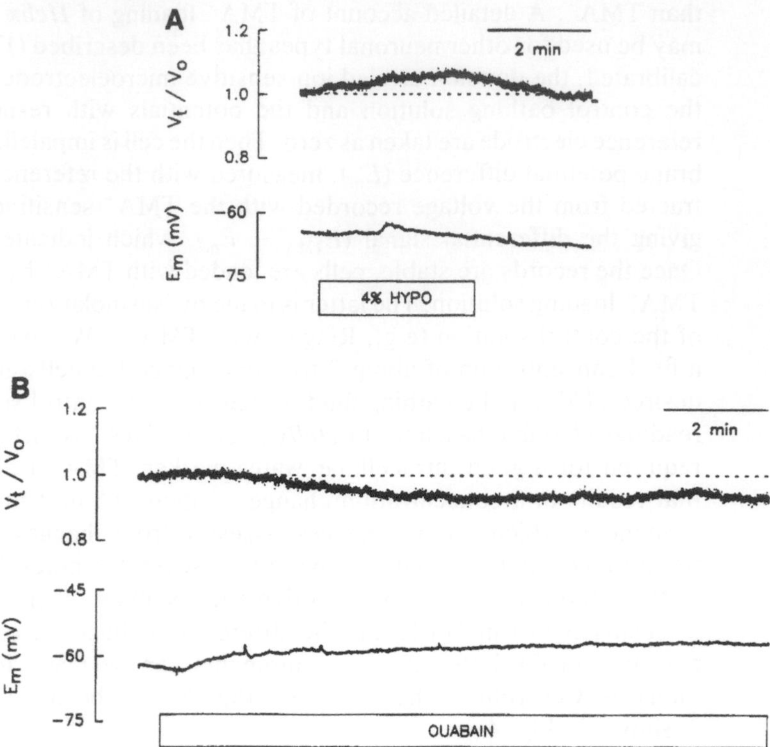


FIG. 4 Shrinkage of a nerve cell from *Helix aspersa* upon inhibition of the sodium pump with ouabain (1 m*M*). On each panel the upper trace corresponds to the normalized cell water volume ($V_t/V_0$) and the lower trace corresponds to the transmembrane potential ($E_m$). (A) exposure of the neuron (cell 77F) to a hypo-osmotic solution (4%), showing that the electrode was able to respond to as little as 4% increases in cell volume. (B) Ouabain-induced decrease in cell water volume. The cell shrank to 7% below its initial water volume (modified from Ref. 17).

where $V_t$ is the cell water volume at time $t$, $V_0$ is the initial cell water volume, $[TMA^+]_0$ is the intracellular $[TMA^+]$ at time $t = 0$, and $[TMA^+]_t$ is the intracellular $[TMA^+]$ at time $t$.

## *Pitfalls and Limitations of the TMA⁺ Technique*

### *Microelectrodes Are Invasive and Change Their Ion Selectivity*

The $TMA^+$ technique is invasive and is limited to large and robust cells that can tolerate puncture with double-barreled microelectrodes having tips of 1.5 $\mu$m or more. Microelectrodes with tips smaller than 1.5 $\mu$m are sensitive to $K^+$ but not to $TMA^+$ (23). Moreover, even large microelectrodes often lose their response to $TMA^+$ upon impalement.

*$TMA^+$ May Alter $K^+$ Conductances*

$TMA^+$ may interact with $K^+$ channels and thereby modify the normal cell physiology or some volume regulatory mechanisms. This is unlikely at least for $Ca^{2+}$-activated $K^+$ channels because they are blocked with a $K_d$ of about 65 m*M*, well above the intracellular concentration required for the technique (27).

*Possible $TMA^+$ Compartmentalization*

One of the assumptions of the technique is that $TMA^+$ is uniformly distributed in the cytosol and is not taken up or released by organelles upon osmotic perturbations. This implies that $TMA^+$ should not be compartmentalized within the cells. There is no final proof for this. Moreover, if $TMA^+$ is able to permeate the plasma membrane in many cell types there is no reason to believe that it does not permeate organelle membranes. To test the latter hypothesis, *Helix* neurons were permeabilized by exposing them to nystatin (100 $\mu$g $ml^{-1}$) and then were loaded with $TMA^+$. The kinetics of $TMA^+$ exit from the cell in the continued presence of nystatin was studied. The efflux of $TMA^+$, in the range from 10 m*M* down to 0.1 m*M*, followed a single exponential time course with a mean rate constant of 1.11 ± 0.01 $min^{-1}$ ($n$ = 3). This suggests that, at least in permeabilized and perhaps intact *Helix* neurons, the intracellular water volume in which $TMA^+$ is diluted behaves as a single compartment (17).

*Changes in $TMA^+$ Activity Coefficient*

Changes in the activity coefficient of intracellular $TMA^+$ during experimental manipulations may lead to errors in the estimation of $[TMA^+]_i$. For instance, dilution of intracellular electrolytes during osmotic swelling reduces the ionic strength so that the $[TMA^+]_i$ is overestimated because of a change in activity coefficient. We found (17) that, in calibration solutions mimicking the change in intracellular ionic strength produced by a large hypo-osmotic challenge, decreasing [KCl] from 100 to 60 m*M*, at a constant concentration (6 m*M*) of $TMA^+$, changes the $TMA^+$ electrode signal by 1.5 mV, yielding an error of 0.3 m*M* in the estimate of $[TMA^+]$.

## Use of Fluorescent Probes to Measure Volume Changes in Single Cells

Noninvasive techniques based on the measurement of changes in the concentration of intracellularly trapped fluorescent dyes, such as 2′,7′-bis(carboxyethyl)-5,6-carboxyfluorescein (BCECF) and fura-2, have been recently introduced to measure CWV changes in cultured cells (13, 28). The changes in CWV are inferred from readings of the fluorescence intensity recorded at the isosbestic wavelengths of BCECF or fura-2, which are insensitive to $pH_i$ or $[Ca^{2+}]_i$, respectively. Changes in fluorescence intensity that are proportional to changes in dye concentration are measured from a restricted plane of the cells (see below). The principle of the technique is that the dye concentration is inversely proportional to CWV. Hence, measurements of changes in fluorescence intensity at the isosbestic point report changes in CWV. The advantage of this technique is that it is possible to measure simultaneously CWV and either $[Ca^{2+}]_i$ or $pH_i$ changes in single cells with unequaled time resolution (0.1–1 Hz).

We have recently developed another technique for measuring CWV, using calcein as the fluorescent probe (29). Calcein, an intensely fluorescent dye derivative of fluorescein, can be easily loaded into single cells (see below). This new technique allows for continuous measurements of CWV changes to within 2%. Calcein is commercially available as acetoxymethyl (AM) ester (Molecular Probes, Eugene, Oregon, catalog #C-1430) and therefore it can be loaded into cells in a noninvasive manner (Fig. 5). The fluorescence intensity of free calcein is directly proportional to its concentration and,

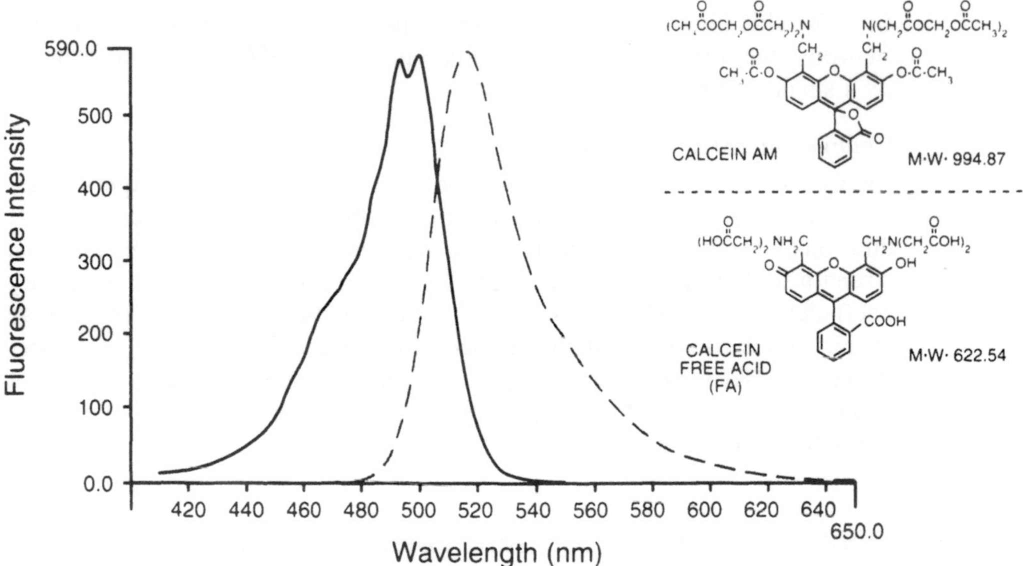


FIG. 5 Fluorescence excitation (solid curve, left) and emission (dashed curve, right) spectra of calcein free acid after hydrolysis of the acetoxymethyl (AM) form. The maximum excitation peak occurred at 497 nm wavelength and the peak emission was at 516 nm.

therefore, inversely proportional to cell volume; it is also independent of changes in the concentation of native intracellular ions, including $Ca^{2+}$, $Mg^{2+}$, and $H^+$, within the physiological range. Because calcein is two to three times more fluorescent than other fluorophores (such as BCECF or fura-2) and can be used at its peak excitation (497 nm) and emission (516 nm) wavelengths (Fig. 5), it has a better signal-to-noise ratio and baseline stability compared to the other dyes. Because of the possibility of reducing the intensity of the excitation light, measurements can be performed with minimal photodynamic damage. The technique can be applied to any cell type that can be grown or affixed to a coverslip. Because calcein properties are different from those of fura-2 or BCECF, this technique allows for independent measurements of CWV.

Calcein shares with other fluorophores a number of advantages over other available methods for measuring the volume of cells attached to rigid substrates: (a) It can be loaded into cells via its permeant hydrolyzable ester without disruption of the plasma membrane. Thus the fluorescence techniques can be used in cells too small or delicate to tolerate puncture. (b) The fluorescence readings can be calibrated with test solutions to yield semiquantitative volume measurements. (c) They are ideal for studying single cells, whereas other methods, such as light-scattering (30) or electrical resistance measurements (31), report average changes in volume of cell populations that may or may not be homogeneous or respond synchronously to osmotic challenges. (d) They have a fast response time. (e) All dyes and optical components needed for any of these techniques are commercially available (see below).

## *Fluorescence Measurements and Optical Setup*

Total fluorescence from a small region of fluorophore-loaded single cells is measured with a microspectrophotometry system, which includes an inverted, epifluorescence microscope (Diaphot-TMD; Nikon, Tokyo, Japan), equipped with a fluor oil-immersion objective, 40×, 1.3 NA lens (CF Fluor; Nikon, Tokyo, Japan). Figure 6 shows a diagram (numbered 1–12) of the setup. The excitation light source (1 in Fig. 6) we use is a 150-W xenon arc lamp (model 66006; Oriel, Stratford, Connecticut). To reduce infrared radiation the light from the excitation light source is passed through a water filter (2 in Fig. 6; model 61940; Oriel) and then it is divided by means of a beam splitter (3 in Fig. 6; parts 78150 and 38106, Oriel). Each beam passes through a spacer tube (4 in Fig. 6; part 7129, Oriel) and a computer-controlled high-speed shutter (5 in Fig. 6; Uniblitz; Vincent Associates, Rochester, New York). Attached to each shutter is a manual multiple filter wheel (6 in Fig. 6; model 77370, Oriel) holding a selection of excitation filters (7 in Fig.

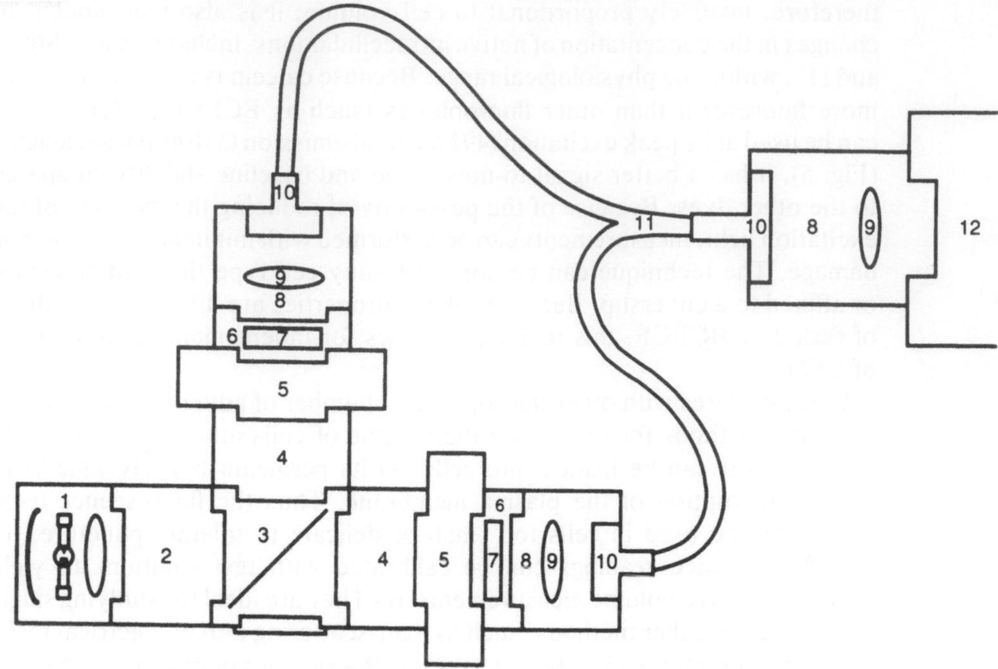


FIG. 6 Diagram of the setup for fluorescence measurements of cell water volume changes. See text for details.

6). Beam 1 passes through a filter centered at 495 ± 10 nm (Omega 495 DF 20; Omega Optical, Brattleboro, Vermont) for exciting calcein at its peak excitation wavelength. When measurements with BCECF are required, beam 2 passes through a filter centered at 440 nm (Omega 440 DF 20) for exciting BCECF at its isosbestic wavelength, and beam 1 passes through the filter centered at 495 nm. For measurements with fura-2, beam 1 passes through a filter centered at 380 ± 6 nm (Omega) and, for exciting fura-2, at its isosbestic wavelength, beam 2 passes through a filter centered at either 362 or 358 ± 5 nm (customized by Omega). When dual-excitation wavelength measurements are done, the filtered light beams are recombined by means of a bifurcated fused-silica fiber bundle (11 in Fig. 6; model 77565, Oriel). The combined excitation light beam is then collimated and coupled to the microscope epiilluminator (8–10 and 12 in Fig. 6).

Figure 7 shows the optical light path when using calcein as the fluorescent probe. Except for the filters and the use of dual-excitation wavelength, the setup and the light path are the same for fura-2 or BCECF measurements. To minimize photobleaching and photodynamic damage, the intensity of the excitation light beam is attenuated with a 10% transmission neutral-density filter (1.0 optical density, metal-coated, UV-grade quartz, Omega Optical).

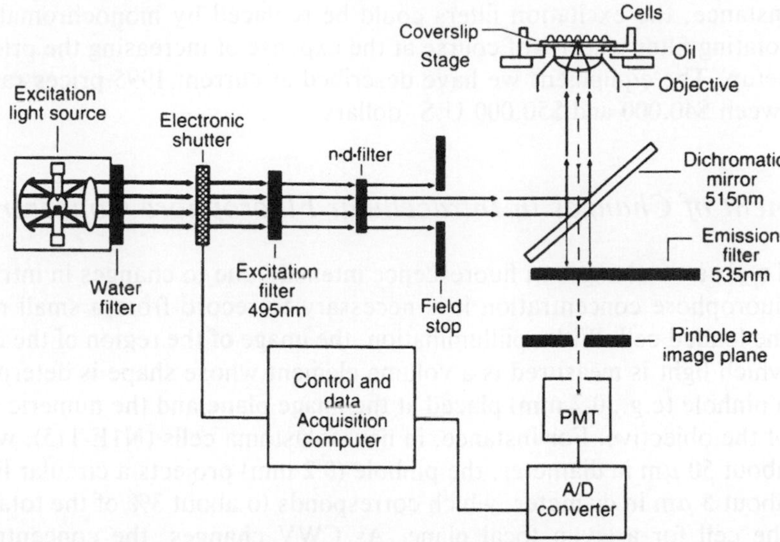


FIG. 7 Optical light path of the fluorescence system when using calcein-loaded cells to measure changes in cell water volume.

The field diaphragm is closed to its smallest opening to limit the excitation area to a minimum. This allows focusing the beam to a chosen cell. A cube holding a dichromatic mirror (515 nm) and a 535 ± 13-nm emission filter is positioned underneath the objective lens in the filter cassette holder of the microscope. The latter filters are used for either calcein or BCECF. For fura-2, a dichromatic mirror (400 nm) and a 510 ± 20-nm emission filter are used instead. The emitted light passes through a 1× relay lens coupled to a turret diaphragm accessory (Microflex PFX; Nikon, Tokyo, Japan) containing seven circular pinhole openings of varying diameter (0.1–10 mm). The whole ensemble is attached to the side port of the microscope.

The emitted light exiting through a selected pinhole is measured with a photometer system. The latter is composed of a photomultiplier (PMT) tube (model 77346, Oriel), a power supply (model 70705, Oriel) and a current preamplifier (model 70710, Oriel). Once amplified, the PMT signal is digitized with an analog-to-digital converter (Lab Master DMA; Scientific Solutions, Solon, Ohio), and displayed and stored on an IBM-compatible PC using customized software. The fluorescence data are collected at a frequency of either 0.1 or 0.2 Hz during shutter openings of 130-msec duration. Subsequent data analysis and presentation are performed using commercial software [e.g., Sigma Plot 5 and TableCurve 3.04 (Jandel, California) and Excel 3.1 (Microsoft, Redmond, Washington)].

The above described system is the one we have designed and built, but alternative components may be used instead of the ones suggested. For instance, the excitation filters could be replaced by monochromators or a rotating filter wheel—of course at the expense of increasing the price of the setup. The equipment we have described at current 1995 prices ranges between $40,000 and $50,000 U.S. dollars.

## *Measurement of Changes in Intracellular Fluorophore Concentration*

To observe changes in fluorescence intensity due to changes in intracellular fluorophore concentration it is necessary to record from a small region of the loaded cell. With epiillumination, the image of the region of the cell from which light is measured is a volume element whose shape is determined by a pinhole (e.g., 0.2 mm) placed at the image plane and the numeric aperture of the objective. For instance, in neuroblastoma cells (N1E-115), which are about 50 $\mu$m in diameter, the pinhole (0.2 mm) projects a circular light spot about 5 $\mu$m in diameter, which corresponds to about 3% of the total area of the cell for a given focal plane. As CWV changes, the concentration of the fluorophore molecules changes in inverse proportion. Thus, changes in fluorophore concentration result in changes in fluorescence intensity. Provided that the concentration of the fluorophore and the pathlength are not too large (see below), the emitted fluorescence is directly proportional to the concentration of the fluorophore, and the following simplified expression (32) can be derived from the Beer–Lambert exponential law:

$$F = 2.3\phi r\varepsilon cd, \tag{3}$$

where $F$ is the fluorescence intensity for a constant light input, $\phi$ is the quantum efficiency of the fluorophore, $r$ is a constant that represents optical instrumental factors, $\varepsilon$ is the extinction coefficient, $c$ is the concentration of the fluorophore, and $d$ is the thickness of the sample, i.e., the pathlength. Equation (3) is valid to within 4% provided that the product $\varepsilon cd$ is $<0.17$ (33).

## *Cell Loading*

Fluorochrome ester-loading procedures vary depending on fluorochrome, cell type, and the presence of intracellular esterases (34). Our experience is derived from some murine cell lines maintained in culture, particularly neuroblastoma X glioma NG108-15 cells, and neuroblastoma N1E-115 cells. Some neuronal cell types do not retain the fluorochrome, probably because of the lack of esterases, e.g., *Helix* neurons. In these cases the free acid of each fluorochrome may be directly microinjected (35). We grow the cells in culture dishes containing 25-mm-diameter glass coverslips (Bellco Glass Inc., 1943-22222). The strategy we use for loading calcein-AM, fura-2-AM, or BCECF is as follows. Each coverslip containing the cells is mounted in a Leiden dish (Medical Systems Corp., Greenvale, New York) and placed on the stage of the epifluorescence inverted microscope. Cells are loaded with calcein by incubation for 10 to 30 min in a standard extracellular solution (e.g., Krebs) containing 2 $\mu M$ calcein-AM (Molecular Probes, Eugene, Oregon). To prepare this loading solution we use a stock containing 6.28 m$M$ calcein [50 $\mu$g calcein-AM in 4 $\mu$l dimethyl sulfoxide (DMSO) plus 4 $\mu$l Pluronic (10% w/w in DMSO)]. Cell loading is monitored fluorometrically. The loading solution is washed out and replaced with standard extracellular solution (SES) when fluorescence of the cell reaches 30 to 40 times the autofluorescence level (i.e., the cell fluorescence without any dye). This usually takes 15 to 30 min. In the SES the fluorescence of the cell continues to rise, probably until all the dye is cleaved. The final cell fluorescence at this stage should be about 100 times the autofluorescence level. The time elapsed between the wash-out of the loading solution and the beginning of an experiment is about 1 hr. In experiments using fura-2 or BCECF, cells are also loaded with the acetoxymethyl ester derivatives, dissolved in standard extracellular solution at a final concentration of 5 $\mu M$. BCECF is added to the loading solution from a 10 m$M$ stock solution in which the fluorochrome is dissolved in DMSO (50 $\mu$g BCECF-AM in ~6 $\mu$l DMSO). Fura-2 is added to the loading solution from a 6.23 m$M$ stock solution containing 50 $\mu$g fura-2 dissolved in 4 $\mu$l DMSO plus 4 $\mu$l Pluronic (10% w/w) in DMSO. The loading time for either probe is 30 to 60 min. Cell loading with the above fluorophores is performed at room temperature. After loading with any of the fluorophores, all solutions are perfused at a rate of 3 ml/min (fluid is exchanged in our current chamber with a time constant of 3.6 ± 0.3 sec).

## *Relation between Relative Fluorescence and External Osmolality of Fluorophore-Loaded Cells*

The technique for measuring relative CWV changes based on changes in calcein fluorescence has been validated using three different murine cell lines: $C_6$ glioma cells, neuroblastoma-glioma NG108-15 cells, and neuroblastoma N1E-115 cells. Here we will outline the method we have followed to calculate CWV changes from the measured changes in fluorescence. We will illustrate the use of the method only for the case of calcein, but similar procedures should be followed for fura-2 or BCECF.

First, to calibrate the system it is important to start by characterizing the relation between relative fluorescence and external osmolality of fluorophore-loaded cells. The osmotic behavior of fluorophore-loaded cells may be investigated by measuring the changes in relative fluorescence resulting from brief (5–10 min) exposure to anisosmotic solutions having osmolalities in the range between −20 and +20% relative to that of the control solution (i.e., the changes in osmolalities should range between 80 and 120% of that of the control solution). Within this range of external osmolalities and exposure times, the cell types we have tested exhibit ideal osmometric behavior. This should be tested for other cell types because there are reports (36) of cells whose volume regulatory mechanisms start to be activated with small increases in cell volume (i.e., 5–7%). However, from our own experience, even when the osmolality is altered to such a degree that cells regulate their volume, the peak of the volume response often falls within the value predicted for an osmometric response (Fig. 9 inset). This is probably due to the fact that with a sudden change in osmolality, cells change their volume to the value predicted by an osmometric response and the regulatory mechanisms are activated after some delay.

An example of the relation between external osmolality and changes in fluorescence is shown for the case of $C_6$ glioma cells. This cell type constitutes a particularly useful example because it allows comparisons to be made with published data obtained with independent techniques (Fig. 8). The data obtained with calcein fluorescence (Fig. 8B) is consistent with published measurements (37) in which cell volume was inferred from light scattering or measured with optical sectioning methods (Fig. 8A), thereby further validating the calcein technique.

Figure 8B shows the osmotic behavior of calcein-loaded single $C_6$ glioma cells investigated by measuring changes in relative fluorescence ($F_t/F_0$) resulting from exposure to anisosmotic solutions having measured osmolalities ranging between −21% (hypo-osmotic) and +21% (hyperosmotic) relative to that of the control solution. Exposure to these solutions resulted in changes in $F_t/F_0$ that reached an apparent steady state within 5 min after the onset of the osmotic challenge (Fig. 8B, inset). As expected (17, 30), within the exposure times and range of osmolalities tested, the cells did not show regulatory volume responses. On returning to the isosmotic solution, the fluorescence signal returned to its initial value. The reciprocal of the apparent steady-state changes in fluorescence ($F_0/F_t$) plotted as a function of the reciprocal of the relative osmotic pressure of the medium ($\pi_0/\pi_t$) are shown in Fig. 8B. The linear relationship between these two variables (correlation coefficient = 0.99) suggests that the changes in fluorescence reflect changes in intracellular concentration of calcein, which in turn reflect changes in cell volume. The slope of the regression line fitted to the data points was 0.34, instead of 1, the slope expected for ideal osmometric behavior (dotted line). The latter line was calculated from Eq. (4):

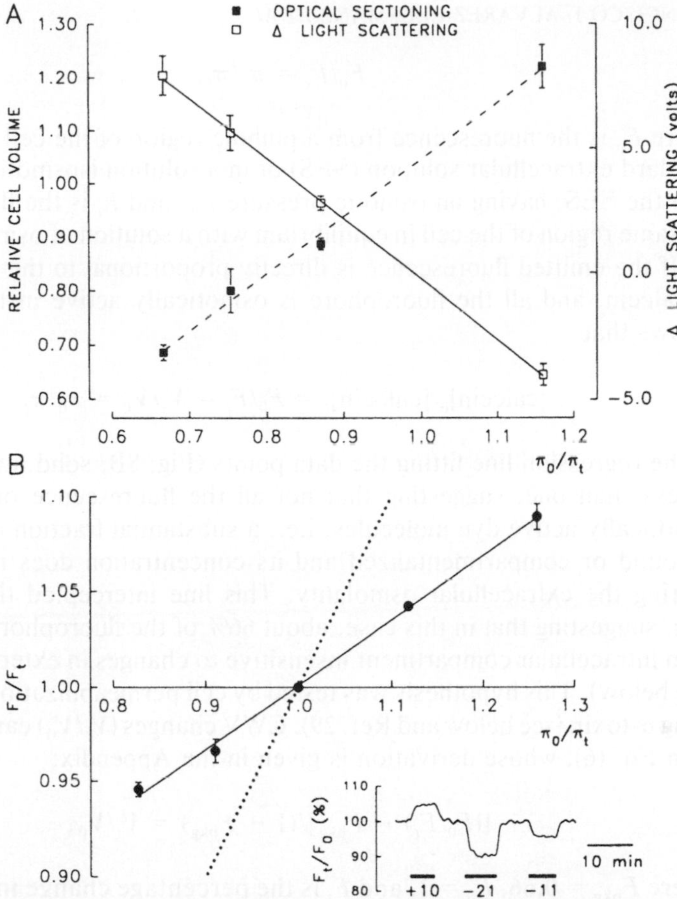


FIG. 8 Osmotic behavior of $C_6$ glioma cells in response to anisosmotic challenges, measured with three different techniques. (A) Relationship between relative cell volume (■, dashed line), maximal change in light-scattering signal (□, solid line), and the reciprocal of the external osmotic pressure $\pi_0/\pi_t$. Values are means ± SE; $n = 3$–11. Data kindly provided by Dr. Kevin Strange (see Ref. 37). Absolute changes in cell volume were determined by optical sectioning methods (37). Note that there is a linear relation between relative cell volume and $\pi_0/\pi_t$. The dashed line has a unity slope because the data were corrected for the volume of solids. Light scattered by a cell is directly proportional to the concentration of intracellular scatterers (30), and thus cell shrinkage or swelling causes increases or decreases, respectively, in the light-scattering signal, which is also linearly related to $\pi_0/\pi_t$, as predicted for an osmometer. (B) Relationship between the reciprocal of the steady-state changes in fluorescence ($F_0/F_t$) and $\pi_0/\pi_t$ in $C_6$ glioma cells loaded with calcein. The slope of the regression line through the solid circles was 0.34. Correcting for the background fluorescence ($F_{bkg} = 0.66$) according to Eq. (A14) yielded the dotted line having unity slope. The line predicting the behavior of a perfect osmometer according to Eq. 4 overlapped with the dotted line in which data were corrected for the bound dye. Inset shows sample records of changes in percentage relative fluorescence ($F_t/F_0$) in a cell exposed to a series of solutions having relative osmolalities as indicated in each bar.

$$F_0/F_t = \pi_0/\pi_t, \tag{4}$$

where $F_0$ is the fluorescence from a pinhole region of the cell in an isotonic standard extracellular solution (SES) or in a solution isosmotic and isotonic with the SES, having an osmotic pressure $\pi_0$, and $F_t$ is the fluorescence of the same region of the cell in equilibrium with a solution of osmotic pressure, $\pi_t$. If the emitted fluorescence is directly proportional to the concentration of calcein, and all the fluorophore is osmotically active in the cytosol, it follows that

$$[\text{calcein}]_0/[\text{calcein}]_t = F_0/F_t = V_t/V_0 = \pi_0/\pi_t. \tag{5}$$

The regression line fitting the data points (Fig. 8B, solid line) has a slope of less than one, suggesting that not all the fluorescence originates from osmotically active dye molecules, i.e., a substantial fraction of fluorophore is bound or compartmentalized and its concentration does not change by altering the extracellular osmolality. This line intercepted the ordinate at 0.66, suggesting that in this case, about 66% of the fluorophore was trapped in an intracellular compartment insensitive to changes in external osmolality (see below). This hypothesis was tested by cell permeabilization experiments using α-toxin (see below and Ref. 29). CWV changes ($V_t/V_0$) can be computed from Eq. (6), whose derivation is given in the Appendix:

$$[(F_0/F_t) - F_{\text{bkg}}]/(1 - F_{\text{bkg}}) = V_t/V_0, \tag{6}$$

where $F_{\text{bkg}} = 0.66$, $F_0 = 1$, and $F_t$ is the percentage change in fluorescence at time $t$, divided by 100. Correcting the data for the fraction of the dye insensitive to changes in external osmolality, the straight line fitting the data overlaps with the line describing ideal osmometric behavior in Fig. 8B.

For microspectrofluorimetric measurement of cell water volume using fura-2 or BCECF, the procedure is the same as for calcein, except that the change in dye concentration is measured at the respective isosbestic wavelengths.

## *Assessment of Trapped Fluorescence*

Esterified dyes are sequestered or bound to more than one intracellular compartment (29, 38, 39). The nonideal osmometric responses on exposure of a cell to anisosmotic calibration solutions in the range between +20 and −20% relative to that of the control solution (e.g., Fig. 8B, solid line), the stepped efflux of the dye upon cell permeabilization, and the histological images (29) all support the hypothesis that a substantial fraction of fluorophore is intracellularly bound or sequestered. This fraction of the dye does not behave as though it were dissolved in the cytosolic water space and is not osmotically active, i.e., its concentration does not change with the external osmolality. Changes in relative fluorescence of osmotically active calcein (or other fluorophores) can be used to assess quantitatively changes in cell water volume using Eq. (6), only if (a) the contribution to the total fluorescence of the osmotically inactive fraction of the fluorophore (background fluorescence) is known and (b) if it is assumed that changes in the fluorescence signal only reflect changes in dye concentration in the osmotically active cytosolic compartment.

Dye compartmentalization is usually assessed by permeabilizing the plasmalemma of cells with detergents such as digitonin. However, cell permeabilization with digitonin (or other detergents) is known to be difficult to control. Even at low concentrations (e.g., 10 $\mu M$), digitonin permeabilizes not only the plasma membrane, but also internal membranes and may even lead to release of intracellular organelles or their contents (40, 41). We have found that the background fluorescence assessed in cells permeabilized with digitonin is underestimated (29).

A better alternative to digitonin for selective plasma membrane permeabilization is $\alpha$-toxin (40, 41), a protein secreted by *Staphylococcus aureus*. This toxin forms hydrophilic pores 1–3 nm in diameter in the plasma membrane of different cell types (42). The molecular mass cut-off of material passing through the pore is approximately 2–4 kDa. Because of its large molecular mass (33 kDa) and the small size of the pores it forms (1–3 nm), under controlled conditions it initially permeabilizes only the plasma membrane, without damage to intracellular organelles. Cells become permeabilized to small molecules, whereas large enzymes, carbohydrates, and organelles are excluded. Therefore $\alpha$-toxin (List Biological Laboratories, Inc., Campbell, California) is a better plasmalemmal permeabilizing agent than digitonin for selective release of the free cytosolic fraction of the free acid forms of calcein, fura-2, or BCECF, whose molecular masses are 623, 832, and 520 Da, respectively.

We have found that the fraction of trapped or bound fluorophore assessed with $\alpha$-toxin is fairly constant between cells of the same type and corresponds to the $y$ intercept of a plot of $F_0/F_t$ versus $\pi_0/\pi_t$ (see Appendix and Fig. 8B). However, each time a new cell type is to be studied, we recommend assessing the fluorescence of the dye bound by permeabilization experiments and comparing the results with those obtained graphically.

In N1E-115 we have estimated that only about 33% of the intracellular calcein behaves as osmotically active and that 67% is trapped. Part of this trapped dye can be released by plasma membrane disruption, leaving a fraction, corresponding to about 22% of the total fluorescence, that is irreversibly bound or trapped in a vesicular-like compartment. Our estimates are in accord with those reported for other fluorophores (38, 43, 44).

## *Regulatory Volume Decrease Monitored in Single Calcein-Loaded Cells*

On exposure to hyposmotic media, neuronal cells, like most animal cells, initially swell and then return to their initial volume, a phenomenon termed regulatory volume decrease (RVD) (17, 45, 46). The mechanisms underlying RVD in neuronal cells have not been studied in detail. In most cells RVD is accomplished by the efflux of osmotically active solutes that fall within two categories: the inorganic ions, $K^+$ and $Cl^-$, and organic molecules, such as free amino acids, polyols, betaine, and other amines (1, 47–50). We have tested the effect of a 40% hypo-osmotic solution on water volume of calcein- or fura-2-loaded NG108-15 and N1E-115 cells. Both cell types respond in an identical manner with a typical RVD. An example is shown in Fig. 9. The upper trace shows the monitored values of $F_t/F_0$ and the lower trace shows the relative CWV, $V_t/V_0$. The bar indicates the time during which a 40% hypo-osmotic solution was applied. The cell swelled to a maximum of 60% above its control CWV. The maximum swelling occurred at 4.8 min; thereafter, RVD ensued at an initial rate of $-2.2\%$ min$^{-1}$ with partial (90%) recovery at 37 min. Then the bathing solution was switched back to the control isosmotic medium and, as expected, the cell shrank, recovering its initial volume in more than 60 min after returning to the isosmotic solution. Before testing the effect of the 40% hypo-osmotic solution, the cell was exposed to a hyperosmotic (13%) and a hypo-osmotic (8%) calibration test solutions. The corresponding changes in $V_t/V_0$ are shown in the inset, which also includes a plot of the relationship between the steady-state values of $V_t/V_0$ and the reciprocal of the relative osmotic pressure of the medium ($\pi_0/\pi_t$). The line denotes the predicted behavior of a perfect osmometer. The filled circles correspond to the osmotic calibration pulses. Note that the measurements fall close to the theoretical line for a perfect osmometer. The open circle corresponds to the peak amplitude of the osmotic response produced by the 40% hypo-osmotic solution. Note that the peak amplitude of this response also falls within the values expected for a perfect osmometer, before measurable RVD ensues.

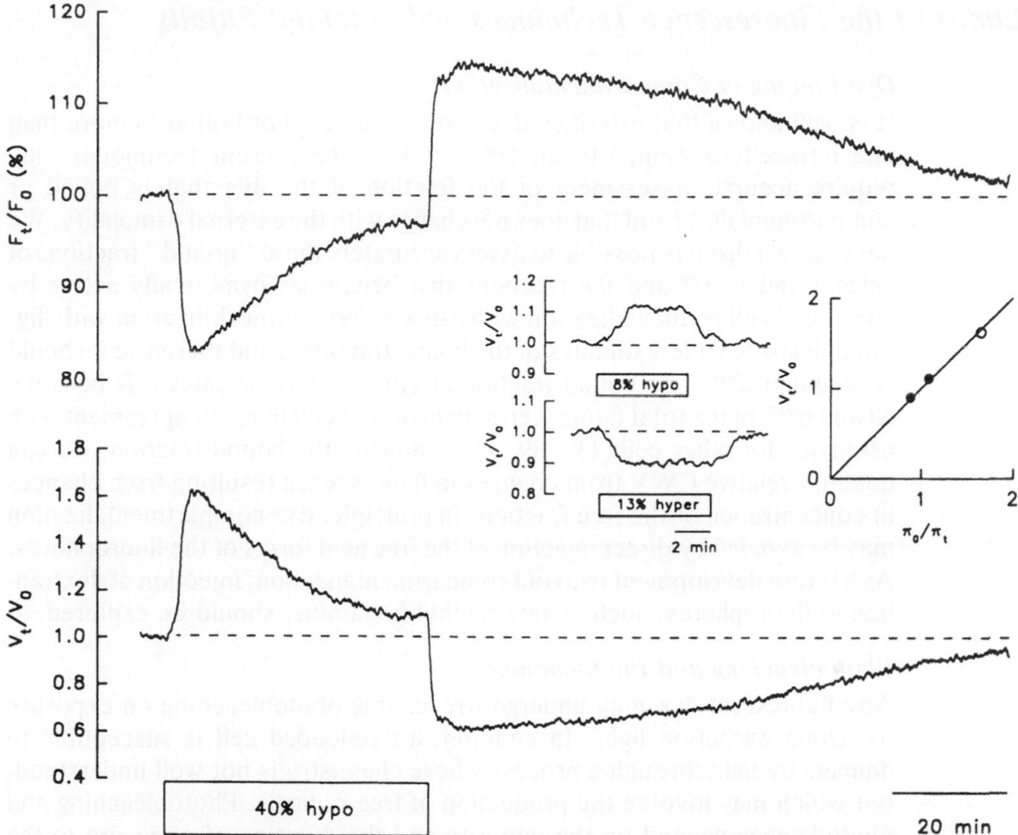


FIG. 9 Regulatory volume decrease monitored in a single murine neuroblastoma X glioma hybrid NG108-15 cell loaded with calcein. Upper trace: percentage relative fluorescence ($F_t/F_0$). Lower trace, changes in cell water volume ($V_t/V_0$) computed after background correction using Eq. (6); $F_{bkg} = 0.7$. Inset: the cell was maintained in isosmotic solution and two calibration pulses were applied before the 40% hypo-osmotic solution. Calibration pulses were 8% hypo-osmotic and 13% hyperosmotic. The relationships between the steady-state relative cell water volume ($V_t/V_0$) and the reciprocal of the relative osmotic pressure of the medium ($\pi_0/\pi_t$) for the calibration pulses are plotted in the insert graph. The line denotes the predicted behavior of a perfect osmometer. The filled circles correspond to the osmotic calibration pulses after background subtraction. The values of $\pi_0/\pi_t$ were 1.09 (for the 8% hypo-osmotic solution) and 0.88 (for the 13% hyperosmotic solution). The open circle corresponds to the peak amplitude of the osmotic response produced by the 40% hypo-osmotic ($\pi_0/\pi_t$ for the 40% hypo-osmotic solution was 1.66). Note that the measurements fall close to the theoretical line for a perfect osmometer.

## *Limits of the Fluorescence Techniques and Potential Pitfalls*

### *Dye Binding or Compartmentalization*

It is well known that esterified dyes are sequestered or bound to more than one intracellular compartment (38, 39, 44). The present techniques may require accurate assessment of the fraction of the dye that is bound or compartmentalized and that does not change with the external osmolality. We have shown that it is possible to assess accurately these "bound" fractions of calcein and fura-2 and the fractions that behave as osmotically active by controlled cell permeabilization with $\alpha$-toxin (29). Permeabilization with digitonin leads to underestimates of the bound fractions, and therefore it should be avoided (29). The bound fraction of either fura-2 or calcein is between 60 and 68% of the total fluorescence in neuronal cell lines, in agreement with estimates for other cells (38, 39, 44). Knowing the bound fraction, we can quantify relative CWV from changes in fluorescence resulting from changes in concentration of the free fraction. In principle, dye compartmentalization may be avoided by direct injection of the free acid forms of the fluorophores. As a future development to avoid compartmentalization, injection of dextran-bound fluorophores, such as tetramethylrhodamine, should be explored.

### *Photobleaching and Photodamage*

Any fluorescent dye may undergo irreversible photobleaching on exposure to strong excitation light. In addition, a dye-loaded cell is susceptible to damage by light through a process whose chemistry is not well understood, but which may involve the production of free radicals. Photobleaching and photodamage depend on the intensity and the duration of exposure to the excitation light, and the concentration of intracellular dye. The only practical way to avoid these problems is to use low-intensity light levels and to keep exposure as short as possible. This can be achieved with appropriate neutral-density filters and a pulsed-fluorescence system equipped with shutters, like the one described in this chapter. One of the virtues of calcein, compared with fura-2, is that it is two or three times more fluorescent than BCECF or fura-2, and it is used at its peak excitation and emission wavelengths and hence it is possible to reduce the intensity of the excitation light producing minimal photodynamic damage and quenching. We have found that the baseline for drift for calcein is not altered by increasing the sampling rate between 0.1 and 1 Hz. Therefore, baseline drift, when present during the time course of an experiment, is attributed mainly to dye leakage.

### *Dye Leakage*

This may produce undesirable baseline drifts. It is not an important problem for calcein, which seems to be well retained, but it may be significant with fura-2. A possible solution is using dyes such as fura-PE3 (TEFLABS, Austin, Texas), which is claimed to have all the properties of fura-2 except that it resists leakage and compartmentalization.

*Cell Movement and Shape Changes*

Single-wavelength measurements like those made with calcein or at the isosbestic point of fura-2 or BCECF are sensitive to cell shape changes. To minimize this problem the microscope is focused at a focal plane where the fluorescence intensity is maximal. This focal plane corresponds or is close to the axial center of the cell. Confocal optical sections of cells loaded with calcein show this assumption to be correct (29). Because the volume of a spherical cell changes with the cube of the cell radius, changes in pathlength due to changes in cell volume are negligible. For a spherical cell having a radius of 15 $\mu$m, an increase in volume as large as 50% would produce an increase in axial pathlength of about 2 $\mu$m. Changes in pathlength due to decreases in cell volume could, in principle, be overestimated. This is because the top surface of the cell might drop below the top surface of the optical image of the volume element formed by the combination of the pinhole with the objective lens of the microscope. We think this is unlikely for relatively large cells because the volume element, which contains the light that is measured by the PMT, is contained within the cell boundary. Confocal optical sections showed (29) that the height of N1E-115 cells varies between 35 and 45 $\mu$m.

*Changes in the Dissociation Constant of Fura-2 for $Ca^{2+}$*

When using fura-2 to monitor changes in $[Ca^{2+}]_i$ and cell volume it is important to bear in mind that changes in the intracellular ionic strength, such as those that could occur for large CWV changes, may alter the $K_d$ for fura-2 (51, 52), leading to an overestimation of changes in $[Ca^{2+}]_i$. A linear logarithmic relationship between $K_d$ and ionic strength exists, and hence it is possible to correct for this problem when necessary.

*Changes in Fluorescence with Medium Microviscosity*

Fura-2 fluorescence, as that of other fluorophores, is enhanced by the microviscosity of the solvent (53). The intracellular fluid microviscosity may change during osmotically induced changes in cell water. However, if this were the case, a decrease in internal microviscosity would decrease the fluorescence intensity of fura-2, preferentially at larger wavelengths, producing an artifactual increase in the 340/380 ratio, mimicking an increase in $[Ca^{2+}]_i$ (39, 53). Measurements of the microviscosity of fluid-phase cytoplasm of mammalian cells (e.g., cultured fibroblasts) show that it is <1.5 cP, greater than that of water (1 cP) (54, 55). The effects of changes in viscosity on the ratio 340/380 at free $[Ca^{2+}] < 1\ \mu M$ are negligibly small at microviscosities in the range between 1 and 3 cP (39). Hence the microviscosity effect does not seem to affect $Ca^{2+}$ readings. Moreover, the effect of viscosity is less important if ratios of 340/362 or 340/358 are employed (39). A way of testing that the swelling-induced $Ca^{2+}$ response is not spurious is to increase the buffer capacity for $Ca^{2+}$, say with BAPTA. Buffering should abolish a genuine increase in $[Ca^{2+}]_i$.

*Spectral Shifts*

Intracellular spectral shifts (51) may displace the isosbestic point of fura-2 or BCECF to regions that are no longer insensitive to $Ca^{2+}$ or $H^+$, respectively. We have empirically observed that the isosbestic wavelength for fura-2 intracellularly is slightly shifted from 360 or 362 nm in calibration solutions to 358 nm in the cytosol.

## Appendix

### *Subtraction of the Background Fluorescence*

Here the assumptions underlying derivation of Eq. (6) are discussed. The first assumption is that there is a linear relation between the amount of dye inside the cell and the total fluorescence within the field of view, irrespective of whether the fluorophore is bound, trapped, or free. This assumption implies that all the fluorescence comes from the dye and that there is no self-quenching. With this assumption,

$$F = KQ\nu/V^{T}, \tag{A1}$$

where $K$ is the proportionality constant relating the amount of dye ($Q$) and the total fluorescence intensity ($F$); $\nu$ is the volume element contained within the field of view, which is determined by the pinhole and the numeric aperture of the objective lens; $V^T$ is the total cell volume in which the dye is bound or diluted, from which $\nu$ is a sample. With a homogeneous cell, such as a red blood cell, $V^T$ would be the whole cell volume. With a nucleated cell having a homogeneous cytoplasm, because the field of view is located in the cytoplasm, $V^T$ would be the cytoplasmic volume (i.e., cytoplasmic ground substance plus small organelles). If the cell has large vacuoles or inclusion bodies, not likely to be represented in the field of view, but otherwise has a homogeneous cytoplasm, $V^T$ is still the cytoplasmic volume. If the cytoplasm is not homogeneous, then it is unlikely that $\nu$ will be representative of the entire cytoplasmic volume and the precise identification of $V^T$ is not possible. Ordinarily, $V^T$ must be considered to be a hypothetical portion of the cytoplasm of which $\nu$ is representative. This definition of $V^T$ may not be very satisfying, but it may be the best that can be done, and from the practical point of view it works. The second assumption is that $Q$ is constant and that there is no redistribution of the dye within intracellular compartments.

When we apply an osmotic challenge to a fluorophore-loaded cell, we monitor changes in relative fluorescence ($F_t/F_0$). However, to work with linear relations between fluorescence and external osmotic pressure it is convenient to plot the reciprocal of the recorded changes in fluorescence expressed as the ratio $F_0/F_t$, where $F_0$ is the fluorescence intensity for the control condition and $F_t$ is the fluorescence reading corresponding to the experimental condition; from Eq. (A1) it follows that

$$F_0/F_t = (KQ\nu/V_0^T)/(KQ\nu/V_t^T). \tag{A2}$$

Because $Q$, $\nu$, and $K$ are all constants,

$$F_0/F_t = V_t^T/V_0^T. \tag{A3}$$

However, $V^T$ is the sum of two volume fractions, namely, the cell water volume fraction in which the dye is diluted in the cytosol, denoted as $V^{os}$, which is the fraction that changes upon alterations in the external osmolality, and a fraction denoted $V^b$, which does not change with the external osmolality, and represents a volume in which the dye is bound or trapped. That is to say, the total fluorescence $F$ has a fraction that comes from $V^{os}$, which varies with external osmolality, plus a fraction that comes from $V^b$, which is constant. Because we are interested in measuring changes in $V^{os}$ it is necessary to devise means to subtract the fluorescence that comes from $V^b$.

Given that

$$V^T = V^{os} + V^b, \tag{A4}$$

it follows that

$$V_0^T = V_0^{os} + V^b \tag{A4.1}$$

and

$$V_t^T = V_t^{os} + V^b, \tag{A4.2}$$

where $V_0^{os}$ is the cell water volume in the control condition and $V_t^{os}$ is the steady-state cell water volume reached after an osmotic challenge.

Substituting Eqs. (A4.1) and (A4.2) into Eq. (A3), we obtain

$$F_0/F_t = (V^b + V_t^{os})/(V^b + V_0^{os}). \tag{A5}$$

For small changes in the external osmolality, $V^{os}$ behaves ideally, i.e., in accordance with the Boyle–van't Hoff equation:

$$V_t^{os} = V_0^{os}(\pi_0/\pi_t), \tag{A6}$$

where $\pi_0$ is the osmotic pressure of the control extracellular solution and $\pi_t$ is the osmotic pressure of the test solution.

Substituting Eq. (A6) into Eq. (A5) we get

$$F_0/F_t = \{V^b + [V_0^{os}(\pi_0/\pi_t)]\}/(V^b + V_0^{os}); \tag{A7}$$

rearranging, we obtain

$$F_0/F_t = [V^b/(V^b + V_0^{os})] + [(V_0^{os}/V^b + V_0^{os})(\pi_0/\pi_t)], \tag{A8}$$

and substituting Eq. (A4.1) into Eq. (A8) and rearranging, we obtain

$$F_0/F_t = [(V_0^{os}/V_0^T)(\pi_0/\pi_t)] + (V^b/V_0^T). \tag{A9}$$

Equation (A9) is a straight line and the fluorescence produced by the fraction $(V^b/V_0^T)$ is the background fluorescence $F_{bkg}$. Consequently, $F_{bkg}$ can be determined from the intercept of the plot of $F_0/F_t$ versus $\pi_0/\pi_t$ and $V_0^{os}/V_0^T$ will be the slope.

Equation (A9) can be rewritten as follows:

$$\pi_0/\pi_t = [(F_0/F_t) - (V^b/V_0^T)]/(V_0^{os}/V_0^T), \tag{A10}$$

and substituting Eq. (A4.1) into Eq. (A10) and rearranging, we obtain

$$\pi_0/\pi_t = [(F_0/F_t) - (V^b/V_0^T)]/(1 - V^b/V_0^T). \tag{A11}$$

To normalize, we can make $V_0^T = 1$, and Eq. (A11) becomes

$$\pi_0/\pi_t = [(F_0/F_t) - V^b]/(1 - V^b), \tag{A12}$$

and according to Eq. (A1), the background fluorescence $F_{bkg}$ will come from $V^b$, which is constant; hence we can write

$$\pi_0/\pi_t = [(F_0/F_t) - F_{\mathrm{bkg}}]/(1 - F_{\mathrm{bkg}}). \tag{A13}$$

We have found that $F_{\mathrm{bkg}}$, determined from the intercept of the plot of $F_0/F_t$ versus $\pi_0/\pi_t$, corresponds to the value determined by controlled permeabilization with $\alpha$-toxin, as described previously (29).

The change in the osmotically active fraction of $V^{\mathrm{T}}$, which is $V_t^{\mathrm{os}}/V_0^{\mathrm{os}}$, can be calculated from the following equation [see Eq. (5)]:

$$[(F_0/F_t) - F_{\mathrm{bkg}}]/(1 - F_{\mathrm{bkg}}) = V_t^{\mathrm{os}}/V_0^{\mathrm{os}} = \pi_0/\pi_t. \tag{A14}$$

Equation (A14) is equivalent to Eq. (6).

## Acknowledgments

We are grateful to David Baker and Malcolm Brodwick for many critical and stimulating discussions, and to José Rodolfo Fernández, Sergio Márquez-Baltazar, and Carrie Preite for skilled technical assistance. This work was supported by the National Institute of Neurological Disorders and Stroke Grant NS-29227 to Dr. Francisco Javier Alvarez-Leefmans.

## References

1. E. K. Hoffmann and L. O. Simonsen, *Physiol. Rev.* **69,** 315 (1989).
2. F. Lang, M. Ritter, H. Völkl, and D. Häussinger, *Adv. Comp. Environ. Physiol.* **14,** 1 (1993).
3. W. J. Mergner and R. T. Jones, *in* "Cell Death. Mechanisms of Acute and Lethal Cell Injury" (W. J. Mergner, R. T. Jones, and B. Trump, eds.), Vol. 1, p. 15. Field and Wood, Medical Publ., New York, 1990.
4. C. Nicholson, *Can. J. Physiol. Pharmacol.* **70,** S314 (1992).
5. K. Strange, *J. Am. Soc. Nephrol.* **3,** 12 (1992).
6. S. R. Gullans and J. G. Verbalis, *Annu. Rev. Med.* **44,** 289 (1993).
7. R. O. Law, *J. Exp. Zool.* **268,** 90 (1994).
8. M. L. McManus and K. B. Churchwell, *in* "Cellular and Molecular Physiology of Cell Volume Regulation" (K. Strange, ed.), p. 63. CRC Press Boca Raton, Florida, 1994.
9. I. Klatzo, *Acta Neuropathol.* **72,** 236 (1987).
10. H. K. Kimelberg, *J. Neurotrauma* **9,** S71 (1992).
11. H. K. Kimelberg, E. R. O'Connor, P. Sankar, and C. Keese, *Can. J. Physiol. Pharmacol.* **70,** S323 (1992).
12. L. Reuss, *Proc. Natl. Acad. Sci. U.S.A.* **82,** 6014 (1985).
13. M. Tauc, S. Le Maout, and P. Poujeol, *Biochim. Biophys. Acta* **1052,** 278 (1990).

14. B.-E. Persson and K. R. Spring, *J. Gen. Physiol.* **79,** 481 (1982).
15. J. K. Foskett, *in* "Optical Microscopy: Emerging Methods and Applications" (B. Herman and J. Lemasters, eds.), p. 237. Academic Press, San Diego, 1993.
16. K. Strange and K. R. Spring, *Kidney Int.* **30,** 192 (1986).
17. F. J. Alvarez-Leefmans, S. M. Gamiño, and L. Reuss, *J. Physiol.* (*London*) **458,** 603 (1992).
18. G. Serve, W. Endres, and P. Grafe, *Pfluegers Arch.* **411,** 410 (1988).
19. K. Ballanyi, M. Strupp, and P. Grafe, *in* "Practical Electrophysiological Methods" (H. Kettenmann and R. Grantyn, eds.), p. 363. Wiley-Liss, New York, 1992.
20. C. U. Cotton, A. M. Weinstein, and L. Reuss, *J. Gen. Physiol.* **93,** 649 (1989).
21. J. S. Adorante and S. S. Miller, *J. Gen. Physiol.* **96,** 1153 (1990).
22. K. Ballanyi, P. Grafe, G. Serve, and W.-R. Schlue, *Glia* **3,** 151 (1990).
23. R. K. Orkand, I. Dietzel, and J. A. Coles, *Neurosci. Lett.* **45,** 273 (1984).
24. W. J. Lederer, A. J. Spindler, and D. A. Eisner, *Pfluegers Arch.* **381,** 287 (1979).
25. F. J. Alvarez-Leefmans, *in* "Practical Electrophysiological Methods" (H. Kettenmann and R. Grantyn, eds.), p. 171. Wiley-Liss, New York, 1992.
26. D. A. T. Dick, *in* "Mechanisms of Osmoregulation in Animals" (R. Gilles, ed.), p. 3. Wiley, New York, 1979.
27. A.Villarroel, O. Alvarez, A. Oberhauser, and R. Latorre, *Pfluegers Arch.* **413,** 118 (1988).
28. S. Muallem, B.-X. Zhang, P. A. Loessberg, and R. A. Star, *J. Biol. Chem.* **267,** 17658 (1992).
29. W. E. Crowe, J. Altamirano, L. Huerto, and F. J. Alvarez-Leefmans, *Neuroscience* (*Oxford*) in press (1995).
30. M. Mcmanus, J. Fischbarg, A. Sun, S. Hebert, and K. Strange, *Am. J. Physiol.* **265,** C562 (1993).
31. E. R. O'Connor, H. K. Kimelberg, C. R. Keese, and I. Giaever, *Am. J. Physiol.* **264,** C471 (1993).
32. H. H. Willard, L. L. Merritt, J. A. Dean, and F. Settle, "Instrumental Methods of Analysis." Wadsworth, Belmont, California, 1981.
33. N. Gains and A. P. Dawson, *Analyst* (*London*) **104,** 481 (1979).
34. J. Slavik, J. "Fluorescent Probes in Cellular and Molecular Biology." CRC Press, Boca Raton, Florida, 1994.
35. F. J. Alvarez-Leefmans, H. Cruzblanca, S. M. Gamiño, J. Altamirano, A. Nani, and L. Reuss, *J. Neurophysiol.* **71,** 1787 (1994).
36. R. J. MacLeod, *in* "Cellular and Molecular Physiology of Cell Volume Regulation" (K. Strange, ed.), p. 191. CRC Press, Boca Raton, Florida, 1994.
37. K. Strange and R. Morrison, *Am. J. Physiol.* **263,** C412 (1992).
38. L. A. Blatter and W. G. Wier, *Biophys. J.* **58,** 1491 (1990).
39. M. W. Roe, J. J. Lemasters, and B. Herman, *Cell Calcium* **11,** 63 (1990).
40. G. Ahnert-Hilger and M. Gratzl, *Trends Pharmacol. Sci.* **9,** 195 (1988).
41. I. Schulz, *in* "Methods in Enzymology" (S. Fleischer and B. Fleischer, eds.), Vol. 192, p. 280. Academic Press, San Diego, 1990.
42. S. Bhakdi and J. Tranum-Jensen, *Microbiol. Rev.* **55,** 733 (1991).
43. C. S. Chew and M. Ljungstrom, *in* "Optical Microscopy: Emerging Methods and Applications" (B. Herman and J. J. Lemasters, eds.), p. 133. Academic Press, New York, 1993.

44. W. Almers and E. Neher, *FEBS Lett.* **192,** 13 (1985).
45. L. C. Falke and S. Misler, *Proc. Natl. Acad. Sci. U.S.A.* **86,** 3919 (1989).
46. H. Pasantes-Morales, T. E. Maar, and J. Morán, *J. Neurosci. Res.* **34,** 219 (1993).
47. M. E. Chamberlin and K. Strange, *Am. J. Physiol.* **257,** C159 (1989).
48. P. H. Yancey, *in* "Cellular and Molecular Physiology of Cell Volume Regulation" (K. Strange, ed.), p. 181. CRC Press, Boca Raton, Florida, 1994.
49. J. Dragolovich and S. K. Pierce, *in* "Cellular and Molecular Physiology of Cell Volume Regulation" (K. Strange, ed.), p. 123. CRC Press, Boca Raton, Florida, 1994.
50. J. M. Sands, *in* "Cellular and Molecular Physiology of Cell Volume Regulation" (K. Strange, ed.), p. 133. CRC Press, Boca Raton, Florida, 1994.
51. A. Uto, H. Arai, and Y. Ogawa, *Cell Calcium* **12,** 29 (1991).
52. D. A. Williams and F. S. Fay, *Cell Calcium* **11,** 75 (1990).
53. M. Poenie, *Cell Calcium* **11,** 85 (1990).
54. K. Fushimi and A. S. Verkman, *J. Cell Biol.* **112,** 719 (1991).
55. N. Periasamy, M. Armijo, and A. S. Verkman, *Biochemistry* **30,** 11836 (1991).

# Index

Printed and bound by CPI Group (UK) Ltd, Croydon, CR0 4YY

Printed and bound by CPI Group (UK) Ltd, Croydon, CR0 4YY
03/10/2024
01040317-0010